I0823750

Confederate Generals in the Western Theater

Confederate Generals in the Western Theater

Volume 1

Classic Essays on America's Civil War

Edited by Lawrence Lee Hewitt and Arthur W. Bergeron Jr.
With a Foreword by Terrence J. Winschel

The Western Theater in the Civil War • Gary D. Joiner, Series Editor

The University of Tennessee Press / Knoxville

The Western Theater in the Civil War series seeks to emphasize an emerging trend in the historiography of the nation's greatest conflict: a more general recognition that events in the West, far from being a sideshow to storied campaigns in the East, were, in many ways, even more decisive in the outcome of the war. Among the works that will be produced are scholarly monographs, biographies of leaders who need reconsideration, and edited collections that present up-to-date scholarship in this rapidly developing field.

First Edition.

The paper in this book meets the requirements of American National Standards Institute / National Information Standards Organization specification Z39.48-1992 (Permanence of Paper). It contains 30 percent post-consumer waste and is certified by the Forest Stewardship Council.

Library of Congress Cataloging-in-Publication Data

Confederate generals in the western theater: classic essays on America's Civil War / edited by Lawrence Lee Hewitt and Arthur W. Bergeron Jr.; with a foreword by Terrence J. Winschel. — 1st ed.
p. cm. — (The western theater in the Civil War)
Includes bibliographical references and index.

ISBN-13: 978-1-57233-700-8 (hardcover)
ISBN-10: 1-57233-700-1 (hardcover)

1. Generals—Confederate States of America—History.
2. Generals—Confederate States of America—Biography.
3. Mississippi River Valley—History—Civil War, 1861–1865—Campaigns.
4. United States—History—Civil War, 1861–1865—Campaigns.
5. Command of troops—History—19th century.
6. Military art and science—Confederate States of America—History.
I. Hewitt, Lawrence L.
II. Bergeron, Arthur W.

E467.C773 2010
973.7092'2—dc22
2009051917

For Thomas Lawrence Connelly

Thomas Lawrence Connelly.
Courtesy of Heather Connelly
Lefkowitz.

Contents

Illustrations

Photographs

Maps

Series Editor's Foreword

As the centennial of the Civil War neared, a dedicated group of historians prepared to explain America's defining conflict in new ways. Steeped in the lore of the war and armed with massive amounts of documents, personal accounts, and manuscripts, they set out to explain the conflict in its true context. Some of these scholars were only two generations removed from the participants. Some, younger, were the students of the older masters. Together they created the core of modern scholarship of the American Civil War. Many of their works have reached iconic status. Modern students often refer to their volumes as authoritative sources on campaigns and commanders. What sets these historians apart from those who came before them?

The decades after the conclusion of hostilities brought many memoirs by surviving commanders, upper-level officers, and veterans of lower rank. Participants on both sides penned articles in publications such as *Confederate Veteran, Galaxy,* and the *Century Magazine*. Many of these pieces were later published in compendia, particularly *Battles and Leaders of the Civil War.* The years from the end of World War I through the mid-1930s saw a revival of interest as the old veterans gathered for reunions before they faded into history. Their accounts of events tended to be limited to their own views and experiences and were naturally skewed by their previous alliances and personal biases. The 1950s and 1960s brought forth a renaissance of writing and analysis from a new generation of scholars. For the first time, the public received more objective views of the war and accepted these new interpretations as America embraced its past.

The published output of the mid–twentieth century was dominated by accounts of the war in the Eastern Theater, where much of the action had occurred. The largest battles were fought there. Most of the heavily populated areas involved in the war were there. However, the campaigns that occurred west of the Appalachian Mountains were the cradles and crucibles for the Union commanders who eventually came to the fore in the East. For the Confederates, this vast region of mountains and valleys, of forests and rivers, was the foundation for their continued survival as a nation. This great storehouse of men and materiel, food, and moral support was vital to the fledgling country. The war in the West is once more receiving the attention of fine scholars as evidenced by the new series of which this volume is a part.

Many of the great historians who focused on the West wrote some of their best analyses as articles in journals or publications with limited distribution. Most of the reading public never got the chance to see this body of work, a wrong that can now be rectified. This volume, *Confederate Generals in the Western Theater: Classic Essays on America's Civil War,* portrays the work of these pioneers and other leading historians on this subject. The editors, Lawrence Lee Hewitt and Arthur W. Bergeron Jr., are both noted historians in their own right. Their selection of essays for this book provides a wonderful chorus of views on this now-emerging area of scholarship. This is a fitting volume with which to begin the new University of Tennessee Press series, The Western Theater in the Civil War. It is appropriately dedicated to the late Thomas Lawrence Connelly, a scion of research on the war in the West and the major authority on the Confederate Army of Tennessee. Many of the essays in this collection were written by his friends and colleagues.

It is my pleasure to reintroduce the lives and exploits of some of the Confederacy's most noted leaders, as well as some of its more obscure commanders, with the hope that readers will savor the best research and writing on this complex theater of war.

Gary D. Joiner
Louisiana State University in Shreveport

Foreword

There have been more books written about the Civil War than any other aspect of American history. As the sesquicentennial of that struggle fast approaches, it is interesting to note that books on "our" war are being written and published at a faster rate than at any point in time since Lee and Grant met at Appomattox, signaling the end of the bloodiest conflict in American history. Coupled with this phenomenon is the intense passion among collectors for Civil War photographs, documents, and artifacts, the level of which reaches new heights with each successive generation. The thirst for such varied materials associated with the Civil War seems almost insatiable. And it is not just Americans who crave the gamut of scholarly literature and ephemera related to the bloodshed that engulfed this nation from 1861 to 1865. Worldwide interest in the Civil War continues to grow as evidenced by the rising number of Civil War Round Tables in Europe and other continents, which even host large-scale battle reenactments.

The breadth and scope of literature on the Civil War is vast and comprehensive. Counted among its treasury of titles are some of the most well known, highly respected, and influential works in American literature. Certainly the Civil War has its share of "classics," those works that by definition have lasting significance or worth. The contributions of these works transcend generations and are indispensable to a study of the "brother's war."

Dominant among the "classics" are works by the veterans themselves who had served in the armies and navies of the North and South. Chief among these works are titles such as *The Personal Memoirs of Ulysses S. Grant,* Joshua Chamberlain's *Passing of the Armies,* E. Porter Alexander's *Military Memoirs of a Confederate,* John Billings's *Hardtack and Coffee,* John B. Jones's *A Rebel War Clerk's Diary at the Confederate States Capital,* Frank A. Haskell's *Battle of Gettysburg,* and the compilation of writings that first appeared as "The Century War Series" in the *Century Magazine* and is better known today as the four-volume *Battles and Leaders of the Civil War.*

Successive generations of historians writing on the Civil War have produced classics of their own. Included among the modern classics are works such as Douglas Southall Freeman's *R. E. Lee* and *Lee's Lieutenants,* Bell Wiley's *Life of Johnny Reb* and *Life of Billy Yank,* and the numerous works by Bruce Catton, which include his trilogy on the Army of the Potomac—*Mr. Lincoln's Army, Glory Road,* and *A Stillness at Appomattox*—and his trilogy on the centennial history of the Civil War, *The Coming Fury, Terrible Swift Sword,* and *Never Call Retreat.* Among the classics of even more recent vintage are titles by James McPherson, such as *Battle Cry of Freedom,* and *The Civil War: A Narrative* by Shelby Foote. And recent publications that are destined to be called classics are John Waugh's *Class of 1846* and Michael Kaufman's *American Brutus,* which deals with John Wilkes Booth and the assassination of Abraham Lincoln.

Even novels such as Margaret Mitchell's epic, *Gone with the Wind,* are considered classics. So too are Stephen Crane's *Red Badge of Courage,* Donald McCaig's *Jacob's Ladder,* Michael Sharra's *Killer Angels,* and his son Jeff's *Gods and Generals.* (Although novels by nature take certain liberties with historical facts, they nonetheless bring the conflict to life for thousands of readers and generate a level of interest in the Civil War as inspired by few writers of nonfiction.)

In addition to these works, any student of the Civil War can rattle off their own list of personal favorites—works they in essence consider classics. This holds true for essays, articles, pamphlets, and the myriad of other publications on the Civil War. The content, analysis, or style of such publications is timeless and will remain as much a benefit to current and future students of the Civil War as they were when first published.

Due to their timeless contribution and the constant demand for these titles, many of the more popular books are reprinted on a regular basis. Sadly, however, the same cannot be said of essays; and many of these truly significant classics of Civil War literature are currently out of print or appeared in various publications that do not enjoy wide circulation. Thus many of these scholarly pieces are known only to professional historians or the most avid students of the Civil War, and their benefit to the study and understanding of the defining event in American history is limited as a consequence.

Confederate Generals in the Western Theater: Classic Essays on America's Civil War is a collection of such essays. Resurrected so they might reach a wider audience and attain their rightful stature in the ever-growing field of literature on the Civil War, these essays come from the pens of the leading Civil War historians of the mid- to late twentieth century, people such as Grady McWhiney, Charles P. Roland, T. Harry Williams, Frank E. Vandiver, Archer Jones, and Edwin C. Bearss. These men, who were the educators and mentors of the current authorities in Civil War historiography, have powerfully, passionately, and persuasively chronicled the people and events that comprised the War Between the States across the spectrum of its complexities. Their writings and speeches have helped forge modern perceptions of the war and its impact on the continuing evolution of the United States, and their influence will be felt for generations to come. You will find that these essays represent some of their best work.

With exacting standards of scholarship, editors Lawrence Lee Hewitt and Arthur W. Bergeron Jr. have carefully, analytically, and skillfully crafted this collection in honor of Professor Thomas Lawrence Connelly, who was himself one of the leading historians of his generation on the Civil War in the Western Theater. Best known for his works on the Confederate Army of Tennessee—*Army of the Heartland, Autumn of Glory,* and *Five Tragic Hours: The Battle of Franklin,* the last of which James Lee McDonough coauthored—Connelly's work was, and remains, exceptional.

It is a fitting tribute to Connelly that this collection of essays is dedicated to him. The Nashville native, who received his Ph.D. under the tutelage of the great Frank Vandiver, almost single-handedly revived interest in the war fought west of the Appalachian Mountains. His writings helped to focus national attention on what was in fact the crucial theater of combat. Recognized by politicians and soldiers alike as such during the war, the military operations along the Mississippi River and in the heartland of the Confederacy proved to be the decisive actions that ended the Civil War. Yet in the aftermath of the war, the significance of the Western Theater was soon eclipsed due to voluminous writings on battles and leaders associated with the Eastern Theater. These writings grew more popular over time and have for generations now dominated the mindset of Americans.

But the pendulum of popularity is slowly swinging to a more historically correct and detailed assessment of the war. Thanks largely to the efforts of the cadre of historians—spearheaded by Connelly—whose writings are contained in this collection, a growing number of historians have come to acknowledge the crucial role of events that transpired in the Western Theater. Even the general public, getting saturated on a steady diet of Gettysburg, is now being drawn to places such as Fort Donelson, Shiloh, Corinth, Vicksburg, Port Hudson, Murfreesboro, Chickamauga, and Atlanta and coming to know the likes of men who are the subjects of these essays.

Connelly would be honored by this collection, especially by the authors of these essays, for he knew most of them personally and greatly admired their work. Those who read these essays will quickly appreciate that most of these men were only a couple of generations removed from the conflict and had a greater degree of intimacy with those who fought or survived America's great struggle than those who currently write on the Civil War. In some cases they had even met and spoken with veterans of the Civil War and thus literally touched the men whose actions forged a nation in the crucible of war. Equally important, many of these authors were/are veterans of World

War II and experienced combat in all its horror on the battlefields of Europe or in the Pacific. Thus they have a common bond with the men in blue and gray, and their writings reflect the emotional ties of a shared experience that bind soldiers across the continuum of time. Their writing evokes the deep emotions of the battlefield that are known only to those who have experienced combat and will instill in you, the reader, a greater understanding and appreciation of the American soldier at all levels of responsibility.

Rounding out and, indeed, complementing this collection are essays from today's leading authorities on the war in the West penned by Richard M. McMurry, Nathaniel C. Hughes Jr., Craig L. Symonds, and Steven E. Woodworth, along with contributions by the editors of this volume. Each author brings a unique and highly focused perspective to his subject based on years of tenacious, painstaking, and thorough research and analysis. Such dogged scholarship has made these authors the acknowledged authorities on the men about whom they write.

Seldom has a collection of essays contained such a lineup of historians. These are the true "power hitters" of their generation in their prime and at their very best. So prepare to enjoy Civil War historiography at its finest. Immerse yourself in these classics. Learn from these masters, for they will have a profound influence on how you view the generals who led Confederate armies in the Western Theater when a divided people struggled to define a nation.

Terrence J. Winschel
Vicksburg National Military Park

Acknowledgments

SEVERAL INDIVIDUALS, ORGANIZATIONS, AND INSTITUTIONS HAVE ASSISTED US IN THIS project in numerous ways. Foremost among them is Charles Elliott, who, though inadvertently, got us started on this project. While attending the Southern Historical Association's annual conference in Fort Worth in 1991, we decided to do an anthology dealing with the military activities that occurred in Louisiana during the Civil War. As part of that endeavor, we asked Charles to write an article on the Battle of Baton Rouge. For more than a decade we patiently waited for him to overcome a case of the "slows" that only George B. McClellan could understand. When we finally received Charles's manuscript, we discovered that instead of delivering what we had asked for, he had written an insightful analysis of Earl Van Dorn as a military strategist. Though it did not belong in the volume we had originally envisioned, it merited publication—and in a book that would have a greater audience than one focused only on Louisiana. Turning lemons into lemonade, we opted to do a collection of writings dealing with Confederate generals who served in the Western Theater. Because of that project's expansion, it is Charles who is now waiting. His essay will appear in *Confederate Generals in the Western Theater,* Volume 2.

Having originally agreed to work on the Louisiana volume, Terry Winschel enthusiastically embraced the new project. In addition to writing the foreword, it was Terry, along with Brian Wills, Gary D. Joiner, and Scot Danforth, who recognized that the readers' experience would be enhanced by separating the previously published articles from those newly written into different volumes. We are grateful to you all. We also wish to thank Craig L. Symonds and Nathaniel C. Hughes Jr. for revising their earlier writings; Charles P. Roland, Richard M. McMurry, Daniel E. Sutherland, Donald S. Frazier, Stacy Allen, Judith Hallock, Wiley Sword, Robert W. Sledge, and especially Bianca J. Rowlett for assisting us with the annotation; Matthew Rutherford, Kevin Dyke, and Annie Authier of the Newberry Library and Janalee Croegaert and Laquita Brown of Northwestern University Library for helping us in our search for extant copies of the articles; Heather Connelly Lefkowitz for providing us with a photograph of her father and Theresa Walling of the University of South Carolina for putting us in contact with Heather; and the staff of the University of Tennessee Press for all their labors on our behalf.

We are grateful to the following individuals and organizations for allowing us to reproduce the articles: Donald S. Frazier and the McWhiney Foundation Press; Stephen M. Wrinn, Craig R. Wilkie, and the University Press of Kentucky; William Underwood, William Blair, and the Kent State University Press; Nicole Riley, Denise D. Logan, and White Mane Publishing Company, Inc.; Ann Toplovich and the Tennessee Historical Society; Susan K. Schott, Fred M. Woodward, and the University Press of Kansas; Beverly Jarrett, Lori Hall Jones, and the University of Missouri Press; Theodore P. Savas and Savas Publishing Company; Richard Hall, Patricia Burgess, and the Southern Historical Association; Wendy Vandervort and the *Journal of Military History* (formerly *Military Affairs*); Anne Bailey, Stan Deaton, and the Georgia Historical Society; David E. Roth, *Blue & Gray Magazine,* and Blue & Gray Enterprises, Inc.; and Patrick G. Williams and the Arkansas Historical Association.

We must acknowledge the friendship and support of our mentors Harry Williams, Bill Cooper, and Charlie Roland, and colleagues Frank Vandiver, Grady McWhiney, Richard McMurry, Ed Bearss, Craig Symonds, Nat Hughes, Mike Ballard, Steve Woodworth, Dan Sutherland, and Archer Jones. Sadly, some of these individuals are no longer with us except in spirit, but the influence of their personalities, words, and deeds will be with us for as long as we live.

Introduction

WHAT HISTORIAN WOULD CHOOSE TO LABOR ON SUCH VOLUMES AS *BRAGG'S BACKSTABBERS, Johnston's Jug Heads,* and *Hood's Hopeless*? The equivalent of Douglas Southall Freeman's epical three-volume *Lee's Lieutenants* will never be written about the Army of Tennessee's high command. We knew there was a dearth of material on the Western Theater in general and on the Confederate side in particular. Would others look favorably upon a volume of essays that dealt with the generals in gray who served outside of Virginia?

We tested the idea with several fellow historians and were buoyed by their enthusiastic responses. Their many and generous offers to contribute to the effort convinced us to do multiple volumes. Some of them, as well as some publishers, advised us to begin the project with a collection of previously published articles. *Confederate Generals in the Western Theater: Classic Essays on America's Civil War* is that volume.

Though they outnumbered their eastern counterparts, far less has been written about the Confederate generals who served between Virginia and the Mississippi River. Volunteers for service aboard the ill-fated CSS *Hunley* were more prevalent than articles suitable for inclusion in this volume. Though not entirely forgotten, these generals deserve better. To correct this injustice nearly fifty historians are currently working with us on five additional volumes—three on the Western Theater and two on the Trans-Mississippi—and we hope to do more.

That said, the essays that follow are not carbon copies of the original. Annotation has been made uniform throughout and provided for those articles that lacked it. Typographical errors were corrected. Nathaniel C. Hughes Jr. availed himself of our offer to allow him to revise his article, and Craig L. Symonds chose to rewrite his earlier account of Patrick Cleburne at the Battle of Chattanooga.

We used several criteria in making our selections. As the first of several volumes, and especially for those less familiar with the subject, we wanted to include an article on every commander of the principal Confederate army in the Western Theater as well as that army's prominent corps commanders. The preeminent division commander in the region, Cleburne, merited an article. Whereas Virginia had only one major military force within its borders for most of the war, that was not the case in the Western Theater. We also included individuals holding significant commands in the region outside the Army of Tennessee, such as the unfairly maligned Mansfield Lovell. Not wishing to ignore the most numerous group of general officers, we wanted to include at least one brigadier.

Geographically we wanted to touch on as much of the war in the West as possible. Some readers might be surprised to learn that according to the American Battlefield Protection Program everything south and west of Virginia and east of the Mississippi is in the Western Theater—even the port of Wilmington, North Carolina, which was the source of sustenance for the Army of Northern Virginia until January 1865. Most veterans of the Civil War would have agreed with this assessment. In this area in particular we were more hampered by what was available than by the limitations of space, and as a consequence certain campaigns, particularly those in the Carolinas and southern Alabama, receive scant attention in this volume.

To introduce younger readers to the writings of different historians of previous generations, we decided to have only one article per author. This prohibited the inclusion of some articles that dealt with our subject, especially by the prolific Steven E. Woodworth. Certain historians needed to be represented. Heading that list was Thomas Lawrence Connelly. Though better known for *The Marble Man: Robert E. Lee and His Image in American Society,* his earlier works included a two-volume history of the Army of Tennessee. Unfortunately, Tom was one of our first hurdles. While his body of work qualifies him as the foremost historian of the Western Theater, he never authored an essay suitable for this publication. For that reason this volume is dedicated to him. That same reason excluded others, some of whom, including Terry Winschel, Brian Wills, and Wiley Sword, are writing articles for our forthcoming volumes.

Finally, the essays selected had to have stood the test of time and merit a wider audience than they likely received in the past because of their original venue of publication. For this reason we gave preference to articles published in state historical journals as opposed to those that appeared in *Civil War Times Illustrated* and those that were part of an anthology not limited to Confederate operations in the Western Theater. Even so, space prevented us from including all that merit reprinting, as did our decision not to include more than one article that focused only on a particular general.

Turning to the fifteen essays in this volume, they are, for the most part, arranged chronologically to parallel the course of the war. Those that are complete biographies appear in the order of their subject's coming to the forefront, if you will. The exceptions are the last two articles, which deal with individuals who never achieved the level of fame—or infamy—of the other thirteen.

The first general to make a significant impact in the Western Theater was Leonidas Polk. Grady McWhiney, in "A Bishop as General," covers Polk's entire career in the Confederate army. Unwilling to finish his biography of the "detestable" Braxton Bragg, it is apropos that McWhiney wrote about the most prominent of that general's critics. Though no fan of Bragg, he concluded that Polk was the worse general of the two. McWhiney implies that the bishop did more to cost the Confederacy its independence than even the reviled Bragg.

Polk's successor to the command of what became the Army of Tennessee was Albert Sidney Johnston. A lifelong student of Johnston, Charles P. Roland presents the general's turbulent but brief career in "Albert Sidney Johnston and the Defense of the Confederate West." We believe Roland's arguments in this recently written essay are more persuasive than those in his book-length biography of Johnston published in 1964. After they have read this essay, fans of Joe Johnston and other western generals will have reason to reconsider their favorites. Though many readers will disagree with Roland, his conclusion that "of all Confederate generals in the West, Albert Sidney Johnston alone seems to have held the potential of a Lee" appears sound.

Probably the most familiar article in this volume is T. Harry Williams's "Beauregard at Shiloh," which was later published almost verbatim in his biography of the Creole general. Nonetheless, it is the best article on Beauregard in the West and focuses on the most controversial week of the war in that theater. Coupling the Roland and Williams articles, the reader is offered probably the most insightful, balanced account of the Confederate plans prior to the Battle of Shiloh ever to appear in a single volume.

Moving away from the Tennessee-Kentucky border and the principal army in the Western Theater, we turn to the lower Mississippi Valley. Arthur W. Bergeron Jr. relates the Confederate career of "Mansfield Lovell," a general whose talents were largely wasted by the Confederacy after the Davis administration made him the scapegoat for the fall of the Crescent City. Davis reluctantly granted Lovell a court of inquiry but refused to give him another command, even though the tribunal exonerated the general's conduct. Lovell did return to duty in 1865, after direction of the war had been wrested from Davis and turned over to the recently appointed General-in-Chief Robert E. Lee and Secretary of War John C. Breckinridge.

Beauregard's brief tenure as an army commander in the West ended with Davis's replacement of him with Bragg. In "Braxton Bragg and the Confederate Invasion of Kentucky in 1862," Lawrence Lee Hewitt analyzes the general's handling of the tremendous burden unexpectedly thrust upon him.

Remembered more for what occurred after 1862, Bragg is considered by many to be the worst of the Confederate generals. Hewitt contends that Bragg has been unfairly maligned.

Following the death of Sidney Johnston in April 1862, Davis began dividing up Department No. 2, which was commonly referred to as the Western Department. That trend ended in December with the appointment of Joseph E. Johnston to coordinate the operations of two of the newly created departments. In "Tennessee and Mississippi, Joe Johnston's Strategic Problem," Archer Jones presents a proactive, even innovative, Johnston that will surprise critics of the general. Unfortunately for Johnston and the Confederacy, Van Dorn's success at Holly Springs only postponed the fall of Vicksburg.

Michael B. Ballard chronicles the last Confederate commander of that city in "Misused Merit: The Tragedy of John C. Pemberton." Undoubtedly, the doomed Pemberton had skills. But bureaucratic genius can't secure public support or inspire troops, two things any general would require to even have a chance of defeating an opponent such as Ulysses S. Grant.

In "Soldier with a Blunted Sword: Braxton Bragg and His Lieutenants in the Chickamauga Campaign," Steven E. Woodworth demonstrates how divisive Bragg's subordinates had become less than a year after his withdrawal from Kentucky. Echoing Hewitt, Woodworth demonstrates that responsibility for the Army of Tennessee's lackluster performance extended beyond Bragg.

One of the few Confederate successes in the Western Theater occurred on the northern end of Missionary Ridge on November 25, 1863. Craig L. Symonds describes this action in "Patrick Cleburne's Defense of Tunnel Hill Revisited." Symonds demonstrates that this engagement was so embarrassing for Union generals Ulysses S. Grant and William T. Sherman that both men lied about what happened to their dying day.

Grant's all-out offensive to win the war in 1864 was far more successful beyond the boundaries of Virginia than within it. The most significant exception to that pattern is recounted by Edwin C. Bearss in "Bedford Forrest and His 'Critter' Cavalry at Brice's Cross Roads." Some might claim that Forrest's greatest tactical victory of the war was still a strategic victory for the Union because it kept the Confederate cavalryman from attacking Sherman's supply line. Nonetheless, Bearss demonstrates that no victorious Civil War general ever equaled Forrest's pursuit of a defeated foe.

Few actions by President Davis failed to draw criticism, and possibly none more so than his replacement of Joe Johnston with John Bell Hood. William J. Cooper Jr. contends in "A Reassessment of Jefferson Davis as War Leader: The Case from Atlanta to Nashville" that Davis was out of options. Similarly, no matter how hopeless it seemed even at the time, Davis's support of Hood's operations in Tennessee was valid. Consequently, it would seem that Hood deserves higher marks than he has heretofore received.

The native Kentuckian turned Texan doesn't get them from Frank E. Vandiver. In "General Hood as Logistician," the mentor of Tom Connelly fascinates the reader with his analysis of Hood's use of his manpower pool. Supply operations are generally forgotten except by those dependent upon them. Nevertheless, Hood failed to realize that for every man he pulled out of a factory to serve in the ranks he was reducing his ability to sustain his army.

Unfortunately for those interested in the war in the West, several of the historians mentioned above turned their interest away from the Western Theater to write about subjects such as John J. Pershing (Vandiver), Huey Long (Williams), or Robert E. Lee (Connelly). Thankfully, Nathaniel C. Hughes Jr. has not defected. His "Hardee's Defense of Savannah" depicts the war as it is entering its final stage, with the major operations in the Western Theater now taking place in the Carolinas.

One of the generals who surrendered in the Carolinas was James Patton Anderson. Another steadfast historian of the Western Theater, Richard M. McMurry chronicles the life of this citizen-soldier in "Patton Anderson: Major General, C.S.A." All but forgotten, Anderson and his military service took him from Pensacola to Perryville and included command of the District of Florida in 1864. He typifies so many of the Confederate generals who need to be rescued from oblivion.

Better known for *Seasons of War: The Ordeal of the Confederate Community, 1861–1865* than for his military writing, Daniel E. Sutherland's "No Better Officer in the Confederacy: The Wartime Career of Daniel C. Govan" is one of the best article-length biographies of any Confederate brigadier. Originally published in the *Arkansas Historical Quarterly,* we are confident that its inclusion will assure it a wider audience, one befitting both the author and his heroic subject. Govan's is an enthralling story, told here by an award-winning historian. Govan's escape unscathed from the maelstrom of Franklin is unbelievable.

In the preface to his history of the Army of Tennessee, Thomas Lawrence Connelly gave several reasons why these men have been forgotten. They have "always been associated with the roughness of the Old Southwest." They rarely won, and "Southerners like a winner." They left "much to the imagination. Many of the western generals had a certain dullness and absence of color which fail to inspire writers. Yet, the basic cause . . . is a paucity of good historical writing." We are confident that this volume would meet with Tom's approval.

Lieutenant General Leonidas Polk. Library of Congress.

A Bishop as General

Grady McWhiney

One of President Jefferson Davis's worst appointments was that of Episcopal bishop Leonidas Polk as a major general in the Confederate army. Polk, who graduated from West Point a year after his friend Jefferson Davis, had spent thirty years in church activities and agriculture before the Civil War. But he knew next to nothing about military affairs or the art of war. In just two months, while commanding the Confederacy's Western Department until Albert Sidney Johnston arrived from the West Coast to assume command, Polk had botched his first military assignment and managed to create problems that would plague the Army of Tennessee for the entire war.

General Polk feuded with his subordinates, exhibited what historian Thomas Connelly called a "sullen aloofness" that would hamper cooperation with his superiors, and planted "seeds of bitter personality and command conflict." He not only failed to establish an effective defense line across northwestern Tennessee but also violated Kentucky's neutrality by seizing Columbus a week before Johnston took command. Polk's actions deprived the unguarded Tennessee border of the slim but important protection that neutrality had provided. Moreover, Polk violated orders by invading Kentucky and then dodged responsibility for the dishonesty of his actions by altering his correspondence with President Davis and Secretary of War Leroy P. Walker, forwarding to Richmond "recopied" correspondence for the War Department files.[1]

All this was only a preview of the damage the apparently humble, sacrificing bishop would inflict on the Confederacy. To most observers, Polk seemed an asset to the Southern cause. "Wonderfully charming" in conversation, over six feet tall, with broad shoulders, blue eyes, and a white lock overhanging his forehead, he possessed "all the manners and affability" of a nobleman. His troops, who never saw the darker side of his personality—his stubborn, childish, quarrelsome nature—adored him. As an aloof bishop he had learned to lead, but as a soldier he would never learn to follow. His treatment of all his commanding officers bordered on insubordination, but usually Polk got away with it.[2]

Throughout his army career, he maintained the remarkable ability to evade the blame for situations he created. He accomplished this early in the war by imposing on his friendship with Sidney Johnston and Jefferson Davis. Polk knew how to flatter and manipulate Davis to protect himself from such army commanders as Braxton Bragg and Joseph E. Johnston. Three modern historians—Thomas Connelly, Steven E. Woodworth, and Judith L. Hallock—all have contributed to a revised view of Polk. Connelly insisted that the bishop, with his "amazing abilities to escape responsibility, was the most dangerous man in the Army of Tennessee."[3]

Bragg, who never liked Polk, believed that just before the Battle of Shiloh, the bishop contributed to General P. G. T. Beauregard's illness, distress, and worry. "Every interview with Genl Polk turns [Beauregard's recovery] . . . back a week," Bragg admitted to his wife. "But for my arrival here to aid him, I do not believe he would now be living."[4]

Bragg also considered the troops that Polk commanded in northern Mississippi hopelessly undisciplined. "I thought my Mobile Army was a *Mob,*" admitted Bragg, "but it is as far superior to Polk's . . . as the one at Pensacola was to it. . . . Such has been the outrageous conduct of our troops [here in northern Mississippi] that the people prefer seeing the enemy." Bragg claimed that Polk did "nothing to correct this. Indeed the good Bishop sets the example, by taking whatever he wishes."[5]

Polk continued this policy. Later in 1862, during the Kentucky Campaign, he ordered an officer to dismount three captured Federal cavalry officers, "turn their horses over to the quartermaster," and make sure their excellent saddles were "specially reserved for [Polk] and his staff."[6]

At Shiloh in April 1862, Polk did nothing brilliant but neither did he do anything stupid. His command of one of the four corps into which General Sidney Johnston divided his army appeared to have been satisfactory, although a contemporary called Polk's battle report the "Romance of Shiloh," implying that much of what the bishop reported never happened.

Polk's troops and baggage train did delay the Confederate march from Corinth, Mississippi, to Shiloh by jamming the streets. Major General William J. Hardee, whose corps was scheduled to get underway by noon on April 3, found his route blocked by Polk's divisions and failed to get his men started until late that afternoon. Polk's inability to clear his men from Hardee's route may have been embarrassing, but it certainly was not the only mistake made by the Confederates during the Shiloh Campaign.

If President Davis and General Sidney Johnston tended to overlook Polk's little and big mistakes during 1861 and early 1862, the situation changed after Bragg became commander of the Army of the Mississippi in June 1862. Bragg considered some of the army's highest ranking officers incompetent; he also

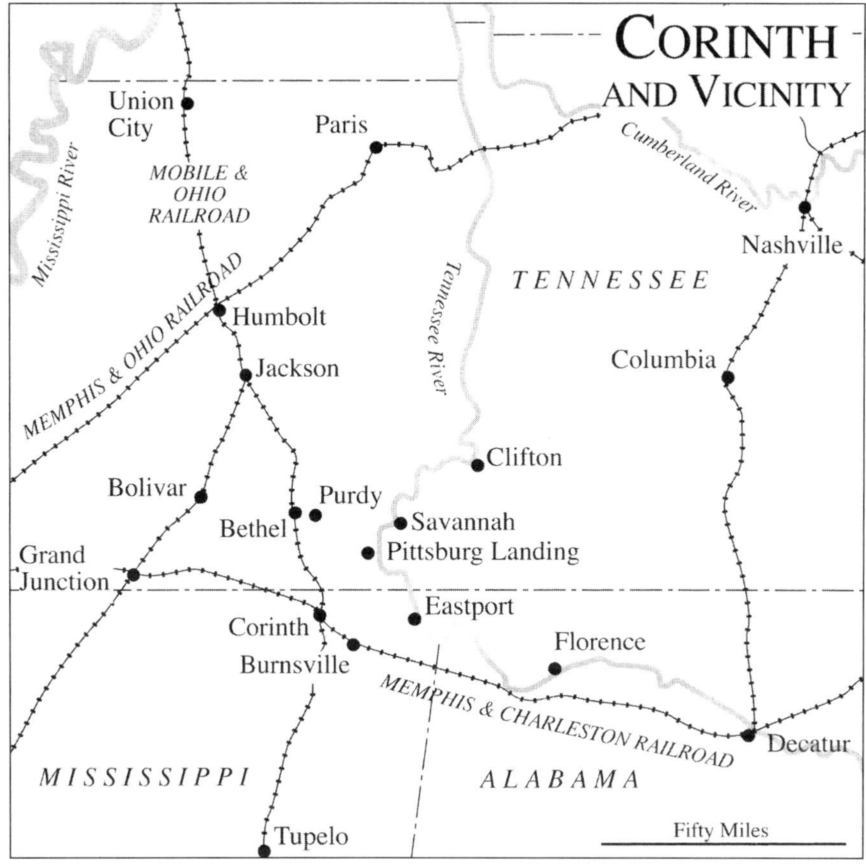

Courtesy of the Grady McWhiney Reseach Foundation.

believed that too many promising young officers had been killed at Shiloh or driven from the service by the rights granted Confederate soldiers to elect their own officers. To improve the quality of his officers, Bragg established promotion boards and stipulated that if elected officers were not accredited by those boards, then the "elective franchise would be considered exhausted."

Bragg, who wanted to place some of his highest ranking subordinates before such examining boards, reported in July 1862 that Hardee was the only "suitable" major general "now present." Such a remark by the army commander was a deliberate slap at Polk, whose military ability Bragg distrusted. "Among the junior brigadiers we have some excellent material," Bragg informed the government, "but it is comparatively useless, being overshadowed. Could the [War] Department by any wholesome exercise of power or policy relieve this army from a part of this dead-weight it would surely give confidence to the troops and add much to our efficiency."[7]

Major General Richard Taylor, who was the son of Bragg's old Mexican War commander and who visited the Army of the Mississippi while on his way from Virginia to Louisiana, heard Bragg call one of his senior generals "an old woman, utterly worthless." Taylor failed to name the "old woman," but apparently it was Polk.[8]

During his Kentucky Campaign in 1862, Bragg became even more disgusted with Polk. On September 28 Bragg left his army temporarily under Polk's command at Bardstown, Kentucky, while he rushed to the state capital at Frankfort to inaugurate a pro-Confederate governor and to consult with Major General Edmund Kirby Smith.

Bragg knew that leaving Polk in command of the army was a mistake, but he had no choice. The army commander had warned Richmond that Polk was incapable of holding the rank or exercising the responsibilities assigned to him. If the president refused to listen and insisted on retaining Polk, what could Bragg do? He must either use Polk in the position entitled by his rank and hope for the best or resign as commander of the Army of the Mississippi, which Bragg was unwilling to do.

When reports indicated that the Federals were advancing from Louisville toward Frankfort, Bragg decided that "Polk [must] move on the enemy & assail the Federals in flank & rear, whilst Smith [attacked] directly on his front." "The enemy is certainly advancing on Frankfort," Bragg informed Polk at 1:00 P.M. on October 2. "Put your whole available force in motion & strike him in flank, and rear. If we can combine our movements he is certainly lost."[9]

Bragg's plan of a Napoleonic concentration of forces on the battlefield was bold but based upon inadequate information. He had only vague reports on the exact location of the Federals and misunderstood Union Major General Don Carlos Buell's strategy. Union troops actually had advanced from Louisville on four roads over a sixty-mile front. "The plan of my movement," Buell explained after the war, "was to force the enemy's left back and prevent him from moving upon my left flank and rear."

But on October 3 Bragg was confident that the Federals were where he thought they were and that Polk, as ordered, was on his way to Frankfort to cooperate with Smith. "The impression strongly prevails that the great Battle will be fought in this vicinity," wrote a staff officer. "With Smith in front and our gallant army on the flank," Bragg boasted, "I see no hope for Buell."[10]

Actually, Buell was in no danger from Polk. The bishop, still at Bardstown, had no intention of marching to Frankfort or attacking Buell; indeed, Polk had sent neither aid nor orders to his subordinate Brigadier General Patrick Cleburne at Shelbyville, although Polk had known for two days that Federal forces threatened the Confederates there. Having done nothing to instruct or to reinforce Cleburne, Polk had the temerity to write Bragg, "It seems to me we are too much scattered."[11]

He was correct, but Polk was about to scatter even further by retreating toward Danville in defiance of Bragg's orders. Believing the Federals were converging on Bardstown, Polk asked and received the support of his ranking generals for his extraordinary decision. Although Major Generals Hardee and Benjamin F. Cheatham later refused to reveal what they had advised, Brigadier General Sterling A. M. Wood readily admitted counseling Polk to disregard Bragg's order and to retreat. Apparently, only Brigadier General J. Patton Anderson, the junior general present at the meeting, favored obeying orders. Anderson pointed out that if Polk failed to march to Frankfort and cooperate with Smith, the results might be disastrous for the Confederates.

Polk explained his reasoning on October 3 in a dispatch to Bragg. "The last twenty-four hours have developed a condition on my front and left flank which makes compliance with [your] order not only eminently inexpedient, but impracticable," the bishop announced. "I have called a council of wing and division commanders, and find that they indorse my views. I shall therefore pursue a different course, assured that when the facts are submitted to you[,] you will justify my decision." Later, Polk would claim that he disobeyed no order; Bragg, he said, directed him to move with all his "available" force, but none of his troops were available for a march to Frankfort. This, of course, was the argument of a bishop, not a general; if his command was available for a retreat, it also was available for an advance to Frankfort.[12]

Polk and some of his supporters insisted that Bragg's order was unclear, but Patton Anderson disagreed. "On the 3rd of October, 1862, I was present at General Polk's quarters in Bardstown," he informed Bragg.

> Your dispatch from Frankfort of date 1 P.M. Oct 2nd was read after an interchange of views in regard to our military condition, as junior officer present, I was called upon by General Polk to give my views as to what was best to be done. I hesitated to do so, whereupon General Polk enquired as to the cause of my reluctance to advise. . . . And I replied, that your order just read did not seem to admit of any other course than that of compliance, and that if any other alternative than that of obedience to the order was adopted, it might involve you and the forces with you near Frankfort in great embarrassment if not defeat—that in your dispatch you distinctly stated that Genl. Kirby Smith would attack the enemy then in your front, and that *we* must move upon him and "strike him in flank and rear."[13]

The decision of Polk to retreat demonstrated more than mere cautiousness; it evidenced a distrust of his commander's judgment totally unsuitable for one considered second in command. Older than Bragg and closer to the president, Polk had been a bishop too long to be a successful subordinate. "With all his ability, energy and zeal, General Polk, by education and habit,

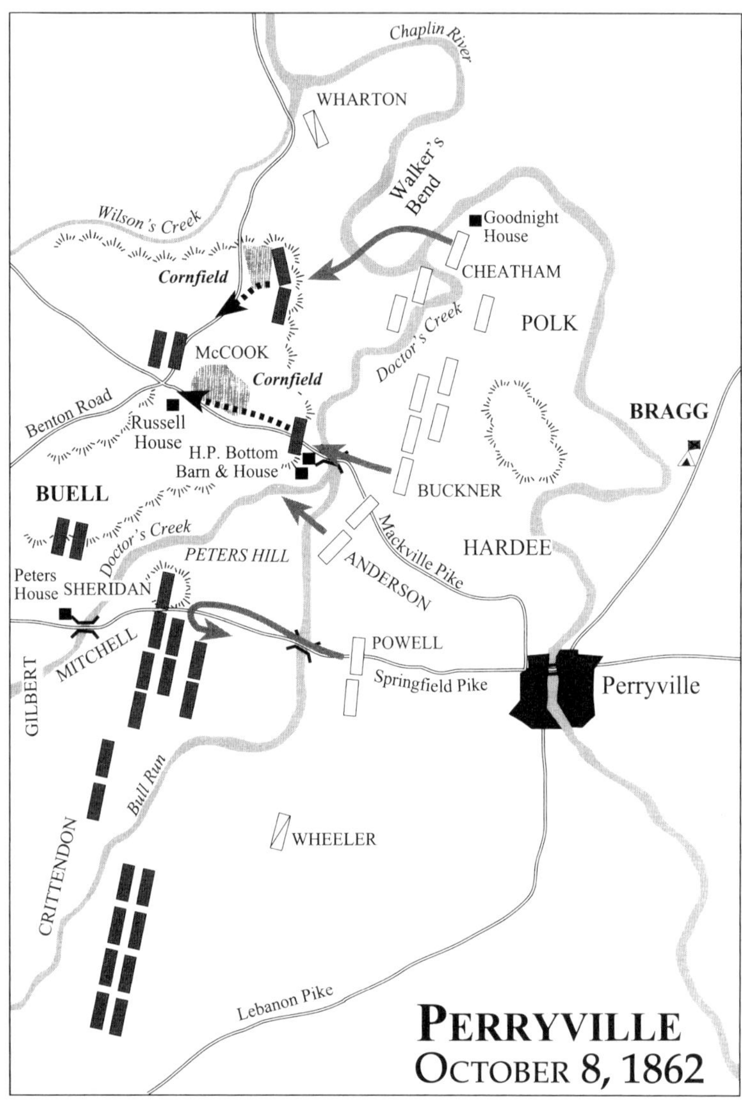

Courtesy of the Grady McWhiney Reseach Foundation.

is unfitted for executing the orders of others," Bragg informed President Davis. "He will convince himself his own views are better, and will follow them without reflecting on the consequences."[14]

During the Kentucky Campaign, Polk made yet another mistake by misarranging troops at the Battle of Perryville. Three roads fanned westward from the town: one went northwest, another due west, and the third southwest. Each crossed an almost dry stream called Doctor's Creek, which trickled northeast into Chaplin River about two miles above Perryville. Between the town and Doctor's Creek, Polk had placed his fifteen thousand Confederates in a concave line facing west. "This position was not so good," observed a staff officer. "Genl Polk's line was weak, his right if outflanked by the enemy would have cut us off from Genl Smith."

Bragg was furious when he rode over the field and saw what Polk had done. At once he ordered Cheatham's division to move up and protect the exposed Confederate flank. "The distance to be marched [by Cheatham's men] was about two miles," recalled an officer, "and the interval which elapsed until their arrival, was a momentous one." Enemy skirmishers had already occupied high ground on the Confederate right, but before additional Federals advanced Cheatham's men arrived and secured the Confederate flank.[15]

Late that afternoon, after a day of hard fighting, Polk believed that a Confederate unit was firing on Brigadier General St. John R. Liddell's brigade of Arkansans as they advanced in the twilight. "Dear me, this is very sad, and must be stopped," Polk muttered as he cantered over to ask the erring commander why he was shooting his friends. "I don't think there can be any mistake about it," replied the surprised officer. "I am sure they are the enemy." "Enemy," exclaimed the bishop, "why I have only just left them myself—cease firing, sir; what is your name, sir?"

After giving his name and *Union* regimental affiliation, the officer asked, "and pray, sir, who are you?" Polk, finally realizing his mistake but seeing nothing else to do, brazenly shook his fist in the Federal officer's face, saying: "I'll soon show you who I am. Cease firing at once."

Without waiting for an answer, Polk turned his horse and rode rapidly back to join Liddell. "They are enemies; fire upon them," the bishop shouted as he reached the Confederate line, "every mother's son of them are Yankees. I saw the colonel commanding the brigade." Liddell's men quickly fired deadly volleys into the startled Federals, but it was now dark and the battle was about over. Polk and Liddell rode forward to see the ground covered with dead and dying Yankees.[16]

General Liddell considered the bishop "a good man, lofty in sentiment, gallant and enthusiastic in the cause," who "meant well and was a true patriot" but unfortunately saw things "with other eyes." Liddell even believed that Polk might have become a fair soldier if he had abandoned his "theatrical manner" and overcome his ignorance of strategy and tactics. A staff surgeon was less generous. "Genl Polk threatened wonders," wrote Dr. David W. Yandell. "He was positively ferocious. But he can't be relied on. . . . He is great at talk, but is monstrous uncertain. I saw enough of the old grey beard at Shiloh & Perryville to cause me to place no great confidence in him. He will prevaricate. He did say he was going to do this and going to do that, but the old man forgets; [unless he is] transferred to [noncombat] house duties, we will all go to the Devil out here."[17]

Following the Kentucky Campaign, Polk rushed to Richmond to denounce Bragg to the president and to anyone else who would listen. The bishop-general not only criticized Bragg's leadership but also boasted that if he "had been in chief command, the strategy of the [Kentucky] campaign would have been very different—and the practical operations, as to tactics would have been very different [too]."

The strategy that Polk claimed he would have adopted revealed both his shortcomings as a general as well as how quickly he had forgotten what actually took place. The bishop said that he had favored "a more rapid and energetic campaign." Instead of the long detour of four hundred miles around by Chattanooga, he would have adopted a more direct and vigorous plan of falling first on one corps of the enemy and then on another, thus destroying the Federals in detail. Instead of marching on a line nearly parallel with that of Buell, Polk favored a cross march, falling on the Federal flank and crushing it before moving on to capture Louisville. And instead of fighting the Battle of Perryville with divided forces, Polk said that he had favored concentrating the Confederate and "then falling on one corps of the enemy, and after crushing that, crushing the other corps—all of which," he insisted, "could have been easily done."[18]

Polk's generalship got no better after the Kentucky Campaign; indeed, his actions there may have been the highlight of his career. At Murfreesboro Polk made the mistake of sending his troops into action piecemeal and, at a critical point in the battle, announcing in a meeting with his commander that Bragg had only "three brigades that are at all reliable" and should immediately retreat. Weeks later the bishop admitted that he had meant to say that only three divisions, not just three brigades, were reliable. During the Confederate withdrawal from Tullahoma in the summer of 1863, Polk demonstrated how he could frustrate his commanding officer without actually disobeying orders. When Bragg ordered him to move his corps to Chattanooga "without delay," the bishop directed his division commanders not to move until they could make their "marches as expeditious as you can with convenience." At the Battle of Chickamauga, Polk's actions and inactions were even more exasperating. At one point he failed to attack when ordered; at another, he failed to locate and notify Lieutenant General Daniel Harvey Hill that Hill was expected to lead an attack. The person responsible for this poorly organized assault, which began nearly four hours later than Bragg had ordered, was the bishop, not Hill. Muddling and lack of concern by Polk had again disrupted Bragg's plans. As historian Judith Lee Hallock notes, Polk's generalship at Chickamauga was "his poorest in a long list of sorry performances."[19]

The bishop's military ability failed to improve after Joseph E. Johnston took command of the Army of Tennessee. Polk apparently got along better with Johnston than he had with Bragg, but their relationship lacked cordiality. Friction between them never surfaced, but a certain lack of confidence in each other prevailed throughout their association. Polk discussed what he considered Johnston's deficiencies as a general with Hardee; in fact, Hardee wrote his friend Polk, "Johnston is wanting in all those particulars in which you feared he was deficient." Perhaps the bishop's major shortcoming was his failure to attempt to smooth the strained relationship, to ease some of the distrust and uncertainty that existed between his army commander and his close friend, the Confederacy's chief executive.[20]

Polk made what some consider his greatest military contribution on June 14, 1864, when, while carelessly inspecting the enemy line from the summit of Pine Mountain, Georgia, the bishop moved a few feet nearer the crest for a final view. As he stood, arms folded, "a cannon-shot crashed through his chest."[21]

His friends and some soldiers considered the death of Polk a great blow to the Confederacy. "He was every inch a gentleman, . . . simple and innocent, yet dignified and imposing," declared a staff officer. "Every private soldier loved him," remembered a Tennessean. "When I saw him here dead, I felt that I had lost a friend whom I had ever loved and respected, and that the South had lost one of her best and greatest generals."[22]

These were pleasant and respectful words, to be sure, but not too helpful in evaluating bishop-general Leonidas Polk and his contribution to the Confederacy. Other contemporaries offered more realistic views. Polk "is a slow coach—one who cannot be depended on," admitted Colonel Taylor Beattie. Major General Samuel G. French wrote kindly of the bishop as "a gentleman and a high Church dignitary" but remembered him as a soldier who was "more theoretical than practical." Bragg described Polk as not just irresponsible and insubordinate but also "luxurious in his habits, rises late, moves slowly, and always conceives his own plans the best." Bragg concluded that he had been injured by Polk "on every field where I have been associated with him."[23]

It may be time for historians of the Confederacy to admit that Bragg was not the Confederacy's worst general. Leonidas Polk may not be able to claim that title either, but he certainly was a bad general. Ignorant of military matters, quarrelsome, at times careless, lazy, insubordinate, and conspiratorial, he was dangerous not only because he knew so little about the art of warfare but also because he delighted in using his religious influence as well as his friendship with high government officials, especially the president, to promote himself and to undermine his enemies. Untrustworthy, he hurt every command and every commander with whom he served. The appointment of Polk to high military rank was one of Jefferson Davis's most serious mistakes.

Notes

"A Bishop as General" was originally published in Grady McWhiney's *Confederate Crackers and Cavaliers* (Abilene, TX, 2002), 209–21, and is reprinted with permission of the McWhiney Foundation Press.

1. Thomas Lawrence Connelly, *Army of the Heartland: The Army of Tennessee, 1861–1862* (Baton Rouge, LA, 1967), 46. For more on Polk's "recopied" correspondence, see Steven E. Woodworth's *No Band of Brothers: Problems in the Rebel High Command* (Columbia, MO, 1999), 12–18.
2. John Fulton, "The Church in the Confederate States," in *The History of the American Episcopal Church, 1587–1883,* 2 vols., ed. William Stevens Perry (Boston, 1885), 2:2563; Arthur James Lyon Fremantle, *The Fremantle Diary; Being the Journal of Lieutenant Colonel Arthur James Lyon Fremantle . . . ,* ed. Walter Lord (New York, 1954), 111.
3. Connelly, *Army of the Heartland,* 47; Steven E. Woodworth, *Jefferson Davis and His Generals: The Failure of Confederate Command in the West* (Lawrence, KS, 1990); Judith Lee Hallock, *Braxton Bragg and Confederate Defeat,* vol. 2 (Tuscaloosa , 1991).
4. Entry of Feb. 19, 1862, Thomas Bragg Diary, Thomas Bragg Papers, 1861–1862, Southern Historical Collection, Wilson Library, Univ. of North Carolina at Chapel Hill (hereinafter cited as Bragg Diary); Bragg to Elise, Mar. 20, 1862, Braxton Bragg Papers, 1847–1869, Rare Book, Manuscript, and Special Collections Library, Duke Univ., Durham, NC.
5. Bragg to Elise, Mar. 25, 1862, Braxton Bragg Papers, William K. Bixby Collection, Missouri Historical Society, St. Louis.
6. St. John Richardson Liddell, *Liddell's Record: St. John Richardson Liddell, Brigadier General . . . ,* ed. Nathaniel Cheairs Hughes Jr. (1866; reprint, Dayton, OH, 1985), 85.
7. U.S. War Department, *The War of the Rebellion: A Compilation of the Official Records of the Union and Confederate Armies,* 128 vols. (Washington, DC, 1880–1901), series 1, vol. 17, pt. 2:647–48. Hereinafter cited as *OR.* All references are to series 1 unless otherwise indicated.
8. Richard Taylor, *Destruction and Reconstruction: Personal Experiences of the Late War,* ed. Richard B. Harwell (New York, 1955), 117.
9. Colonel W. G. M. Davis to Bragg, Oct. 2, 1862, telegram, Braxton Bragg Papers, William P. Palmer Collection, Western Reserve Historical Society, Cleveland, Ohio (hereinafter cited as Bragg Papers, Palmer Collection); George W. Brent Diary, Oct. 3, 1862, Bragg Papers, Palmer Collection (hereinafter cited as Brent Diary); *OR,* vol. 16, pt. 2:896–97.
10. *OR,* vol. 16, pt. 2:896–97; Robert U. Johnson and Clarence C. Buel, eds., *Battles and Leaders of the Civil War,* 4 vols. (New York, 1884–88), 3:47; J. Stoddard Johnston, "Bragg's Campaign in Kentucky: From Munfordville to Frankfort," J. Stoddard

Johnston Military Papers, Filson Club, Louisville, KY; Brent Diary, Oct. 3, 1862, and George W. Brent's Memoranda, Oct. 3, 1862, both in Bragg Papers, Palmer Collection (hereinafter cited as Brent's Memoranda).

11. Cleburne to Polk, Oct. 1, 1862, Polk to Bragg, 10 A.M., Oct. 2, 1862, and Hardee to Bragg, 10 A.M., Oct. 2, 1862, all in Bragg Papers, Palmer Collection.
12. *OR*, vol. 16, pt 1:1094–95, 1101, 1107; Wood to Bragg, Apr. 13, 1863, and Anderson to Bragg, Apr. 15, 1863, Bragg Papers, Palmer Collection; William M. Polk, *Leonidas Polk: Bishop and General*, 2 vols. (New York, 1915), 2:140.
13. Anderson to Bragg, Apr. 15, 1863, Bragg Papers, Palmer Collection.
14. *OR*, vol. 16, pt. 2:566; *OR*, vol. 16, pt. 1:1091; Bragg to Davis, May 22, 1863 (copy), Bragg Papers, Palmer Collection.
15. Brent's Memoranda, Oct. 8, 1862, Bragg Papers, Palmer Collection; J. Stoddard Johnston, "Battle of Perryville," J. Stoddard Johnston Military Papers, Filson Club, Louisville, KY.
16. *OR*, vol. 16, pt. 1:1157–60; Polk, *Leonidas Polk* 2:161–62; Johnston, "Battle of Perryville."
17. David W. Yandell to William Preston Johnston, Nov. 8, 1862, Albert Sidney and William Preston Johnston Papers, Mrs. Mason Barret Collection, Manuscripts Division, Howard-Tilton Memorial Library, Tulane Univ., New Orleans; Liddell, *Liddell's Record*, 100–101, 119.
18. Thomas Ruffin, *The Papers of Thomas Ruffin*, 4 vols., ed. J. G. de Roulhac Hamilton (Raleigh, NC, 1918–1920), 3:270–71.
19. *OR*, vol. 20, pt. 1:700–701; *OR*, vol. 23, pt. 2:900–901; Judith Lee Hallock, "Braxton Bragg and Confederate Defeat," vol. 2 (Ph.D. diss., Stony Brook Univ., New York, Dec. 1989), 117.
20. Hardee to Polk, July 27, 1863, in Leonidas Polk Papers, Univ. of the South, Sewanee, TN.
21. Joseph H. Parks, *General Leonidas Polk, C.S.A.: The Fighting Bishop* (Baton Rouge, LA, 1962), 382.
22. Statement by Henry Watterson, quoted in ibid., 374–75; Sam R. Watkins, *"Co. Aytch," Maury Grays, First Tennessee Regiment of a Side Show of the Big Show* (Jackson, TN, 1952), 154.
23. Taylor Beatty Diary, Oct. 20, 1863, Taylor Beatty Papers, 1780-1917, Southern Historical Collection, Univ. of North Carolina at Chapel Hill; Samuel G. French, *Two Wars: An Autobiography of Gen. Samuel G. French . . .* (1901; reprint, Huntington, WV, 1999), 202; Bragg to James A. Seddon, Sept. 25, 1863, James A. Seddon Papers, Duke Univ., Durham, NC.

General Albert Sidney Johnston. Library of Congress.

Albert Sidney Johnston and the Defense of the Confederate West

Charles P. Roland

According to family tradition, Confederate President Jefferson Davis lay ill in bed one day in late August 1861, when he heard familiar footsteps on the hallway below. "That is Sidney Johnston's step," he said. "Bring him up." A few moments later Johnston was ushered into the president's room.[1]

This was doubtless an emotional occasion for both men. They had been friends since their cadet days together at West Point, where Johnston, who graduated in 1826, was two years ahead of Davis. Johnston held the coveted position of adjutant of the corps, and Davis had admired him deeply. Years later they had served together gallantly in the Battle of Monterrey in the Mexican War. In the 1850s, when Davis was United States secretary of war, he had played a key role in appointing Johnston colonel and commanding officer of the newly created 2nd Cavalry Regiment, an elite organization that included in its roster such subordinates as Lieutenant Colonel Robert E. Lee, second in command, and Majors William J. Hardee and George H. Thomas, along with others who achieved distinction in the Civil War.

Later in the decade, the war department selected Johnston to command an expedition to quell a threatening rebellion by the Mormons in Utah Territory. He achieved the rank of brevet brigadier general in this operation. On the eve of the Civil War, Johnston was assigned to command the Pacific Department, with headquarters in San Francisco. He resigned his commission when his adopted state, Texas, seceded, and he made a Herculean journey across the southwestern region of the nation to join the Confederacy.

Davis was overjoyed with Johnston's appearance in Richmond, not only because the two were dear friends, but also because Johnston was perhaps the most distinguished field officer of the United States Army at that time. A description of him published in the *Harper's Weekly* magazine during the Utah operation captured him precisely:

> Johnston is now in the matured vigor of manhood. He is about six feet in height, strongly and powerfully formed, with a grave, dignified, and commanding presence. His features are strongly marked, showing his Scottish lineage, and denote great resolution and composure of character. His complexion, naturally fair, is, from exposure, a deep brown. His habits are abstemious and temperate, and no excess has impaired his powerful constitution. His mind is clear, strong, and well cultivated. His manner is courteous, but rather grave and silent. He has many devoted friends, but they have been won and secured rather by the native dignity and nobility of his character than by his powers of address. He is a man of strong will and ardent temper, but his whole bearing testifies to the self-control he has acquired.[2]

At the time of his promotion during the Utah expedition, Lieutenant General Winfield Scott, commanding general of the army, wrote to President James Buchanan, "[Johnston] is more than a good officer. He is a God send to the country through the army." Lieutenant General Ulysses S. Grant wrote in his memoirs that the Union officers at the beginning of the Civil War considered Johnston to be the most formidable opponent they would meet in the conflict.[3]

At the moment, Davis was urgently in need of someone to command the far-flung Western Theater of the Confederacy. Johnston seemed to be the perfect choice; he was a native of Kentucky and had spent most of his career in Texas and other parts of the West. On September 10, Davis issued the order assigning Johnston to the command of Confederate Department No. 2, a vast area spreading from the Appalachian Mountains on the east to, and including, Indian Territory [now the state of Oklahoma] on the west. He was assigned the rank of full general, the highest field officer of the Confederacy. He proceeded at once to Nashville and took over the active command of his department.

He faced a grave decision—what to do about the state of Kentucky. The Confederacy considered Kentucky to be a Confederate state. The Kentucky government had declared the state neutral, and the Lincoln administration had sent in no military formations, though Union troops were being recruited, armed, and trained in the state. Ten days prior to Johnston's arrival in Nashville, Confederate Major General Leonidas Polk occupied Columbus, Kentucky, a key strategic point located on the Mississippi River at the northern terminus of the Mobile & Ohio Railroad. Should Johnston order Polk to withdraw into Tennessee, along with a small contingent of Confederate troops under Brigadier General Felix K. Zollicoffer, who were located at Mill Springs, guarding the Cumberland Gap in eastern Kentucky, or should Johnston occupy and defend Kentucky for the Confederacy?

Urged by Governor Isham Harris of Tennessee, Johnston took the bolder course and ordered a contingent of 4,000 troops under the command of Brigadier General Simon Bolivar Buckner from Nashville to Bowling Green, Kentucky. Johnston pointed out to the Richmond authorities that otherwise, the Confederacy would lose Kentucky and invite an immediate invasion of Tennessee. But the move into Kentucky was motivated primarily by political considerations, and it was based on the hope and belief that Kentuckians would rally in great numbers to the Confederate cause. If they failed to do so, Johnston would find his situation precarious.

In some ways, Johnston's strategic views were circumscribed by those of Davis, whose policy was that of territorial defense—voluntarily yielding no part of the Confederacy to the enemy. This dictated a cordon or line form of defense and prohibited an early concentration of Confederate forces. Consequently, Johnston's small army (initially, some 30,000) was spread across a Kentucky front of approximately 500 miles with wide gaps along the line. It was anchored on the west by Polk's troops at Columbus on the Mississippi, a position that Polk at once began to fortify in order to deny the use of the river to the Federals. The eastern flank was held by Zollicoffer's force at Mill Springs on the Cumberland. The center lay at Bowling Green, with Major General William J. Hardee in immediate command. Bowling Green was also the site of Johnston's headquarters, as well as the Confederate government of Kentucky.

Painfully aware of the weakness of his line, Johnston worked diligently to strengthen it. His hope of receiving thousands of volunteers from Kentucky soon vanished. He appealed for troops to the governors of the other states within his department, but their response was disappointing. He then sought reinforcements from the Confederate authorities, suggesting that these troops be sent from other parts of the Confederacy. In January 1862, he wrote to the Confederate adjutant general: "All the resources of the Confederacy are now needed for the defence [*sic*] of Tennessee." His efforts to strengthen his line were all in vain, and he faced a showdown against a Union force of approximately twice his own.[4]

The most vulnerable place in this attenuated line lay where the Tennessee and Cumberland rivers ran through it in their course to the Ohio. Small forts guarded these points: Fort Henry on the Tennessee and Fort Donelson on the Cumberland. The two were only eleven miles apart. Johnston appreciated the weakness of these forts, but he was unable to remedy it; perhaps because of his preoccupation with the overall strategic situation, he did not pay sufficient heed to the vulnerability of the forts.

The opening attacks on the Confederate line came at its extremities. In early November, Brigadier General Ulysses S. Grant led a force from his base at Cairo, Illinois, and struck a small Confederate outpost at Belmont, Missouri, across the Mississippi from Columbus. Polk's troops drove off the attackers by ferrying reinforcements across the river. In mid-January 1862, Johnston's eastern force, now commanded by Major General George B. Crittenden, attacked Union troops under Brigadier General George H, Thomas north of the Cumberland River in the Battle of Mill Springs (also known as the Battle of Fishing Creek or the Battle of Logan's Cross Roads). The Confederates were defeated and driven back across the river.

These actions were preliminary to a far more serious operation, an attack by Grant in the western segment of the Confederate front. It came at the points where the Tennessee and Cumberland rivers penetrated the Confederate line. On February 6, a joint army-navy force of 17,000 troops under Grant and a flotilla of seven gunboats commanded by Flag Officer Andrew H. Foote moved against Fort Henry. The gunboats alone quickly overpowered the fort's batteries and the commander, Brigadier General Lloyd Tilghman, surrendered the position. Grant and Foote now prepared to take Fort Donelson.

The loss of Fort Henry opened an irreparable breach in Johnston's line. The Tennessee River was navigable by Union gunboats and transports all the way to the Muscle Shoals in northern Alabama. Two-thirds of Johnston's army in Kentucky was subject to being trapped within the great arc of the river.

On February 7, Johnston assembled his ranking subordinates to adopt a strategy for meeting this desperate situation. The group included Hardee and General P. G. T. Beauregard, who had been sent to Kentucky from Virginia to serve as Johnston's second-in-command. Beauregard was the lowest ranking of the five full generals of the Confederacy. At the Bowling Green meeting, Johnston proposed a momentous plan to abandon his advanced line and withdraw to northern Mississippi outside the loop of the Tennessee River. The wings of his army were temporarily to act as independent forces until they could be united at some point there. He would command the eastern wing; Beauregard, the western wing (primarily Polk's force).

This plan assumed that Fort Donelson, on the Cumberland, was untenable, and Nashville, the capital of Tennessee and an important industrial and commercial city, would temporarily be yielded to the enemy. After the Civil War, Beauregard claimed to have opposed the entire plan and to have urged Johnston to hasten by rail and boat with his Bowling Green force to confront Grant at Fort Donelson, but there is no convincing evidence that Beauregard offered such an alternative to Johnston's strategy. Both Johnston and Hardee wrote at the time that the three generals had agreed unanimously that the Kentucky line must be abandoned.

Should Johnston, nevertheless, have fashioned and adopted such a plan as Beauregard later suggested? Possibly it would have worked. But in addition to Grant's troops at Fort Donelson, Johnston's Bowling Green troops also faced another army of more than twice their strength. It was commanded by

Major General Don Carlos Buell and was advancing against Johnston from northern Kentucky. Johnston, with good reason, feared that fighting Grant and Buell together in Kentucky would expose his army to destruction or capture there. Grant wrote in his memoirs that Johnston should have taken the risk; that the outcome could not have been worse for the Confederacy if Johnston and all his army had been captured. This is a fatuous argument. It ignores the vital role played by Johnston and, after his death, by the army throughout the remainder of the war in the West.

The wisdom of Johnston's decision is debatable. That he exhibited immense moral courage in making it is beyond cavil, in my judgment. He well understood the seriousness of giving up, even temporarily, two Confederate states. He was also keenly aware of the public furor that would be aroused by this action. An outcry of indignation greeted his retreat; a delegation from the Tennessee Legislature appealed to Davis, demanding that Johnston be removed. But Davis rejected the demand and said, "If [Johnston] is not a general, we had better give up the war, for we have no general." Despite the uproar, Johnston took the action that he considered imperative in order to save his army.[5]

But Johnston made a monumental mistake after deciding that Fort Donelson could not be held. He sent some eleven or twelve thousand additional troops into the fort. When, on February 15, Grant made his attack on it, the ranking generals there, Brigadier General John Floyd and Brigadier General Gideon Pillow, in one of the most flagrantly mishandled operations of the entire war, surrendered the fort and its troops, including those who had just arrived. Only with the greatest difficulty did Johnston gets his Bowling Green army across the Cumberland River at Nashville and avoid being trapped north of that stream as well as north of the Tennessee.

Now began a race for Johnston and Beauregard to get their troops into Mississippi ahead of the Federals. Johnston first had to decide where the wings of his army were to be brought together. A glance at the map indicated that the easiest place was the town of Corinth, in northeastern Mississippi. It lay where the two major railroads of the western Confederacy—the Mobile & Ohio, running north and south, and the Memphis & Charleston, running east and west—intersected. But neither Johnston nor Beauregard mentioned Corinth in their early conversations and correspondence. They probably considered it too close to the Tennessee River, only twenty miles from Pittsburg Landing, and thus likely to be captured before they could reach it.

But for various reasons, the Federals were slow in moving a force up the Tennessee. Not until almost a month after the fall of Fort Donelson did they establish a camp at Pittsburg Landing. On March 13, Major General Charles F. Smith began debarking an army of five divisions there. Soon a sixth division arrived, a total of approximately 40,000 men. Following Smith's death a few days later from an infection, Grant took command of the force.

Meanwhile, both Johnston and Beauregard decided independently to unite their wings of the Confederate army at Corinth. This was a relatively easy task for Beauregard; he simply transported them on the Mobile & Ohio Railroad and posted them at key points along the way. Johnston's task was extremely difficult and hazardous. He ran the risk of being trapped above the Tennessee River. Nevertheless, after a grueling march by foot from Murfreesboro, southeast of Nashville, he crossed them over the river on the Memphis & Charleston bridge at Decatur, Alabama, then shuttled them by rail to Corinth. By late March he had the two wings of his army there, plus some 10,000 additional troops under Major General Braxton Bragg brought by rail from Pensacola and Mobile, and another 5,000 under Brigadier General Daniel Ruggles brought by rail and river from Louisiana.

Johnston now planned a mighty blow at Grant's army, which was still encamped at Pittsburg Landing. The night of April 3, Johnston received word that Buell's Union army was moving rapidly to join Grant. Johnston immediately issued orders for his force to march from Corinth and deliver the attack before Buell's arrival. He wired Jefferson Davis of his intention and indicated that his assault would be made with his three main corps, those of Hardee, Polk, and Bragg abreast, with Bragg's corps, the largest of the three, on the right, followed by a reserve force under Brigadier General John C. Breckinridge, who had replaced Crittenden at Corinth.

But Johnston then delegated to Beauregard the duty of issuing the march and attack order, and Beauregard planned a different formation from the one Johnston had indicated to Davis. Beauregard's order called for an attack in successive waves by various corps, with the main effort on the right to shear the Federals from their base at the Landing. This arrangement would have a profound effect on the outcome of the battle. Apparently, Johnston did not learn of the change of formation until the column was on the march and at a time he considered too late to make a change.

After innumerable delays and obstacles, by late afternoon of April 5, the Confederates were deployed for the attack. Johnston suddenly found himself facing the moment of truth, a time the great Prussian military theorist Carl von Clausewitz stated usually precedes a great engagement: the breakdown of the will of the army subordinates; their loss of nerve for the undertaking. At this point, said Clausewitz, only the will and character of the commander keeps the operations alive. Beauregard suddenly began to argue that all hope of taking the Federals by surprise was lost; that they would "be entrenched to the eyes," and that the assault would be bloodily repulsed. Most, if not all, of the corps commanders agreed. Johnston listened courteously, then said quietly that he still expected to find the Federals unprepared. Then, "We shall attack at daylight tomorrow." This was the supreme command decision of the entire campaign.[6]

The Confederates struck at dawn on April 6 in the vicinity of a small, country Methodist church named "Shiloh," which is the name they gave the battle. They found their enemies unprepared; the surprise was complete; Johnston's judgment was fully vindicated.

How was he able to sense the situation so accurately? The renowned nineteenth-century British military theorist Colonel G. F. R. Henderson answered this question generically in discussing the attributes of great generals: "If in appearance great risks [are] run, it [is] with the full knowledge that the enemy's character or his apprehensions would prevent him from taking those simple precautions by which the critics point out that the whole enterprise might easily [be] ruined. [Great generals] penetrate . . . their adversary's brain!" Grant was convinced that the Confederates lacked the will and purpose to launch a counterattack. Johnston seems to have penetrated his brain on this point.[7]

But the Federals reacted quickly and soon established a line which they defended stubbornly. The Confederate attack formation, Beauregard's formation, gave away much of the advantage of the Confederate surprise. The initial assault was made by Hardee's corps only, about one-third of the Confederate strength, excluding the reserves. Not until more than an hour later was Bragg's corps, the largest of the three attacking corps, committed to action. The bulk of Polk's corps was committed still later. If the full Confederate strength had made the opening assault, with Bragg's heavy corps on the right, as Johnston had originally intended, the attack probably would have overwhelmed the Federals before they could have recovered from their surprise and formed a defense line.

By midmorning, all troops on both sides, except for the Confederate reserve, were engaged, and the entire front of about three miles was ablaze. The fighting was extraordinarily fierce; both Grant and his top division commander, Major General William Tecumseh Sherman, said years later that they had seen no more furious action during the war. Thousands of Union troops abandoned the fight and fled to the river; great numbers of Confederates broke ranks to engage in looting the deserted, well-stocked Union camps. On one occasion, Johnston personally reprimanded a young officer who was doing so, then, to soften the rebuke, he picked up a tin cup and said that it would be his part of the loot.

Johnston left Beauregard to oversee the rear while he (Johnston) commanded at the front. Riding across the line from left to right, he communicated directly with corps, division, and brigade commanders and made adjustments in his formation. By late morning, he was aware that, although the general attack seemed to be going well, his thrust on the right had been stopped by powerful resistance along a wagon road in the woods and the edge of a farm peach orchard. He immediately ordered Breckinridge's reserve brigades into the attack in this sector.

Still the Confederates were unable to break the Union line. At about two o'clock, Johnston personally rode along the front of one of Breckinridge's reluctant regiments, tapping the points of their bayonets with the tin cup that he had said was his part of the loot. Then, wheeling his horse, he cried, "I will lead you," which he proceeded to do, at least for a short distance. The regiment then moved forward and took its objective.[8]

Johnston was elated with his success. He kicked up one foot to show an aide, Governor Isham Harris of Tennessee, where a bullet had ripped up the sole of his boot, and he said gaily, "Governor, they came very near putting me *hors de combat* in that charge." Then he sent Harris to bear instructions to one of the brigade commanders.[9]

When Harris returned a few minutes later, he found Johnston reeling unsteadily in the saddle. Harris inquired anxiously whether Johnston was wounded. Johnston replied slowly, "Yes, and I fear seriously." Harris steadied him on his horse, led him a few yards into a wooded ravine where they were sheltered from the shot and shell, and laid him on the ground. Johnston's staff gathered around and looked frantically for a wound on his upper body. Meanwhile, he bled to death through a rent in the main artery of his right leg. By two-thirty, he was dead. His personal physician, who doubtless could have saved his life, was attending wounded Confederates and Federals, ordered to do so by Johnston. Ironically, this humane move cost him his own life.[10]

Beauregard now assumed command and ordered the attack to be continued. The fighting did continue; the Union troops in the Hornet's Nest were surrounded and forced to surrender at about six o'clock; the Confederate line pressed forward in an effort to consummate a decisive victory. But suddenly Beauregard ordered the attack halted. In his report of the battle, he explained that his troops were exhausted and that he had received news that Buell's army would not reach Pittsburg Landing in time to save Grant's army. Beauregard's plan was to rest his troops and complete the victory the next morning.

This proved to be a fundamental, and near fatal, mistake. Buell's troops reinforced Grant's during the night, and the following day (April 7) the combined armies counterattacked the Confederates and obliged Beauregard to disengage his army and retreat to Corinth. The bloody Battle of Shiloh (Pittsburg Landing, to the Federals) thus ended in a significant tactical and strategic Union victory, leaving a powerful invading army poised to drive down the Mississippi Valley and split the Confederacy in half.

Johnston's death at the most critical point of the battle has left two of the most enduring unanswered questions of the Civil War. First, would he have won a decisive victory at Shiloh if he had not been killed? The trend among historians today is to say that he would not have done so. But this ignores the testimony of certain knowledgeable participants to the effect that a definite lull occurred in the attack, especially on the right, following his death—a lull that enabled the Union commanders to establish a final defense line around Pittsburg Landing.

Confederate General Bragg wrote in his battle report that the delay caused by Johnston's death prevented a complete Confederate victory. Union Brigadier General Stephen A. Hurlbut, who commanded the left flank division in the Hornet's Nest, wrote in his report that at about three o'clock, because his position was no longer under attack, he was able to withdraw his troops from the line in order to place them in the final defensive position near Pittsburg Landing.

The most telling affirmation concerning the result of Johnston's death came from Union General Sherman. He said in his memoirs: "The rebel army, commanded by General Albert Sidney Johnston . . . beyond all question fought skillfully from early morning till about 2 P.M., when their commander-in-chief was killed. . . . There was then a perceptible lull for a couple of hours, when the attack was renewed, but with much less vehemence, and continued up to dark." The break in the fighting occurred at the very point of culmination in the battle. It is a reasonable assumption that without it, the Confederates might have consummated a victory.[11]

Critics of Johnston find fault with the command role he took in the battle. They say he should have remained at the rear, directing the attack through his aides and couriers; that he ceased to be an army commander and became a mere regimental commander; and that he was recklessly careless in exposing himself in such a way as to be killed. Under ordinary circumstances, these criticisms may be valid.

But Johnston was not operating under ordinary circumstances. He was keenly aware that a commander's ability to inspire his troops to superhuman effort in combat is as important as the orders he gives on the field; he was keenly aware that his troops at Shiloh were virtually untrained civilians and that his appearance at the front was critical to their morale and determination. He was also cognizant that victory hung at a delicate balance by early afternoon, and that his personal conduct might tip the scale.

Military history is filled with examples of great generals who have performed acts similar to Johnston's at Shiloh and have exposed themselves to injury or death in doing so. Alexander and Caesar did so in wars of antiquity; Napoleon and Wellington did so in the modern era. Wellington's chief of cavalry was killed by a bursting shell while standing immediately by Wellington's side on the line. It could just as easily have been Wellington. Rommel and Patton conducted themselves in a similar fashion in World War II. Later, a British commission studying the issue of command effectiveness reached the decision that Rommel was the most effective field commander of the conflict. This was so, they said, because he commanded at the front where he could survey the scene for himself, make judgments on first-hand observation, and convey his orders and instructions to his subordinates eye to eye, with no possibility of misunderstanding. His "electric presence" at the front added a weapon to the German attack, they concluded. Every account of the Battle of Shiloh indicates that Johnston's electric presence at the front added a potent weapon to the Confederate attack.[12]

A final question remains: What would have been Johnston's role in the war if he had survived the Battle of Shiloh? My evaluation is that he would have been an incalculable asset to the Confederacy. Whereas such generals as Grant, Sherman, Lee, and Beauregard grew immensely with experience in the war, Johnston was unable to do so because he died in the first truly great encounter of the struggle. But he demonstrated in it that he was a man of towering character, unshakable will, and dauntless spirit; that he was capable of anticipating and outthinking his adversaries; and that he was a troop leader of extraordinary charismatic qualities in battle. Moreover, he enjoyed the complete confidence of his commander-in-chief, Jefferson Davis. Of all Confederate generals in the West, Albert Sidney Johnston alone seems to have held the potential of a Lee.

Notes

"Albert Sidney Johnston and the Defense of the Confederate West" appeared in Charles P. Roland's *History Teaches Us to Hope: Reflections on the Civil War and Southern History*, ed. John David Smith, 163–74, (Lexington, KY, 2007), and is reprinted with permission of the University Press of Kentucky. Most of the material in this essay earlier appeared in Charles P. Roland, *Albert Sidney Johnston: Soldier of Three Republics* (Austin, TX, 1964).

1. Jefferson Davis, *Jefferson Davis, Constitutionalist: His Letters, Papers and Speeches*, 10 vols., ed. Dunbar Rowland (Jackson, MS, 1923), 8:232.
2. *Harper's Weekly*, Jan. 30, 1858.
3. Winfield Scott to William Preston, Feb. 11, 1858, Wyckliffe-Preston Papers, Box 51, Department of Special Collections, Univ. of Kentucky Library, Lexington; Ulysses S. Grant, *Personal Memoirs of U. S. Grant*, ed. E. B. Long (1885; reprint, New York, 1952), 187.
4. Johnston to Samuel Cooper, Jan. 22, 1862, "Headquarters Book of Albert Sidney Johnston" (copy), Louisiana Historical Association Collection, Manuscripts Division, Howard-Tilton Memorial Library, Tulane Univ., New Orleans.
5. Edward W. Munford, "Albert Sidney Johnston," Albert Sidney and William Preston Johnston Papers, Mrs. Mason Barret Collection, Manuscripts Division, Howard-Tilton Memorial Library, Tulane Univ., New Orleans (hereinafter cited as Johnston Papers); Hudson Strode, *Jefferson Davis: Confederate President* (New York, 1959), 221.

6. Carl von Clausewitz, *On War,* 3 vols., trans. Colonel J. J. Graham (London, 1962), 2:55, 192; Alfred Roman, *The Military Operations of General Beauregard in the War between the States, 1861 to 1865,* 2 vols. (New York, 1884), 1:277–79; Braxton Bragg, "General Albert Sidney Johnston and the Battle of Shiloh," Johnston Papers.
7. Colonel G. F. R. Henderson, *The Science of War: A Collection of Essays and Lectures, 1891–1903,* ed. Colonel Neill Malcolm (London, 1933), 175.
8. William Preston Johnston, *The Life of General Albert Sidney Johnston* (New York, 1878), 613–15.
9. Ibid.
10. Ibid. See also Isham G. Harris to Preston, Apr. 6, 1862, written in William Preston Diary, War Department Collection of Confederate Records, Record Group 109, National Archives.
11. William T. Sherman, *Memoirs of General William T. Sherman,* 2 vols. (New York, 1875), 2:245.
12. Ronald Lewin, *Rommel as Military Commander* (New York, 1968), 243.

General Pierre Gustave Toutant Beauregard. Courtesy of the U.S. Army Military History Institute, Carlisle, Pennsylvania.

Beauregard at Shiloh

T. Harry Williams

Shiloh was a crucial battle in the career of P. G. T. Beauregard. For him it was a make or break battle. A victory would have made him one of the greatest military figures of the Confederacy, and he could have traveled the glory road. A defeat or an indecisive result would mean that his military road would continue to lead downhill. He had started on the downward path in 1861 in Virginia. Summoned to Richmond by President Davis, he had come with the renown of Sumter behind him and with the reputation of being one of the finest generals in the South. Then the trouble started. Davis placed him in command of the largest army in Virginia, on the Bull Run line. Soon Beauregard began to demonstrate that he was not ready to command a field army. He produced and pressed upon the government several plans of grand strategy that bordered on the fantastic. They were impossible of execution because they were not based on the realities of available Confederate resources. Beauregard formulated them in a sort of Napoleonic dreamworld; they were too grandiose and complex to be carried out by the kind of military organization the Confederates had. The same weakness appeared in his planning of battlefield strategy at the first Battle of Manassas. His combat schemes failed because they were too elaborate to be completed by the organization at his disposal. Success at Manassas came without his having done much to bring it about and even despite grave errors on his part that might have brought disaster. On the credit side of his military ledger, he was courageous and pugnacious. At Manassas he had handled his men well. Judged by what he had shown up to the summer of 1861, he could command a small army or a corps in a large one. He showed promise, but he needed more seasoning before he could direct a field army.

In the months after Manassas he revealed another weakness. He developed a passion for the use of the pen and became involved in a series of controversies, most of them useless, with Richmond. Finally Davis, fed up with Beauregard's dialectical talents and convinced that he was no field general, arranged to send the general to the Western Department as second in command to Albert Sidney Johnston. Beauregard's friends thought he was being shelved, and they were probably right. Beauregard arrived in the West just before U. S. Grant smashed the center of Johnston's line at Henry and Donelson, forcing the Confederates to loosen their hold on Kentucky and retreat through Tennessee. In the withdrawal Beauregard commanded the left wing of the army, the forces in and around Columbus, Kentucky, while Johnston led the troops retiring from the Bowling Green line. By the last week in March the Confederate fractions were reunited at Corinth in northeast Mississippi. There Johnston and Beauregard discussed the recent disasters and their plans for the future. They knew that Grant's army had moved up the Tennessee and had landed on the west side at Pittsburg Landing, about twenty-five miles from Corinth. It was rumored that D. C. Buell was marching to join him with 25,000 men. Both generals agreed that a blow should be struck at Grant as soon as possible, before Buell arrived. With an offensive in mind, Beauregard drew up a plan to reorganize the army. Johnston was designated as commander, Beauregard second in command, and Braxton Bragg chief of staff. The new organization contained four corps: the first, under Leonidas Polk, 9,136 troops; the second, Bragg, 13,589; the third, William J. Hardee, 6,789; the reserve corps, John C. Breckinridge, 6,439.

Not much time was allowed Johnston and Beauregard to plan an offensive. Late on the night of April 2 a telegram from the commander at Bethel, about twenty miles north of Corinth, was handed to Beauregard. It stated that the Federals were maneuvering in strength on his front. Immediately Beauregard decided that the Union commanders had divided their forces for an advance on Memphis. He wrote on the bottom of the telegram, "Now is the moment to advance, and strike the enemy at Pittsburg Landing," and told his chief of staff, Thomas Jordan, to take it to Johnston. Jordan went to the commander's quarters and gave him the message. Johnston said he would like to discuss it with Bragg; accompanied by Jordan, he crossed the street to Bragg's rooms. Bragg indorsed the proposed move, but Johnston raised objections, the chief one being that the troops needed more training. Jordan, voicing what he knew were Beauregard's views, replied that waiting would only enable the Federals to increase their strength and that an attack now would catch them by surprise. Johnston finally yielded. He authorized Jordan to draft a preparatory order for an advance.

In Bragg's room Jordan wrote a circular order to Polk, Bragg, and Hardee directing them to be ready to move by six the next morning. A similar directive was telegraphed to Breckinridge, commanding the reserve corps east of Corinth. Jordan told an aide to awaken Beauregard at five and tell him that an advance order had been issued. Soon after sunrise Jordan was summoned to Beauregard's quarters. He found the general sitting up in bed writing notes on the backs of telegrams and envelopes. A copy of these Jordan took to his office to use as a basis in framing a directive for the march order to Pittsburg Landing and for the battle order. As he wrote, he also had before him as a model a copy of Napoleon's order for the Battle of Waterloo. Before he finished, he went to Beauregard's quarters to sit in on a conference attended by Johnston, Bragg, and Hardee. Jordan said it would take time to write and distribute the detailed order. So Beauregard, drawing a rough sketch on the top of a camp table, carefully explained to the corps generals the march routes and the battle order. Without waiting for the written order, which would follow later, they were to have their troops moving by noon. It was then about 10 A.M.[1]

From Corinth to Pittsburg Landing was approximately twenty-five miles. To reach the main Federal line, the Confederates would have to march about twenty miles. Two narrow dirt roads led from Corinth to the Landing. One, the Ridge or Bark Road, ran north and then east. The other started east, turned in a northerly direction to a hamlet called Monterey, and joined the Ridge Road about four miles from Pittsburg. From Monterey the Purdy and Savannah roads led north to intersect the Ridge Road. At the point where the Savannah and Ridge roads crossed, about eight miles from the Landing, was a home known as Mickey's or Mickey's House.

Beauregard's written march order directed Hardee to advance on the Ridge-Bark Road, with the head of his column to bivouac that night (the third) at Mickey's. At 3 A.M. on the fourth Hardee was to move on until he approached the enemy position and then deploy in line of battle. Bragg's big corps was to assemble at Monterey and march in two wings on the Purdy and Savannah roads to the Ridge Road. The head of column of the right or Savannah wing was to reach Mickey's before sunset; the head of column of the left or Purdy wing was to reach the intersection at the Ridge Road by night. On the morning of the fourth Bragg was to follow in rear of Hardee on the Ridge Road and form a second battle line. Polk, who had only one division at Corinth (the other was at Bethel), was to leave half an hour after Hardee, bivouac behind the latter that night, and follow his line of march in the morning. At Mickey's Polk was to halt and form as a reserve. His division at Bethel was to move down the Purdy Road and join him. After the march order had been written, somebody realized that Polk and Bragg might get on the Ridge Road at the same time. Accordingly, Polk was instructed to stop at the Purdy intersection until Bragg's left wing had passed. The reserve corps was to assemble at Monterey after Bragg left and move by the best route to Mickey's or wherever the army was by the morning of the fourth. Beauregard's obvious intention, although not clearly indicated, was to place Polk in Bragg's left rear and the reserve corps in his right rear.[2]

Beauregard's battle order would bring the army into action with the corps arranged one behind the other: Hardee, Bragg, Polk, Breckinridge. Instead of each corps having a specific sector of the Federal line to attack, the first two would advance in parallel lines stretching across a three-mile front. Not only was this formation certain to cause a confusion and mingling of units, but it also meant that Hardee and Bragg would have to give their attention to matters on the entire front. The order of battle reads strangely when viewed in the light of Beauregard's avowed strategic objective. The Federal army was between two creeks flowing into the Tennessee. While attacking along the whole Federal line, the Confederates planned to turn the enemy left and drive the Federals away from the river and back on the northern creek where they would have to surrender.[3] This being the objective, the parallel arrangement was faulty. Instead of having a formation in depth on the right, Beauregard's line was equally strong at all points, and his attack was likely to hit each sector of the enemy line with equal strength.

His reasons for the line formation, given after the war, are not quite convincing. He explained that he put Hardee first because that general had the best corps and that he placed Bragg second because many of the latter's troops were recruits and would do better behind Hardee. He also contended that the parallel arrangement was better adapted for the unknown terrain into which the army was advancing.[4] After the war Johnston's partisans charged that Beauregard changed the original order of advance planned by his superior. This accusation they based on a telegram Johnston sent Davis on April 3 announcing that the army was about to advance. Johnston said Polk commanded the left, Bragg the center, and Hardee the right. That is, each corps was assigned a designated sector of the front. Rather than Beauregard altering Johnston's design, it seems probable that Johnston did not know what Beauregard was doing. The hasty planning at headquarters was being done by Beauregard and Jordan, with Johnston acting largely as an onlooker. Johnston did not see the written order until the march started, too late to change it even if he wanted to.[5]

From the language of the written order of April 3 it seems plain that Beauregard intended to concentrate the army around Mickey's that night and to attack the next morning. The order was drafted when he thought that the march could be started at an early hour. But the preparations for the movement—the explanations to the corps generals, the framing of instructions—took time. He had to postpone the jump-off until noon. Even then the march did not start. The streets of Corinth were jammed with wagons and troops. As the hours wore on, they remained there. Beauregard blamed Polk for the delay. He said that Polk, not understanding the verbal order, blocked with his troops and his train the line of Hardee's march. Polk said that he could not move until Hardee did and that Hardee did not receive instructions to move until three. Whatever the facts of the case, which probably were that Beauregard, as at Manassas, forgot to send an order, Hardee did not start until late in the afternoon and Polk until nearly dark. When Polk stopped for the night, he had covered only nine miles.[6] Because of the various delays, Beauregard sometime during the day did the obvious thing. Without preparing new written orders, he recast his time schedule. Now he planned to reach Mickey's by the evening of the fourth and attack on the fifth. In other words, he shoved most of the movements in the written order forward twenty-four hours.[7]

Despite what Beauregard wrote after the war and what historians have written since, the initial delays were not too serious. The movements of Hardee and Polk conformed to the new schedule. Hardee, once started, moved fast. By the morning of the fourth the head of his column was at Mickey's, only four miles from where he was to deploy in battle. Before midday Polk reached the Purdy intersection and waited for Bragg's left column to move by. He waited three hours. Finally he received a dispatch from Bragg saying that he was moving his whole corps on the Savannah road and for Polk not to hold up. Polk then went on to Mickey's, his march that day covering seven miles. Bragg was having all kinds of trouble with his unwieldy corps. Starting late on the afternoon of the third, he did not get the head of his column to Monterey until 11 A.M. on the fourth; his second division arrived late in the afternoon. Although it was only six miles from Monterey to Mickey's, his head of column did not approach the rendezvous until dark. Some of his units were moving in all during the night. Bragg's delays slowed Hardee as well as Polk. He dispatched Hardee in the morning to check his advance until the two corps were closer together. The slowness of his march Bragg ascribed to inefficient guides, the poor condition of the troops, and the improvised organization of the army. Also, he had been held back by his efforts to locate the reserve corps, which had not yet appeared.[8]

On the night of April 4 most of the Confederate army was approximately where it was supposed to be. Tired, bedraggled, and hungry, it was in and around Mickey's and in position to launch an attack on the morrow. Polk's division from Bethel and the reserve corps had not arrived, but they were expected early the next day. As they were to form part of the reserve or third line, the attack could be started in their absence. The worst feature in the situation was the location of Polk's troops. Because of the march mix-up, they were ahead of Bragg. Before Bragg could deploy behind Hardee, he would have to march through Polk.

Soon after midnight a heavy rain started and was still falling at 3 A.M., when Hardee was supposed to move out to form the first battle line. Because of the darkness and the rain, he could not start until dawn. Hardee's corps was not large enough to cover the front between the two creeks. To fill the gap Beauregard had authorized him to place one of Bragg's brigades on his right. By the time Hardee got his own troops and Bragg's unit deployed it was ten o'clock. Then Bragg began to arrange his own line. The hours were slipping by.[9]

Between six and seven Johnston arrived on the field. He and Beauregard had left Corinth on the fourth; that night they slept at Monterey. Before sunrise they were on their horses and riding toward Mickey's.[10] Impatiently Johnston watched Bragg array his line. The left division was not present. Finally Johnston sent a staff officer to ask Bragg where it was. Bragg said it was somewhere in the rear and he was trying to locate it. Johnston contained himself and waited, for over two hours according to one account. At twelve-thirty he looked at his watch and exclaimed "This is perfectly puerile! This is not war!" Riding to the rear, he found the division in the road, its advance blocked by some of Polk's troops. Almost frenzied, Johnston ordered the road cleared. By two it was free, and the last of Bragg's men passed to the front. Bragg now had to deploy the division in line, and Polk had to get his troops up and deployed in Bragg's left rear. Polk, who had been fuming all morning because he could not move till Bragg was out of the way, did not complete his formation until four, at which time his division from Bethel joined him. Soon after the reserve corps arrived. Since morning the Confederates had advanced about two and a half miles. The hour was too late to attack.[11]

As Polk was fixing his line, he was told that Beauregard wanted to see him. He found Beauregard standing in the road talking with Bragg. Speaking with much feeling, Beauregard said, "I am very much disappointed at the delay which has occurred in getting the troops into position." Polk replied that the fault was not his, that he had been held up by the troops ahead of him, meaning Bragg's. Beauregard said that because of the delay the attack would have to be called off. To succeed it had to be a surprise, and with the Confederate army within two miles of the Federal outposts a surprise was impossible at this late hour. The army would have return to Corinth. At this point Johnston and several other officers, perhaps attracted by the loud language, came up. Johnston asked what the matter was. Turning to his superior, Beauregard poured out reasons why the attack must be called off. Twenty-four hours had been lost, their presence was surely known to the Federals. "Now they will be entrenched to the eyes," he cried. He seemed to be unnerved by the miscarriage of his careful plans. Johnston, showing more balance and courage, said that he doubted the Federals knew of their approach and that anyway, having come this far, the army could not turn back. He concluded the informal council by saying, "We shall attack at daylight tomorrow!" As he walked off, he said to a staff officer, "I would fight them if they were a million."[12]

Shortly Beauregard's gloomy feelings were cheered. Hardee asked him to ride in front of his men to encourage them. Beauregard expressed reluctance, but when Johnston joined in the request he agreed. He stipulated that there must be no cheering; the noise might betray the presence of the Confederates to listening enemy outposts. Hardee so directed, but the order had to be repeated as he cantered down the line.[13]

The soldiers had little to cheer about except the presence of the hero of Manassas. That afternoon the rain stopped, and the sun broke through the mists. The night was clear and cold. For security reasons, the only fires allowed were in holes in the ground. In many units the food rations were short. The weary, wet soldiers, most of whom had been under arms since before dawn, slept on the ground. Beauregard forgot to give orders to set up his tent and had to spend the night in an ambulance wagon.[14]

Three creeks formed an important part of the terrain of the Battle of Shiloh. They bounded the area in which the battle was fought. On the south was Lick Creek, which took its rise about twelve miles from the Tennessee, flowed in a northeast direction, and entered the river south of Pittsburg Landing. On the north was Owl Creek, which flowed parallel with Lick and emptied into Snake Creek, which in turn joined the river north of the Landing. Near the river the distance between the streams was five miles; it was three miles at the point where Grant's army was encamped between Owl and Lick. Owl was the stream on which the Confederates intended to drive the Federals and destroy them.

The land between the creeks was a rolling plateau, rising in places to a height one hundred feet above the river. A few farms dotted the area, but most of it was covered with heavy timber and brush and crossed by ravines. The roads were country dirt ones; most of the primary roads ran in an east-west direction, which made for bad communications for the Confederates. About three miles from the Landing and almost in the center of the area was a little log church called Shiloh, from which the battle would take its named.[15]

On the night of April 5 the two largest armies yet to come together in the war slept within a few miles of each other, the Federal army strangely unaware of the presence of its enemy. As is usually the case with Civil War battles, the estimates of the numbers of the contending forces differ and conflict. It seems certain, however, that Johnston approached the field with close to 40,000 men. In his camps between the creeks Grant had about the same number, possibly a few thousand less. At Crump's Landing four miles downstream (north) from Pittsburg he had another division of about 7,000. This latter unit did not participate in the fighting on April 6. In the battle of that day the two armies seem to have been approximately equal in size.[16]

As the morning of April 6, a Sunday, dawned, a heavy white mist hung low over the woods in front of the Confederate positions. Then the sun broke through and dispelled the fog. Excitedly the romantic Confederates passed the word around that it was another sun of Austerlitz. The clear sky, the bracing air, the freshness of the Southern spring day united to remind the soldiers of the land for which they fought. So did an address by Johnston read to each regiment as it formed in line. Remember the precious stake involved in the coming battle, the commanding general exhorted; remember the mothers, wives, and children hanging on the outcome; remember "the fair, broad, abounding land" and the happy homes that would be desolated by defeat; remember above all the women of the South, whose noble devotion had never been exceeded in any age.[17]

While the troops were forming in line, the generals stood around Johnston's campfire. Apparently another argument over the feasibility of an attack started, with Beauregard again raising objections. It was interrupted by the sound of shots as Hardee's skirmish line encountered the first Federals. Johnston said the battle had started and it was too late to change the dispositions.[18] At five o'clock Hardee's whole line moved forward, followed by Bragg 500 yards behind. Johnston and Beauregard stood on a slight eminence watching the men advance. The sound of the firing increased as the Confederates drove through the Federal outposts and then lulled temporarily as the attackers moved toward the enemy camps. Shortly before seven Beauregard met Johnston near the latter's headquarters. The commanding general said that the battle had opened in grand style and that he was going to the front. Mounting his horse, he said to his aides, "Tonight we will water our horses in the Tennessee River."[19]

One would like to know more of what passed between the two generals in their last meeting, particularly of the command function that Johnston assigned to his junior. After the war Beauregard said that Johnston gave him "the general direction" of the battle.[20] Taken at face value, this statement would mean that Beauregard was to control the principle movements of troops all over the field. Obviously Johnston did not intend him to exercise this power—first, because that would have left nothing for Johnston to do, and second, because Beauregard from his position in the rear could not have directed movements at the front. Johnston's purpose is evident from Beauregard's actions during the day. While the commanding general pressed the attack at the front, Beauregard was to command the troops in the rear sector, particularly the two reserve corps, Polk in rear of Bragg and Breckinridge in rear of Hardee. At the right moment he was to commit them to battle and to send forward any other troops to points where they were needed. This function Beauregard performed until Johnston's death, always moving his headquarters forward as the Confederate front line advanced.

In the Confederate records the Battle of Shiloh is a story of headlong attacks, brave fighting, confused and unscheduled advances, and a main objective not attained. As the Confederates drove the surprised Federals before them over the rugged terrain, the lines of the attackers lost their next line formations. Units from one corps inevitably got mixed with those in another, with a single tangled, irregular line resulting. Finally the corps commanders improvised an arrangement whereby each one directed the attack on a specific area of the front. From left to right the Confederate line was now commanded by Hardee, Polk, Bragg, and Breckinridge. The corps generals devoted most of their efforts to leading charges instead of to organizing their masses and feeding them up to the front. At an early hour many Confederate troops began to straggle off to plunder the Federal camps or to make their way to the rear. These latter Beauregard endeavored to stop with cavalry and organize into battalions to be sent forward again.

Partly because of the actions of the corps generals and partly because of the terrain and stiffening Federal resistance, the initial Confederate advance was slowed. The battle tended to develop into a series of frontal assaults conducted more or less independently. From the Confederate viewpoint the advance was fatally off schedule. The left was moving faster than the right. The Federals were being driven back toward their base on the river instead of northward onto Owl Creek. As the Federals on the Confederate right retired, they came to an old sunken road in a heavily wooded area. In this natural trench General Benjamin Prentiss rallied the remnant of his division; other units later formed on his flanks. Whereas in modern war such a strong position would have been bypassed or contained, the Confederates tried to take it. For hour after hour, while on the left the advance was grinding past Shiloh church, the Confederates flung eleven bloody and vain charges at the place they aptly called the "Hornet's Nest."[21]

Johnston saw what was going wrong with his plan. About noon he moved to the right to personally direct the attack on that vital sector. As he rode among the men, he carried a tin cup in one hand. He had taken it earlier from an officer who had come out of a deserted Federal tent carrying some valuable articles, which he had shown to the general. Johnston rebuked the man for plundering, then regretting his words, he took the cup, saying, "Let this be my share of the spoils today." Johnston exhorted the men to go forward. Tapping their bayonets with his cup, he said, "These must do the work." At one point where the soldiers were obviously reluctant to charge, Johnston offered to lead them. Shamed, they sprang forward and drove the Federals back.

Johnston sat on his horse watching the retreating Yankees. Governor Isham Harris of Tennessee, serving as a volunteer aide, galloped up to the general. He saw Johnston reel in the saddle. "General, are you hurt?" cried Harris. "Yes, and I fear seriously," Johnston replied. A bullet had severed the large artery in his right leg. Maybe a retiring Union soldier had paused and drawn a bead on what he thought was an important officer near the front; maybe a stray ball fired at no one in particular just happened to strike Johnston. Guiding the horse away from the line of fire and holding Johnston with one arm, Harris stopped in a ravine and lifted the unconscious general from the saddle. Other officers gathered around. Any person with the most elementary knowledge of first-aid could have stopped the flow of blood and saved Johnston's life. But in the Civil War nobody knew anything about first-aid except the medics, and Johnston had sent his surgeon to look after some prisoners.[22] Johnston died at two-thirty. Governor Harris spurred to the rear and delivered the sad news to Beauregard shortly after three.[23]

After the war Johnston's partisans, particularly his son, liked to say that he died at the moment of victory. They claimed that he had achieved triumph elsewhere on the field and was organizing his right for the final push when the fatal bullet hit him. As a matter of fact, victory had not been won anywhere; the Confederate forces were not fighting under a common direction; and Johnston did not have complete control of his forces on the right, let alone on the whole front. By going from unit to unit at the edge of battle, exhorting the men to charge and offering to lead them, Johnston was performing more like a corps or division general than a commander. At the time of his death, he did not have a single staff officer with him, which indicates that he was not exerting much control over the battle.[24] Whatever general direction was being exercised issued from Beauregard.

Up to the time of Johnston's death, Beauregard had been performing the function assigned him by the commanding general—that of ordering movements in the rear of the battle. Soon after the attack started, he set up field headquarters on a high point between the Pittsburg and Purdy roads. From this point he deployed Polk and Breckinridge in columns of brigades and instructed them to follow Bragg and go in wherever they were called to help; if in doubt where to go, they should move toward the sound of heaviest firing. About the middle of the morning, as the Confederate line advanced, he moved up to within half a mile of the abandoned Federal camps. At two o'clock he established his third headquarters of the day near Shiloh church. Always he had his staff riding over the field collecting reports from the front and rounding up stragglers. Any inactive units that he spotted he directed to the front, sending most of them to Hardee on the left. Just before he received the news of Johnston's death, he was about to shift some troops to the center.[25]

With Johnston dead, Beauregard assumed command of the army. Immediately he acted to keep the impetus of the offensive rolling. Contrary to what his enemies said during the war and later, he was not ignorant of the situation at the front or indifferent to the outcome of the battle. From the reports of his staff he had a fairly accurate picture of how far the attacks had gone. He knew that the Federal right had retired toward the Landing and that the left was still holding. Ordering that the news of Johnston's death be kept from the men, he directed that the advance continue all along the line. To coordinate the attack on the Federal left he ordered Bragg to take charge of the Confederate right and sent General Daniel Ruggles to command the center. The so-called lull of an hour, which followed Beauregard's assumption of control, was not due to any confusion resulting from the change in command, but to the time involved in shifting additional troops toward the fatal sunken road.[26]

While the Federals from the right were constructing a new and powerful defense line on the bluff above the Landing, the Confederates were concentrating for a final effort against the Hornet's Nest. Ruggles collected over sixty pieces of artillery and pounded the position with a merciless fire. Shaken by the barrage, the Federal troops on the right and left of Prentiss withdrew to the Landing. Prentiss, under orders from Grant to hold to the last, fought on with 2,200 men. Although virtually encircled by attackers, he continued to resist until five-thirty, when he surrendered. If any one man saved the Federal army at Shiloh, Prentiss was the man. Even captured, he and his troops were useful to Grant. A Confederate regiment was detailed to watch over the rich bag of prisoners.

After the surrender of Prentiss, the tired Confederates drifted toward the Federal line around the Landing. The Union forces were massed in a semi-circular position with their backs to the river. The line was strongest on the left or the south side. Here the Federals had assembled fifty artillery pieces to meet the expected attack from the Confederate right. In addition, two Federal gunboats stood by in the river ready to throw their shot when the Confederates advanced. On the Federal right the line faced generally west. For the safety of this sector Grant felt little alarm. Opposite was Hardee, whose pecking assaults indicated he was incapable of mounting a dangerous attack. Besides, Grant's division from Crump's Landing, which had been ordered to the field early in the day and whose commander had confused his route, was finally nearing the scene and would shortly join the troops on the right. Only for his left did Grant fear. Only from their right, where most of their troops were massed, could the Confederates possibly deliver a decisive blow.

They did not have much power to do it here. At this critical moment the Confederates had no reserve to put in to clinch victory. The last unit of Breckinridge's corps had been committed to battle by early afternoon. For the final assault of that bloody day, Bragg could marshal only two relatively fresh brigades from his own corps—not really fresh, for they had been through the carnage of the sunken road. One of them was badly short of ammunition. At Bragg's order they charged, bravely but without much dash, and were repulsed.[27] As they advanced, a part of a regiment from Buell's army appeared in the Federal line. On the previous day one of Buell's divisions had reached the west side of the river. Soon after the battle started Grant ordered it to the field. Its vanguard was just now crossing. Even without the presence of these troops Bragg's attack would have failed. The reports of the brigade commanders told the story: their men were too exhausted to fight.[28]

It was now after six o'clock. From his headquarters at Shiloh church two miles in the rear, Beauregard sent his staff officers to the corps generals with instructions to suspend the attacks and retire to the enemy camps for the night. His reasons for withdrawing, as given in his preliminary report, were that the troops were tired and scattered, darkness was coming on, and the Confederates had substantial possession of the field. In short, Beauregard thought that he had the Federals whipped, that he could do nothing more that day, and that after resting his men he could complete his victory on the morrow.[29]

To the end of his days Beauregard would be criticized and condemned for stopping the attack. It would be said that he let slip the great opportunity for victory in the West: one more assault and Grant's army would have been driven into the river and destroyed. Bragg started the criticism, at least officially, when he wrote in his report on April 20 that his troops were starting a final attack with every chance of success when the withdrawal order came. As the years passed, Bragg remembered more and more about the episode, until finally he thought that he had threatened to disobey the order. How Bragg could have imagined, after witnessing what happened to the charge of his two brigades, that he had a chance to seize the Landing defies comprehension. He did not think it two days after the battle. Then he ascribed the Confederate failure to the demoralized and disorganized condition of the troops caused by the want of discipline.[30]

Today Beauregard's decision seems as right as it did to him on the evening of that hard-fought Sunday. He did not know, of course, of the new factor in the battle, the arrival of Buell's troops, that changed the entire situation. Nor, apparently, had he been apprised that the Federal division from Crump's Landing was approaching the field. But he did know that his own men were tired, hungry, and spiritless after thirteen hours of fighting, too exhausted even to cheer when told they had won a victory. He knew that many of the units were scattered, disorganized, and out of control and that the latest attacks had been feebly delivered. These things he had learned from the reports of his and other officers; some of them he had seen with his own eyes. As he rode over the rear area of the field, he saw groups of men resting on their arms, too weary to move; he saw hordes of stragglers plundering the enemy camps. He saw also the sun going down. He wanted to get his army in hand before darkness. As a matter of fact, with the approach of night, as the reports of the brigade and regimental commanders show, many units were retiring from the line without orders. The withdrawal directive merely recognized an action partially in process of execution. Even with an early start, the disorganization of the Confederate forces was so great that some units did not reach their bivouac until eight o'clock. When all the elements in the situation are weighed, it seems obvious that Beauregard had no recourse but to assemble his army for another attempt the next day.[31]

That night Beauregard made his headquarters in Sherman's tent near Shiloh church. There came the corps commanders to discuss the events of the day and plan tomorrow's moves. All felt confident that victory had been achieved and that an attack the next day would complete the destruction of the Federal army. Their optimism was confirmed by the receipt of a dispatch from Colonel Ben Hardin Helm, Lincoln's brother-in-law, in northern Alabama that Buell was not marching toward Pittsburg Landing after all but toward Decatur. Captured General Prentiss, who was Jordan's guest and who was having a wonderful time teasing his hosts with predictions of defeat the next day, unwisely said that the report was untrue. They refused to believe him. At Beauregard's direction Jordan sent a telegram to Richmond announcing the capture of every enemy position and "a complete victory." On the basis of this message, Davis reported to Congress that the Federal army had been practically destroyed.[32]

One Confederate soldier was not so sure the Federals were not being reinforced. Colonel Nathan Bedford Forrest, not yet recognized as a great cavalryman, dressed some of his men in captured Federal coats and sent them into the enemy lines. They reported back that heavy replacements were arriving but that a sudden night attack would push the Federals into the river. Forrest found Hardee and presented his information. The cavalry leader advised an immediate attack or a withdrawal. If the Confederates tried to fight the fresh Federal masses the next day, he said, they would be "whipped like hell." In a rather casual way the corps commander told Forrest to take his intelligence to Beauregard. Forrest was unable to locate the commanding general's headquarters. Once again he sent his scouts to the Federal camps, and again they related that reinforcements were coming in. Once more Forrest sought out Hardee, at two in the morning. This time Hardee, told him to return to his regiment and keep a vigilant watch.[33] In such an offhand manner was vital military intelligence often handled in the Civil War.

Forrest's information and his salty analysis of the fate awaiting the Confederates were both correct. That night 17,000 of Buell's troops were ferried over the river. With these arrivals, the Crump's Landing division, and the hard core of his own army, Grant had at his disposal on the morning of April 7 at least 40,000 men. Determined to seize the initiative, he launched an attack on the Confederates at daylight. The fighting on the second day was almost an exact reversal of that of the day before. The Confederates were surprised, strategically, because they had expected to be the attackers. During the night and in the early morning hours the Confederate generals had not done too good a job of reorganizing their forces. When the Federals struck, some units were several miles in rear of the first line of encampments. The Confederate line of battle was formed slowly and was not completed until after the Federals had rolled past the Hornet's Nest.

Because of the heavy losses of the previous day and the large number of stragglers who had left the field, Beauregard could put in action only something over 20,000 troops. From right to left the Confederate line was commanded by Hardee, Breckinridge, Polk, and Bragg. Something of the confusion attending the withdrawal on the preceding night is seen in the fact that

Hardee and Bragg had exchanged wings, that Hardee commanded two of Bragg's brigades, and that Bragg directed one of Polk's divisions. Polk arrived on the field late with his other division. As the battle swayed back and forth on a fluid front the Confederate units tended to become more mixed and scattered than on Sunday. Several seem to have been fired on by their own troops. Beauregard noticed one group in a woods who appeared to be clad in white uniforms. At first he thought they were Federals, but he saw they were fighting on the Confederate side. Inquiry developed that they were Louisiana troops. They were equipped with blue coats, and on the day before had been fired into by Confederates. To prevent the repetition of this danger, they had turned their coats inside out.[34]

The impact of the Federal attack forced the Confederates back all along the line. Although the Southern troops resisted stubbornly and at points even counterattacked, they could not halt the resistless blue advance.[35] Their failure was not due entirely to inferior numbers or to the lack of a proper reserve. Confederate observers noted that the men seemed to be losing their dash and fire. Even when general officers led them in person to points at the front, they responded feebly. Beauregard himself, on two occasions, seized the colors of slowly advancing regiments and led them forward. When an officer friend reproved him for rashness, he answered, "The order must now be 'follow,' not 'go'!" Sometimes a unit, after being placed in line, would stand a short time and then slowly melt away.[36] It was not just that the men were bone-tired after two days of battle; their spirits were close to being broken by the abrupt reversal of fortune, by the sudden snatching away of apparent victory.

Jordan, who had been observing the demeanor of the men, went to Beauregard shortly after two and said, "General, do you not think our troops are very much in the condition of a lump of sugar thoroughly soaked with water, but yet preserving its original shape, though ready to dissolve? Would it not be judicious to get away with what we have?" Beauregard replied, "I intend to withdraw in a few moments."[37] He too had been studying the soldiers and the situation. Against the fresh, superior Federal forces, the Confederates had no chance of victory. If they remained on the field, they would be pounded to pieces. The only recourse was to get the army away to safety. For an hour Beauregard had been contemplating a withdrawal. Now his mind was made up. Staff officers rode to tell the corps generals to retire but slowly and in good order. In rear of Shiloh church Beauregard posted a strong rear guard with artillery support. He wanted the Federals to know that although he was leaving, he was doing so with dignity and honor and not in rout and disaster. By four o'clock the Confederates had left the field and were on the road to Corinth.[38]

No Federals pursued them that day or attacked them that night when they encamped a few miles from the field. Grant's army was in no shape to pursue; it had been too roughly handled. A heavy rain started after dark, making the roads impracticable for artillery the next day. Without supporting artillery, pursuing infantry could easily be checked by a few enemy guns. The next morning Sherman attempted a sort of half-pursuit. The Confederate rearguard turned him back with its cavalry alone.

Even without the presence of harassing Federals, the Confederate withdrawal to Corinth was a grim journey. The weary, discouraged foot soldiers plodded on over the narrow, muddy, almost impassable road; among them jolted the wagons carrying the thousands of groaning wounded. The way of the march was littered with abandoned supplies. Bragg found that few officers were with their men. "The whole road presents the scene of a rout," he wrote in anger and disgust to Beauregard.[39] The sad job of carrying the wounded to Corinth strained the army's transportation facilities. Shiloh was the first bloody battle of the war. For both sides the casualties were terrific. The Union losses were 1,754 killed, 8,408 wounded, 2,885 captured or missing—a total of 13,047. The Confederate losses totaled 10,699: 1,726 killed, 8,012 wounded, 959 missing.[40]

Shiloh is the most "iffy" battle of the war. Its might-have-beens have fascinated writers. What would have happened if the Confederates had launched their attack on the fifth? What if Johnston had not died? What if Beauregard had made another attack on the sixth? If the Confederates had defeated Grant, would they have smashed Buell and regained the West? Most of the dramatic possibilities of the battle have been exaggerated. If the Confederates had attacked on the fifth, they would have encountered approximately the same size Federal force on that day and the following as they did on the sixth and seventh. One division of Buell's army was on the other side of the Tennessee by midafternoon of the fifth and could have reached the Landing by night. Another could have arrived early the next day. Only if the Confederates could have made their attack on the fourth, as planned in Beauregard's original order, would they have had a real chance to destroy Grant. On the evening of the sixth Beauregard's cause was lost. Even without Buell's troops, Grant could have probably stopped an attack with the aid of the fresh division from Crump's Landing. But what if a last assault had driven the Federals into the river? The Confederate army would have been so shattered that it could not have followed its success. On the following day, even after a so-called victory, it could muster only some 20,000 troops. And over the river would have been Buell with 25,000 fresh soldiers. Beauregard could not have advanced for a long time and without reinforcements. Probably he would have had to retire to Corinth to regroup.

In the West Beauregard showed definite improvement as a field commander. The Napoleonic complex, the penchant for grand planning, the tendency to exaggerate the resources available to him—these characteristics were still a part of him but held in obvious restraint. His battle plan for Shiloh, like the one before Manassas, was whipped up in too short a time. He was yet to learn that a detailed design was not the work of a few hours. The march order from Corinth to the battlefield, while overly optimistic as to the results that could be obtained, was not, as has been sometimes charged, unduly complex. Although his battle arrangement was faulty, it too was innocent of the complexity which had often marred his plans in Virginia. After he took command following Johnston's death, he did all that any general could have done in the circumstances. The withdrawal on the second day was conducted with skill. His one bad mistake was on Sunday night, when he failed to take adequate action to reorganize his army and made no attempt to ascertain the intentions of the enemy. He still tended to overlook an important detail, to assume that the enemy would act as he wanted him to act.

Nevertheless, he gave promise, with continued experience, of developing into a useful field general. The promise was not fulfilled because his career in the field did not continue. Shiloh and its aftermath increased the dislike and distrust which Davis felt for him. Within a few months after the battle the president found a pretext to remove him from field command. Beauregard was relegated to direct the defenses of Charleston and to the role of an engineer. For the rest of the war he held relatively unimportant assignments. Sumter, Manassas, Shiloh—and then the glory was ended.

Notes

"Beauregard at Shiloh" was first published in *Civil War History* 1, no. 1 (Mar. 1955): 17–34, and is reprinted by permission of the Kent State University Press.

1. Alfred Roman, *The Military Operations of General Beauregard in the War between the States, 1861 to 1865,* 2 vols. (New York, 1884), 1:270–72; Robert U. Johnson and Clarence C. Buel, eds., *Battles and Leaders of the Civil War,* 4 vols. (New York, 1884–88), 1:579–81, 594–96.
2. U.S. War Department, *The War of the Rebellion: A Compilation of the Official Records of the Union and Confederate Armies,* 128 vols. (Washington, DC, 1880–1901), series 1, vol. 10, pt. 1:392–95. Hereinafter cited as *OR.* All references are to series 1 unless otherwise indicated.
3. Ibid., vol. 10, pt. 1:397.
4. Johnson and Buel, *Battles and Leaders* 1:581–82; Thomas Jordan and J. P. Pryor, *The Campaigns of Lieut.-General N. B. Forrest . . .* (New Orleans, 1868), 149.
5. *OR,* vol. 10, pt. 2:387; Johnson and Buel, *Battles and Leaders* 1:554.
6. Johnson and Buel, *Battles and Leaders* 1:596; Roman, *Beauregard* 1:275–76; William M. Polk, *Leonidas Polk: Bishop and General,* 2 vols. (New York, 1893), 2:90–93.
7. All accounts of the Battle of Shiloh say that from the first Beauregard planned to reach Mickey's by the evening of the fourth and attack on the fifth. This interpre-

tation can be squared with the written order only by assuming that Beauregard meant to get the army to Mickey's, four miles from the Federal position, on the third; advance early on the fourth and spend the day deploying; and attack on the fifth. That Beauregard would waste a day in deployment, when time was so precious to the Confederates, seems absurd. If it be objected that the Confederates did require a whole day to deploy on the fifth, the answer is that the circumstances were unusual and that they never dreamed it would take that long.

8. William Preston Johnston, *The Life of Gen. Albert Sidney Johnston* (New York, 1878), 564–65; Johnson and Buel, *Battles and Leaders* 1:582; Polk, *Leonidas Polk* 2:93–97; *OR,* vol. 10, pt. 1:463–64; *OR,* vol. 10, pt. 2:390–91.
9. Johnston, *Albert Sidney Johnston,* 560–61; Polk, *Leonidas Polk* 2:97–98; *OR,* vol. 10, pt. 1:567.
10. *OR,* vol. 10, pt. 1:400; Johnson and Buel, *Battles and Leaders* 1:596–97.
11. Johnston, *Albert Sidney Johnston,* 560–63; Polk, *Leonidas Polk* 2:97–99; *OR,* vol. 10, pt. 1:406, 414, 464, 614.
12. *OR,* vol. 10, pt. 1:407; Johnston, *Albert Sidney Johnston,* 567–71; Johnson and Buel, *Battles and Leaders* 1:555, 583–84, 597–98; Roman, *Beauregard* 1:277–79. Beauregard claimed that Johnston called the council, but the evidence shows clearly it was largely accidental.
13. *OR,* vol. 10, pt.1:400; Roman, *Beauregard* 1:530, 533.
14. William G. Stevenson, *Thirteen Months in the Rebel Army . . .* (New York, 1864), 148–49; Roman, *Beauregard* 1:348.
15. Jordan and Pryor, *Forrest,* 117–18; Johnson and Buel, *Battles and Leaders* 1:465–86, 495–98.
16. *OR,* vol. 10, pt. 1:112, 398; Johnson and Buel, *Battles and Leaders* 1:485, 537–39; Johnston, *Albert Sidney Johnston,* 670, 685.
17. Jordan and Pryor, *Forrest,* 121; Johnston, *Albert Sidney Johnston,* 582; Thomas D. Duncan, *Recollections of Thomas a Duncan* (Nashville, 1922), 53; letter of Sergeant A. P. B., Apr. 14, 1862, in *New Orleans Evening Delta,* Apr. 21, 1862, clipping in Arthur W. Hyatt Papers, Louisiana and Lower Mississippi Valley Collection, Hill Memorial Library, Louisiana State Univ., Baton Rouge (hereinafter cited as Hyatt Papers); *OR,* vol. 10, pt. 2:389.
18. Johnston, *Albert Sidney Johnston,* 569; *OR,* vol. 10, pt. 1:464.
19. Johnson and Buel, *Battles and Leaders* 1:557, 599; Roman, *Beauregard* 1:284–85.
20. Johnson and Buel, *Battles and Leaders* 1:586.
21. Ibid., 586–91; Jordan and Pryor, *Forrest,* 121–31; Roman, *Beauregard* 1:283–307; Johnston, *Albert Sidney Johnston,* 587–609; Duncan, *Recollections,* 58–60; Albert Dillahunty, *Shiloh* (Washington, DC, 1951), 9–15; Otto Eisenschiml, *The Story of Shiloh* (Chicago, 1946), 27–49.

22. Johnston, *Albert Sidney Johnston,* 611–15; Roman, *Beauregard* 1:537.

23. Johnson and Buel, *Battles and Leaders* 1:590.

24. Eisenschiml, *Story of Shiloh,* 39.

25. *OR,* vol. 10, pt. 1:401–402; Johnson and Buel, *Battles and Leaders* 1:586–90; Roman, *Beauregard* 1:285, 289, 294–96.

26. Roman, *Beauregard* 1:297–98; Johnson and Buel, *Battles and Leaders* 1:590; Stanley F. Horn, *The Army of Tennessee* (Indianapolis, 1941), 134–35.

27. Johnson and Buel, *Battles and Leaders* 1:590–91; Roman, *Beauregard* 1:301–304; Jordan and Pryor, *Forrest,* 131–34.

28. *OR,* vol. 10, pt. 1:551, 555.

29. Ibid., vol. 10, pt. 1:386–87.

30. Ibid., vol. 10, pt. 1:466–67; Don C. Seitz, *Braxton Bragg: General of the Confederacy* (Columbia, SC, 1924), 111–13.

31. H. M. Stanley, *Autobiography of Henry Morton Stanley* (Boston, 1909), 198; letter of Sergeant A. P. B., Apr. 14, 1862, in *New Orleans Evening Delta,* Apr. 21, 1862, clipping in Hyatt Papers; Roman, *Beauregard* 1:304–305, 547–50.

32. Roman, *Beauregard* 1:305; Jordan and Pryor, *Forrest,* 135–36; Johnson and Buel, *Battles and Leaders* 1:602–603; *OR,* vol. 10, pt. 1:384–85; James D. Richardson, ed., *A Compilation of the Messages and Papers of the Confederacy, including the Diplomatic Correspondence, 1861–1865,* 2 vols.(Nashville, 1905), 1:208. Jordan stated in *Battles and Leaders,* above, that he received a telegram from Helm in the afternoon and gave it to Beauregard after sunset, that is, after Beauregard had ordered the battle stopped. Beauregard said, in his preliminary report, *Official Records,* above, that he received a message saying Buell had been delayed and would not be able to reach Grant in time so save him. It is possible that two messages about Buell were received and that the one Beauregard mentioned may have come in before the battle ended. In such case, Beauregard might have confused them in his report.

33. Jordan and Pryor, *Forrest,* 136–37.

34. Roman, *Beauregard* 1:316; Basil W. Duke, *Morgan's Cavalry* (New York, 1909), 86.

35. Roman, *Beauregard* 1:308–19; Johnson and Buel, *Battles and Leaders* 1:591–93; Horn, *Army of Tennessee,* 139–42; Dillahunty, *Shiloh,* 16–9; Kenneth P. Williams, *Lincoln Finds a General,* 5 vols. (New York, 1949–59), 3:383–88.

36. *OR,* vol. 11, pt. 1:402; Seitz, *Bragg,* 113; Roman, *Beauregard* 1:317; Jordan and Pryor, *Forrest,* 142.

37. Johnson and Buel, *Battles and Leaders* 1:603.

38. Roman, *Beauregard* 1:318–99; Jordan and Pryor, *Forrest,* 143–46; Johnston, *Albert Sidney Johnston,* 651–53; *OR,* vol. 10, pt. 1:388.

39. Stevenson, *Thirteen Months in the Rebel Army,* 170–71; *OR,* vol. 10, pt. 2:400.

40. D. W. Reed, *The Battle of Shiloh* (Washington, DC, 1909), 23; Thomas L. Livermore, *Numbers and Losses in the Civil War in America, 1861–65* (Boston, 1901), 79–80; Johnson and Buel, *Battles and Leaders* 1:485, 537–39; Johnston, *Albert Sidney Johnston,* 656.

Major General Mansfield Lovell. Library of Congress.

Mansfield Lovell

Arthur W. Bergeron Jr.

JOSEPH E. JOHNSTON CALLED HIM ONE "OF THE BEST OFFICERS WHOSE SERVICES WE CAN command" and pronounced him "as fit to command [a] division as any [man] in our service."[1] Judah P. Benjamin described him as "a brilliant, energetic, and accomplished officer."[2] Braxton Bragg remembered him as "one of the best artillery officers in the old service."[3] His old and close friend Gustavus W. Smith praised him with these words: "Possessed of extraordinary physical strength, activity and endurance—with mental ability of a very high order—training and experience in military and civil service as well as in ordinary business pursuits—brave almost to the point of rashness—in conduct and character guided by the highest sense of right."[4]

Anyone who did not already know the subject of this essay might wonder to whom these compliments refer. Certainly, the name of Mansfield Lovell would be one of the last to occur to most Civil War buffs and scholars. The few of them who recognize his name know that he was a Northerner who joined the Confederate army after First Manassas and who tried unsuccessfully to defend the important city of New Orleans. In fact, about the only time his name even appears in writings about the conflict is when authors blame him for the fall of New Orleans. Criticism fell heavily on Lovell both during the war and in the years immediately thereafter. Professor Daniel E. Sutherland wrote in a 1983 article, "Lovell was a man condemned, and subsequent historians were slow to give him justice."[5]

Sutherland's article, entitled "Mansfield Lovell's Quest for Justice: Another Look at the Fall of New Orleans," sparked this author's interest in the general. Sutherland expanded upon work begun by New Orleans newspaperman and historian Charles L. "Pie" Dufour. In his book *The Night the War Was Lost* and an article written the next year, Dufour demonstrated that Lovell had been made the scapegoat for the loss of the Crescent City. He stated that President Jefferson Davis deserved the blame and criticized him for his "utterly unjust treatment of General Lovell."[6] Using sources unknown or unavailable to Dufour, Sutherland delved more deeply into the controversy. He concluded: "It is almost certain—as certain as it can be without a written confession—that Davis actively and consciously sought to make Lovell the 'victim' demanded by the Confederate people and history."[7]

The questions in this author's mind created by Sutherland's work had nothing to do with his or Dufour's assessment of blame for the fall of New Orleans or with Davis's actions in making Lovell a scapegoat. What was the general's background, how did he come to receive command of the defense of the Confederacy's most important city, and what was the rest of his Civil War career like? No historian has ever attempted a biography of Lovell, so such a project became very appealing. This essay was written prior to going very deeply into research on Lovell. The biography remains just one of several of those "irons in the fire." The essay is, hopefully, just the first step toward completion of the project. With that caveat in mind, what follows is a tentative look at this little known Confederate general.

Born in Washington, D.C., on October 20, 1822, Mansfield Lovell was the first son of Dr. Joseph and Margaret E. Lovell. Dr. Lovell was the first surgeon general of the United States Army Medical Department and had received that appointment four years before his son's birth. He came from Massachusetts. His grandfather had served in the Continental Congress and had signed the Articles of Confederation, and his father had fought in the Continental Army during the American Revolution. Mansfield's mother came from an old New York family. Both of his parents died within a few weeks of each other in 1836, and he lived for two years with a guardian in New York state.[8]

Lovell received an appointment to the United States Military Academy in 1838 and graduated ninth in a class of fifty-six on July 1, 1842. His classmates included twenty-one men who became generals during the Civil War. Among them were Earl Van Dorn, James Longstreet, John Pope, William S. Rosecrans, Martin Luther Smith, Abner Doubleday, and Gustavus Woodson Smith. After graduation, Lovell became a second lieutenant in the 4th United States Artillery. He continued to study while on active duty, and his friend Gustavus Smith said "he was soon recognized as one of the most prominent young officers in the artillery" service.[9] The 4th Artillery formed a part of General Zachary Taylor's army when the Mexican War began in 1846. Lovell fought in the Battle of Monterrey and was wounded.[10]

Shortly after this engagement, Lovell became chief of staff to Brigadier General John A. Quitman of Mississippi. That volunteer general was commander of a brigade in Taylor's army. Gustavus Smith claimed that Quitman had asked the younger regular army officers in the army to recommend someone for the position on his staff and that "with great unanimity they named Mansfield Lovell."[11] Regardless of the circumstances of his appointment, Lovell and Quitman soon became very good friends. The latter's biographer wrote that they "established a close intellectual and personal relationship."[12] In the late 1850s, two of Lovell's brothers married Quitman's daughters. This relationship between Lovell and Quitman probably became like that between a father and son. After Quitman's death, Lovell wrote a letter to one of his daughters. He said that the general had "inspired 'especially in his younger friends a confiding reliance and affectionate regard, much akin to the holy feeling that should exist between child and parent.'"[13]

Lovell, now a first lieutenant, accompanied Quitman after he was transferred to the army of General Winfield Scott. Quitman eventually received command of a division, and his men played an important part in the campaign from Vera Cruz to the City of Mexico. In a reconnaissance with Quitman just before the Battle of Chapultepec, Lovell performed bravely. Though several times exposed to heavy enemy fire, he succeeded in pinpointing important enemy positions and in leading to safety a small group of soldiers threatened with capture by the Mexicans. The next day, during the assault on the Belen Gate, Lovell received his second wound of the war. He and engineer Lieutenant Pierre G. T. Beauregard were sent by Quitman to investigate white flags raised over Mexico City. The officers left in charge surrendered the place to them. Lovell signaled the general that all was well, and Quitman's troops marched into the capital.[14]

At the end of the war, Lovell returned to his artillery company. He received a brevet as captain for gallantry at Chapultepec. Because the company's other officers had all been killed or wounded, Lovell became its commander when it was reorganized as a regular light artillery battery. He held this command until 1851. At that time, he was assigned to another company of the 4th Artillery stationed at Fort Hamilton in New York Harbor. Lovell married Emily M. Plimpton, the daughter of an army officer who had served in the War of 1812, Seminole War, and Mexican War. In 1854, Lovell resigned his commission and went to work for Cooper & Hewitt's Iron Works in Trenton, New Jersey.[15]

Lovell's friendship with John A. Quitman continued after the Mexican War ended. The latter became involved in a number of filibustering schemes that had Cuba as their objective, and he brought Lovell into several of them. In 1850, Narciso Lopez planned an invasion of the island. He visited Quitman, then governor of Mississippi, and "offered Quitman command of the filibuster army and rulership over a Cuban republic if the invasion force . . . proved successful."[16] The governor wrote to Lovell and asked if he was interested in participating. Quitman tempted Lovell with the possibility of becoming either a prime minister or secretary of war in the new government. Eventually, Quitman declined the command offered by Lopez but did assist him his preparations.[17]

Several years after the failure of Lopez's campaigns against Cuba, Quitman again began looking toward acquiring the island for the United States. Lovell visited him at his plantation near Natchez in November 1852 and talked about Quitman's plans. In July 1853, Quitman spent several weeks in the North seeking support for an invasion. Not until late 1854 and early 1855 did he begin formalizing his command structure. Quitman's most recent biographer writes: "Exact ranks and assignments within Quitman's invasion hierarchy remain unclear, but important spots went to Mansfield Lovell, . . . Gustavus Woodson Smith, and Jones M. Withers. . . . Lovell and Smith both resigned *their* commissions as of December 18, so that they would be free to go whenever Quitman gave the signal."[18] The opposition of President Franklin Pierce and Spanish reinforcements to Cuba caused Quitman to end his scheme in early 1855.[19]

This long and close relationship between Lovell and Quitman explains in large part the former's support of the South and Southern rights during the 1850s. Gustavus Smith pointed to the "rebel" tradition in the Lovell family as one reason for that support. He wrote that Lovell's "deep seated and firm convictions on [the] subject [of Southern rights] were based upon thorough knowledge of the history of the country from the time of the declaration of independence made by the thirteen British colonies."[20] Yet, since Professor Robert E. May describes Quitman as "the premier secessionist of antebellum Mississippi" and "one of the half dozen or so most prominent radicals and the most strident slavery imperialist in the entire South," his influence had to have been predominant in Lovell's life.[21]

The Lovell-Quitman friendship may also have been a factor in Lovell's future problems with Jefferson Davis. Quitman and Davis were personal and political rivals. This antagonism dated from the Mexican War when Davis had served under Quitman. According to one source, that relationship "provoked each of them to revile each other."[22] As Franklin Pierce's secretary of war, Davis played a role in blunting Quitman's designs on Cuba. Davis undoubtedly knew that Lovell was one of Quitman's protégés and must have held that against him.

Lovell's association with filibustering created friendships with a number of army officers, some of whom would later recommend him for a position in the Confederate army and support him after the fall of New Orleans. One of these was Beauregard, for whom Quitman wrote a letter of recommendation when the Louisianian considered joining William Walker in Nicaragua. Another frustrated former army officer who toyed with the idea of going to Nicaragua was George B. McClellan. In an 1857 letter to Little Mac, Johnson K. Duncan, also one of Lovell's friends, voiced the frustrations many of these men felt after leaving the army: "You, Lovell, G. W. [Smith] and myself are anything but satisfied with our present occupations."[23]

Lovell apparently tried to recruit 4,000 Americans for an army to support Benito Juarez and his liberal party in the Mexican civil war. In February 1858, he wrote McClellan offering him a leadership role in the army and stated: "If we could raise money I would send you a call in less than 3 months."[24] Even Joseph E. Johnston became tempted to participate in this Mexican adventure. He had been friends with Gustavus Smith, Lovell, McClellan, and other younger officers for some years. The group received no encouragement from the Juarez regime, and their plans fizzled out. When Johnston became Quartermaster General in 1860, he received congratulations from Lovell and Smith. He wrote them an acknowledgment and said, "Filibusters are doing better, I think, than Filibustering."[25]

In April 1858, Lovell became Superintendent of Street Improvements for New York City. The following November, his old friend Gustavus Smith named him to the position of Deputy Street Commissioner shortly after Smith assumed the duties as Street Commissioner. During the next year, Lovell received command of a militia company known as the City Guard. It was "composed of about one hundred gentlemen of means and position" in the city.[26] Lovell trained the men in both infantry and artillery drill, making them particularly efficient in the latter. He opposed the election of Abraham Lincoln after the latter's nomination in the summer of 1860. Lovell supported the Crittenden Compromise and other measures that he thought would avert an armed conflict between the sections.[27]

Various Southerners began contacting Lovell and Gustavus Smith about supporting their cause as early as December 1860. In that month, Beauregard visited with them on his way to assume his duties as superintendent at West Point. He told them that if Louisiana seceded from the Union he would leave the army. Lovell and Smith apparently told him "that they would act in the same manner" if similar circumstances occurred.[28] Braxton Bragg wrote to the two after his appointment as commander of the Louisiana State Army in early February 1861. He was seeking "experienced soldiers to command and train his volunteer army" and may have offered Smith and Lovell commissions. Bragg did ask their recommendations on other former army personnel whom he might also contact. Lovell replied to Bragg's letter without naming anyone, but he asked "what might [noncommissioned officers] expect in the way of rank?"[29] Bragg interpreted this letter to mean that Lovell wanted assurance of specific rank before he would join the Southern cause.

Jefferson Davis sent Raphael Semmes to New York in February 1861 to purchase weapons and other supplies for the newly formed Confederate States of America. He instructed Semmes to contact Lovell and Smith while on his mission. Davis wanted him to tell them that the South "would be happy to have their services in our army."[30] Later that month, Davis asked Beauregard to contact some of his friends about becoming bureau heads. One of Beauregard's letters went to Lovell and Smith, and he inquired when the Confederacy could expect them to offer their services. Smith replied that both he and Lovell sympathized with the seceded states though not citizens of any of them. He went on to say that he and Lovell thought that either Davis or his secretary of war would make them a direct offer. Smith stated that if such an offer were made and "if [it were] up to our standard (as we understand it)" they would favorably consider it "and in all probability" accept it.[31]

Some of this correspondence makes it sound as if Lovell and Smith were trying to sell their services to the highest bidder. In fact, Bragg later criticized Lovell for his "mercenary" actions. This criticism was unjustified. Though perhaps a biased account, Smith wrote after Lovell's death that the latter had "no shadow of doubt in his mind in regard to the legal and moral right of the course he pursued" in supporting Southern independence.[32] Smith and Lovell obviously wanted some kind of official assurance that Davis really wanted them in the Confederate army and an idea of what rank they would be offered. Considering Davis's hatred of Quitman and possible dislike of his friends, Smith and Lovell's wariness is understandable. Lovell's sentiments were expressed in a letter to McClellan after the surrender of Fort Sumter. He told Little Mac that he and Smith "had sworn they would never submit to 'throat-cutting' abolitionists who sought to give blacks 'political & civil equality with the white man,' a doctrine [that] . . . the South would resist to the death."[33]

Another criticism leveled at Lovell was that he did not go south until four months after the war started. His detractors doubted his dedication to the cause because he had waited so long to offer his services to the Confederacy. It is interesting that the same attack was not made on Smith. Of course, the latter was a native of Kentucky, and it could be argued that like that state he chose not to take sides until Kentucky's neutrality was violated. Because of illness, Smith could not leave New York City when war broke out, and he did not reach Lexington, Kentucky, until late July or early August. Perhaps Lovell chose not to make his way south until his friend could travel. He left his family in New York City and accompanied Smith to Lexington. The latter wrote after the war that Lovell had hoped that an armed conflict could be avoided and that he did not resign his job as deputy street commissioner until "he became satisfied that the Northern States would use all the resources of the general government . . . to keep the Southern States in the Union at the point of bayonet."[34]

Having learned that his two old friends had reached Lexington, Joe Johnston on August 19 recommended them to President Davis as division commanders in his army around Manassas. Johnston wrote Davis that Lovell and Smith had "always wanted to serve us" but had "not come forward [before], because, not belonging to seceded states, they didn't know how [they] would be received."[35] The two did not get to Richmond until September 11. Confederate forces had violated Kentucky's neutrality by occupying Columbus seven days before and this undoubtedly prompted them to offer their services to the Confederacy. Smith received a commission as major general on the nineteenth and soon was leading a division in Johnston's army.[36] So far no evidence has come to light concerning why Davis did not make Lovell a general at that time or whether Lovell was considered for a command under Johnston.

Yet, Lovell was in the right place at the right time. The government in Richmond began receiving complaints about Major General David E. Twiggs, commander of Department No. 1. That military department consisted of the state of Louisiana as well as the Gulf Coast counties of Mississippi and Alabama and had as its primary mission the defense of New Orleans. Former governor André B. Roman wrote Davis, "In spite of his good intentions the infirmities of General Twiggs, which confine him to his armchair, disqualify him completely for the situation he holds."[37] The seventy-two-year-old general realized that he needed assistance and asked the War Department to send two brigadier generals to the Crescent City as soon as possible. On September 25, Lovell was made a brigadier general and ordered to report to Twiggs. He would have charge of the coastal defenses of the department, an assignment that undoubtedly resulted from his recognized experience as an artillery officer. A week later, the War Department directed Brigadier General Daniel Ruggles to proceed from Pensacola, Florida, to New Orleans as an additional subordinate to Twiggs.[38]

Prominent Louisianians continued to ask Davis and Secretary of War Judah P. Benjamin to assign a competent officer to replace Twiggs. Finally, on October 5, the old general wired Richmond that because of his ill health he wanted someone sent to relieve him in command of Department No. 1. For some reason, Lovell had not yet left the capital. On October 6, he traveled to Fairfax Courthouse to consult with Beauregard on the defense of New Orleans. The next day, the War Department promoted Lovell to the rank of major general and named him as Twiggs's replacement. Twiggs received instructions to remain on duty until Lovell arrived. Lovell left Richmond on October 10 and made a brief stop in Norfolk, where his family had just arrived from New York under a flag of truce.[39]

Secretary of War Benjamin may have played a role in Lovell's assignment. In a letter to Joe Johnston, Benjamin called Lovell a "justly-esteemed officer."[40] Benjamin described Lovell to Governor Thomas O. Moore as "a brilliant, energetic, and accomplished officer" when informing that official of Lovell's assignment.[41] John B. Jones, a clerk in the War Department, intimated in his diary that Benjamin was responsible for Lovell's and Smith's appointments as generals. Late in 1861, Davis considered removing the Mississippi counties along the coast from Department No. 1 and adding them to Bragg's department, but Benjamin persuaded him to leave the area under Lovell.[42]

To date, no evidence has been found of a prewar connection between Lovell and Benjamin that would explain why the latter would have had any reason to favor Lovell. Benjamin was from Louisiana, and he may have seen Lovell's experience as an artillery officer as a good qualification for sending him to take over the defense of New Orleans.

Lovell's assignment greatly angered Braxton Bragg, commander at Pensacola, Florida. He had grown tired of his rather stagnant situation there and desired a more active field, so that he could prove himself as a commander. In a letter to a friend, Bragg stated that Davis had promised him command of the Gulf Coast from Pensacola to New Orleans. When Twiggs made it known that he would retire, Bragg expected the president to "show his sincerity and confer this command on me."[43] Bragg complained to Governor Moore after Lovell's appointment: "The command at New Orleans was rightly mine. I feel myself degraded by the action of the government."[44] He spoke harshly of Lovell as an "eleventh-hour" convert who had been "purchased in the open market by the highest bidder."[45] While Bragg acknowledged Lovell's competence and energy, he told Moore that "but for his inordinate vanity [Lovell] would be a fine soldier."[46]

Thus, Mansfield Lovell received command of one of the most important points in the Confederacy despite his northern birth and recent appointment as a general. Why would Davis have given such a position to a man he obviously had little reason to reward? Lovell's presence in Richmond as complaints about Twiggs reached the capital may account for his initial commission as a brigadier. Yet, the president did not have to order him to take charge of Department No. 1 when Twiggs asked for relief.

Davis had at least two senior generals he could have sent to New Orleans. He might easily have placed Louisiana under Bragg, unifying the defense of the Gulf Coast. Instead, in the same order that had Lovell relieve Twiggs, the War Department extended Bragg's command to include Alabama (thus reducing Lovell's department).[47] This was obviously an effort to placate one of Davis's favorite officers. Two days prior to this order, Beauregard wrote to Benjamin asking for a transfer to his native state. Disgusted by Davis's refusal to reinforce the army at Manassas so that it could go on the offensive and fearful that the enemy would soon attack the Crescent City, Beauregard wished assignment there. The chief executive responded in a letter that said the Creole general "was too valuable in Virginia to be shifted to another theater."[48] One cannot help but wonder whether Beauregard's request had any effect on Davis's decision to promote Lovell and have him replace Twiggs.

That consideration aside, we still have no answer to the question of why Davis sent Lovell to take command at New Orleans. In the absence of evidence in Davis's writings, the author has based his conclusion on the president's actions. It is clear that Davis thought that the forts below New Orleans would stop any enemy naval force that tried to ascend the Mississippi River.

Thus, he felt that the only real threat to the Crescent City would come from the upper Mississippi River. There are good indications of Davis's low regard for Department No. 1 in his choices of Twiggs as its first commander and of Ruggles as one of the generals sent to aid Twiggs. The latter was obviously too old and infirm for his job. Except for leading a division at the Battle of Shiloh, Ruggles never held an important command during the war. Had Davis really feared for the safety of New Orleans, he would have placed it under Bragg's command or sent another experienced general there. Feeling that no significant fighting would occur in Louisiana, Davis saw little possibility of Lovell gaining any recognition as commander of Department No. 1. In short, it was a "safe" assignment.

Lovell reached New Orleans on October 17 by rail from Jackson, Mississippi, and formally assumed command the next day. The New Orleans press noted his arrival and spoke of him in approving terms: "General Lovell comes among us with a high character for energy and great administrative talents, and with an already distinguished military reputation."[49] He at once began trying to rectify the numerous deficiencies he found in the city's defenses. Pie Dufour has provided a fine summary of his activities. He wrote: "General Lovell's correspondence with Richmond in his first week in New Orleans shows how quickly and efficiently he sized up the situation, evaluating needs with the eye and knowledge of an experienced and energetic officer. Lovell's energy indeed was boundless."[50]

Jefferson Davis tied Lovell's hands from the very first, making it virtually impossible for him to successfully defend the city against a strong naval attack. Before he left Richmond, Lovell spoke to both Davis and Benjamin about exercising command over the naval forces at New Orleans and on the lower Mississippi. The president refused to allow Lovell to have any control over the navy. He wrote:

> The fleet maintained at the port of New Orleans and the vicinity is not a part of your command; and the purposes for which it is sent there or removed from there are communicated in orders and letters of a Department with which you have no direct communication. It must, therefore, be obvious to you that you could not assume command of these officers and vessels, coming within the limits of your geographical department, but not placed on duty with you, without serious detriment to discipline and probable injury to the public service.[51]

In January 1862, Lovell wrote Benjamin complaining about the lack of cooperation from the navy. He reminded Benjamin of his appeal the previous October: "I felt satisfied that if the protection of the navigable streams running up into the country was removed from my control it would in all probability not be properly arranged in connection with the land defenses. . . . This is just what has happened."[52] Lovell could not order available gunboats to positions where they could best aid the forts on the lower river in resisting an attack. Neither could he push the completion of ironclads under construction at the Crescent City. Although he gave the navy as much assistance as possible, Lovell received little support from its commanders. Davis would not change his original decision, however. As Dufour and others have noted, "The divided command at New Orleans was a vital factor in its fall."[53]

Despite the obstacles he faced, Lovell succeeded in improving the defensive posture of New Orleans. He had entrenchments constructed around the city and organized some 14,000 troops for its defense. The men mounted several hundred heavy artillery pieces that had arrived during Twiggs's tenure as commander. Powder mills began turning out several tons of gunpowder a day. Lovell accumulated small arms, medical stores, clothing and other supplies for his men. Lovell's friend Gustavus Smith made a correct assessment when he wrote: "Within two months after General Lovell took command the military defenses of New Orleans were in condition to successfully resist any land attack that could probably be brought against the city."[54]

Lovell experienced somewhat less success in strengthening Fort Jackson and Fort St. Philip, the masonry forts ninety miles below New Orleans. He sent more men to the forts and supported his subordinates' efforts to improve the men's drill and discipline. Beauregard had suggested to Lovell that he have obstructions placed in the river between the forts. Lovell had his men construct a raft of cypress logs, schooners and iron cable to block the river. As an artilleryman, he recognized the weakness of the cannons, most of them old smoothbores, mounted in the forts. He attempted unsuccessfully to obtain some columbiads and rifled guns to replace the outdated ordnance. Lovell hoped that with the more effective cannons his men could cause heavy damage to an attacking enemy fleet when it slowed down to try to get past the raft. He felt that such a bombardment, properly supported by the navy, would turn back the Federal vessels.[55]

In early February 1862, Davis and the War Department began stripping New Orleans of men, material and naval vessels. Lovell received orders to send 5,000 men to Columbus, Kentucky. A month later, Davis instructed him to hurry several thousand more men to Beauregard in western Tennessee. To reinforce the naval forces above Memphis, fourteen armed river steamers moved up river from New Orleans. Lovell had only about 3,000 poorly armed and untrained militiamen left in the city. Benjamin informed Lovell that the government expected to defend New Orleans "from above by defeating the enemy" on the upper Mississippi. In early March, Lovell complained to the secretary of war about how his command was being decimated. He wrote: "Persons are found here who assert that I am sending away all troops so that the city may fall an easy prey to the enemy."[56] Several weeks later, Lovell informed Benjamin that, despite the difficulties he faced, "I shall do my duty as I understand it."[57]

All of Flag Officer David G. Farragut's naval squadron had gotten into the Mississippi River from the Gulf of Mexico by early April. In spite of this obvious threat, Davis attempted to order away from New Orleans the ironclad *Louisiana.* That vessel was not yet completed, but Davis wanted it sent north to face the Union ironclads moving toward Memphis. Both Lovell and Governor Moore protested such an action. Lovell asked that if the *Louisiana* had to go he be allowed to keep her rifled guns so he could place them in Fort Jackson. Davis still thought the forts alone would destroy Farragut's squadron. He tried to persuade Governor Moore of the correctness of his strategy: "The Louisiana may be indispensable to check the descent of the iron boats. The purpose is to defend the city [New Orleans] and valley."[58] Davis finally relented and agreed to leave the ironclad at the Crescent City.

Some of the mortar vessels under Farragut began shelling Fort Jackson and Fort St. Philip on April 15. Three days later, the bombardment began in earnest, with all twenty mortar boats participating. The Federals hoped to reduce the two forts with this fire so that the frigates could steam past without having to engage the Confederate guns. After five days, Farragut saw that this strategy would not work. The mortar vessels had caused little damage and few casualties, and only the morale of the forts' defenders suffered. Farragut decided to have his frigates run past the forts before daylight on April 24. One of his captains succeeded in cutting a gap in the obstructions. In a fight that lasted about four hours, all thirteen Federal gunboats succeeded in passing the forts and destroying almost all of the wooden Confederate vessels stationed near the forts. The next day, the advance elements of Farragut's squadron brushed aside some token opposition from batteries south of the city and dropped anchor before New Orleans. Rather than subject the civilian populace to a bombardment, Lovell decided to evacuate the city. He ordered as many troops and supplies as could get out to Camp Moore.[59]

Why had the most important city in the Confederacy fallen to the enemy so easily? In his official report, Lovell gave these three reasons:

> 1st. The want of sufficient number of guns of heavy caliber which every exertion was made to procure without success;
>
> 2d. The unprecedented high water, which swept away the obstructions upon which I mainly relied, in connection with the forts, to prevent the passage of a steam fleet up the river; and
>
> 3d. The failure, through inefficiency, and want of energy of those who had charge of the construction of the iron-clad steamers *Louisiana* and *Mississippi* to have them completed in the time

> specified so as to supply the place of obstruction; and finally the decision of the officers in charge of the *Louisiana* to allow her, though not entirely ready, to be placed as a battery in the position indicated by General Duncan and myself.[60]

Pie Dufour has advanced two main reasons for the fall of New Orleans. First, as quoted before, he said, "The divided command at New Orleans was a vital factor." Dufour went on to write: "Next to the divided command as a primary cause of the fall of the city comes the Richmond government's utter lack of appreciation of the realities of the situation at New Orleans."[61] In assessing the blame for this disaster, Dufour exonerated Lovell and concluded: "The responsibility for the fall of New Orleans rested squarely at the doors of the President of the Confederacy."[62] This author agrees completely with Dufour's conclusions on causes and responsibility. In no respect has this author found any reason to criticize Lovell's performance as commander at New Orleans. He did everything within his power to defend the city and, with proper support from Richmond, might have succeeded in turning back Farragut's attack.

As stated at the beginning of this essay, both Pie Dufour and Daniel Sutherland have adequately covered Davis's efforts to make Lovell the scapegoat for the Union capture of New Orleans. Interested readers should look at their works. It is appropriate to quote Sutherland's conclusion on why Davis went so far in persecuting Lovell: "Davis obviously sought to avoid personal blame for what in retrospect could only be judged a terrible blunder. Yet he was too intelligent, too able a man not to realize that he and his cabinet were at least partly to blame. In this context, Lovell was simply a convenient scapegoat."[63] It is hoped that this essay has demonstrated that Lovell was a perfect candidate for being made a scapegoat. If Bragg, Beauregard, or any other Southern general had been in Lovell's place, Davis probably would not have even tried to lay all the blame on them.

Lovell did not remain inactive after the evacuation of New Orleans. He began immediately to reorganize his troops and to send them to Vicksburg, Mississippi. By fortifying the bluffs near that town, Lovell hoped to prevent the enemy fleet from ascending the river any farther. He had received a proposal for fortifying Vicksburg from a Mississippi militia officer in December 1861 but could not provide any assistance to the project at that time. The Confederate defeat at Shiloh prompted Lovell to contact Beauregard about erecting heavy artillery batteries near the town. Lovell thought that this would serve a double purpose. First, it would protect the river. Second, it would provide an anchor for a defensive line running through Vicksburg to Meridian in case Beauregard's army had to retreat from Corinth. At that time, Lovell had no engineer officer to send to begin the work, so he asked Beauregard to send an engineer from his army. Once batteries had been erected, Lovell planned to transfer twelve to fifteen cannons from his works above New Orleans to be mounted at Vicksburg.[64]

The fortification of Vicksburg proceeded as quickly as Lovell could shift men, artillery and supplies northward from Camp Moore. Additional heavy artillery pieces for the river batteries arrived from Mobile, Richmond and possibly other points. Lovell sent Brigadier General Martin L. Smith, an experienced engineer, to assume command at Vicksburg and push construction of the fortifications. Smith reached the town May 12 and found three batteries completed and a fourth almost ready. He soon had more than 2,500 men under his command. On May 18, the leading ships of Farragut's squadron steamed up to the town. The Federals demanded the surrender of Vicksburg, but Smith refused. Unsure of the strength of the Confederate defenses, the Federals took no action but awaited the arrival of the remainder of the fleet. By May 20, the defenders had eighteen heavy guns mounted in seven batteries. Farragut ordered a bombardment of Vicksburg but decided not to risk an attack on the place. Lovell's promptness in providing defenses for Vicksburg prevented its fall in the spring of 1862.[65]

Lovell went to Vicksburg on May 21 to inspect its defenses. The next day, General Ruggles arrived from Beauregard's army with orders to take command there. Lovell quickly protested to Beauregard and asked him to send Ruggles to northern Mississippi where he could do no harm. Beauregard responded by ordering Ruggles to Grenada. He then wrote Lovell that the War Department had added Vicksburg to his command and removed it from Lovell's. The Creole general decided to allow Lovell to continue supervision of the Vicksburg defenses since all of the troops there came from his department. Lovell established his headquarters at Jackson, Mississippi, from which point he could move by rail to either Vicksburg or Camp Moore as circumstances dictated. He told Beauregard that he would maintain garrisons at both of those points to protect the railroads. Lovell hoped to be able to confine the Federals to a few garrisons along the Mississippi River.[66]

After Beauregard's army evacuated Corinth and retreated to Tupelo, Davis began a shake-up in the Western Theater. As part of the shake-up, he searched for someone to replace Lovell. Davis at first ordered Major General John B. Magruder to Jackson from Virginia, but that general could not go there immediately. Then Davis instructed Bragg to assume temporary command until Magruder could arrive. At that point, Beauregard informed the War Department that he was taking sick leave and that he needed Bragg to remain with the army in his absence. Davis accused Beauregard of abandoning his army and removed him from command. He named Bragg as Beauregard's replacement. Davis then decided to have Major General Earl Van Dorn replace Lovell.[67]

It appears that Davis's older brother, Joseph, had recommended Van Dorn, who was a native of Port Gibson, Mississippi, to take over Lovell's department. Joseph wrote to the president complaining of Lovell's evacuation of New Orleans and describing discontent with Lovell remaining in field command. Several months later, Lovell wrote to his wife that his brother had talked to Joe Davis and learned that the elder Davis had been instrumental in the change of command. Lovell also told his wife that at the time Joe Davis had been misinformed about what had actually happened at New Orleans but had since "seen the error of his ways[,] and would undo the mischief if the opportunity should offer."[68]

Neither Davis nor anyone in the War Department directly notified Lovell of the impending change in command. Lovell had learned by June 20 of Van Dorn's orders to relieve him. He sent a letter to Davis informing him of what he had heard and told the president that Beauregard had offered him "command of a fine corps in his army." He spoke of this opportunity as "a far better position for a soldier than one where he is incessantly striving to accomplish important objects with ridiculously disproportionate means."[69] Lovell also sent a request to the War Department asking for orders: "Would prefer to be sent to the field of immediate action. Will I be allowed a change of position to put an end to unfounded popular clamors?"[70] Though Davis had no intention of doing so immediately, he had the War Department tell Lovell that a court of inquiry would soon convene to look into the fall of New Orleans. For that reason, Van Dorn was being sent to relieve him.[71]

Van Dorn reached Vicksburg on June 27 and assumed command the next day. Lovell remained in the town to assist Van Dorn in the defense of the place. After the Federals broke off their bombardment of Vicksburg in late July, he went to Jackson to await orders for the court of inquiry. In Jackson, he learned that the inquiry had been postponed. He decided to go to Richmond and see what had happened. In a meeting with Davis, Lovell was told that a court would be convened soon. Davis also told him that he should return to Mississippi and serve under Van Dorn until the inquiry began. Lovell had no way of knowing that Davis had lied to him and was delaying as long as possible the court of inquiry.[72]

When Lovell returned to Jackson, Van Dorn assigned him to the command of a division still in the progress of organizing. Bragg had wanted Major General John C. Breckinridge and part of his division with him on his invasion of Kentucky and recommended Lovell as a replacement to command the remnants of the division. Some of Breckinridge's men did eventually serve under Lovell. Bragg had left Van Dorn's and Major General Sterling Price's commands to tie down Union forces in northern Mississippi and western Tennessee so that they could not aid the Federals in Kentucky. Van Dorn planned an offensive into western Tennessee that would involve both his and Price's troops. He asked Davis for permission to organize and equip 12,000–15,000 recently exchanged prisoners so that they could participate in the campaign. He intended these men to form part of Lovell's command.[73]

While Van Dorn traveled to Holly Springs to make preparations for the campaign, Lovell remained in Jackson preparing Van Dorn's troops to move toward the north. He entered into those duties enthusiastically and wrote his wife: "I could not have arranged it better if I had been allowed to fix it for myself." Once Van Dorn's and Price's forces united and were reinforced by the exchanged prisoners, the combined army would be as strong as Sidney Johnston's at Shiloh. Lovell expected to be second in command of the army since he outranked Price. He went on to say to his wife: "Every one of my friends seems to think I have done a first rate thing in getting into position in this manner without any help from the War Dept."[74] Lovell reached Van Dorn's army at Davis' Mill, eighteen miles north of Holly Springs about September 23. There Van Dorn formally organized Lovell's division by assigning to it three infantry brigades and one cavalry brigade. If the returned prisoners were forwarded to the army, they would form a second division under Lovell, who would then become a corps commander.[75]

Van Dorn planned to attack Major General William S. Rosecrans's army at Corinth and capture that place. The Confederate army outnumbered that of Rosecrans (22,000 to 15,000), but the Federals could quickly concentrate another 8,000 men from various outposts as reinforcements. Price cautioned Van Dorn to wait until the exchanged prisoners could join the army. Van Dorn did not want to wait until those men were armed and reorganized. The delay might give Rosecrans time to bring in reinforcements and strengthen his defenses. Lovell met Price for the first time on September 28. He informed his wife of his disappointment in Price's appearance and said, "He may be a very gallant man, but unless I am much mistaken is not a first rate soldier."[76]

Lovell's division led the advance toward Corinth. On the eve of the attack, he wrote that he expected the enemy to evacuate the town rather than try to defend it. He was in excellent spirits: "I feel as fine and hearty as a youngster of twenty and have had a better appetite than I have for the past ten years." He thought that "with good generalship" the army could drive the Federals back to the Ohio River "without the loss of much life."[77] Rosecrans did not retreat without a fight, and Van Dorn's army fought a two-day battle, October 3–4, trying to take Corinth. Lovell apparently supported Van Dorn's decision to attack rather than carry out a war of maneuver designed to force Rosecrans to fall back. When one of his brigade commanders told him that an attack could not succeed, Lovell replied that "if we could not succeed we had better lay down our arms and go home."[78]

Lovell's division played an important role in the first day's fighting. His men routed the Federals on their front and captured an artillery piece. After the battle, he wrote his friend Gustavus Smith that the army had almost succeeded in taking Corinth. He stated that "*had we known accurately the nature of the works and their positions we should have taken the place on Friday beyond all doubt.*" Like Van Dorn, Lovell underestimated the strength of Rosecrans's army, thinking that most of it had fought the Confederates on October 3. Actually, a large portion of the Union army remained in reserve during that day's battle. Lovell's account to Smith went on to read:

> We lost time in "groping" our way, and night came on before we could finish up our work which was in full tide of success. During the night fully 8000 men came in from the different outposts, and the enemy was saved—Two hours of daylight on Friday would have given us the most brilliant victory of the war, and disconcert[ed] all the enemy's plans for his offensive winter campaign. If the returned prisoners had been ready we should have succeeded.[79]

During the second day of fighting, Lovell was to push his division forward on the Confederate right and await word that Price's two divisions had broken through the Union lines. His men were then to join in the assault. The division commander on the left did not attack when scheduled, and then the center division went in prematurely. Both divisions suffered heavy casualties and had to fall back. Lovell awaited orders to advance but never received them. When his men finally began to prepare for an assault, he received word of the defeat of the other two divisions and orders to send a brigade to cover their retreat. This brigade helped check the pursuit by the Federals, and Lovell's division slowly withdrew from the field with the rest of the army. The next day, at Hatchie Bridge, Lovell's men acted as the army's rear guard and threw back an attack by Rosecrans, allowing the army to continue the retreat unmolested.[80]

Though the attack on Corinth had failed, Lovell felt good about his role in the campaign. He wrote his wife:

> If I am to believe all that I hear, my conduct throughout the operations has redounded much to my credit all through the Army. Hundreds of persons have come to me and said that although they began the campaign with prejudices against me, they would now rather serve under me than any one else. . . . I did my duty bravely and nothing more.[81]

Some criticism of Lovell's conduct during the battle has been made. The Missouri troops under Price's command questioned Lovell's failure to attack on October 4. They felt that if Lovell had sent his division into the assault, the battle would have been won.[82] Albert Castel, in his biography of Sterling Price, has picked up on this criticism. He said that Lovell "probably . . . had concluded that an assault on Corinth was hopeless, and therefore decided to have no share in it." Conceding that Lovell's assessment may have been right,

Castel nevertheless pronounced him "guilty of betraying the confidence of [his] commanding general." Castel made this undocumented comment in a footnote: "Van Dorn probably was aware of this attitude on the part of Lovell . . . but dared not make an issue of it because it would reflect on his own ability as a general (it is just possible that Lovell, a member of his class at West Point, had some sort of personal hold over him, too)."[83]

There is absolutely no evidence to support any of this criticism and certainly none to back up Castel's allegations. That Lovell still hoped for victory on October 4 is shown in a conversation he had with a subordinate after he received Van Dorn's order to retreat. He told that officer: "I don't understand this, Colonel. I've got a position here, and I can whip anything that can come out of Corinth or hell and by G-d, I don't want to leave it."[84] Van Dorn praised Lovell's conduct during the battle, writing President Davis: "General Lovell has now the entire confidence of the troops and gained reputation in the late battle."[85] Though members of the same graduating class at West Point, Lovell and Van Dorn had had no contact with each other between 1842 and the Civil War except possibly during a brief period in the Mexican War. It is impossible for me to believe that Lovell could have had any kind of hold on the hot-blooded Van Dorn that would have prevented the latter from criticizing his role in the battle.

The Corinth Campaign was Lovell's last active service in the Confederate army. Van Dorn reorganized his command after the battle and assigned Lovell to command one of his corps. Lovell remained with the army during its operations in northern Mississippi in October and November, but his men did no serious fighting. Davis urged Lieutenant General John C. Pemberton to relieve Lovell of command. Pemberton had assumed command of the newly organized Department of Mississippi and East Louisiana on October 14. Thus he became Van Dorn's superior. The War Department reorganized Pemberton's department on December 7. Those orders named Van Dorn commander of Lovell's corps and relieved Lovell. The orders directed Lovell to "await further orders" since he had not "been assigned to duty by the War Department."[86] When he learned of these orders, Van Dorn wrote Lovell: "Let it be your proud consolation that you have fought gallantly, skillfully at Corinth—persistently and bravely on the long retreat as rear-guard from that unfortunate field to Holly Springs—and subsequently from Holly Springs to Grenada. . . . I am truly sorry to lose you from the Army."[87]

Davis kept Lovell in limbo by delaying the court of inquiry to look into the fall of New Orleans. Joe Johnston asked for Lovell's reinstatement in the army in Mississippi in December 1862, but Davis turned down his request. In January 1863, Lovell traveled to Richmond to testify in a Navy Department inquiry concerning New Orleans. He asked Davis at the end of that investigation to return him to duty, but again Davis refused. Lovell finally persuaded some friends in the Confederate Congress to initiate action that would force the calling of a court of inquiry. The War Department issued orders on February 18 for the court to convene in Jackson, Mississippi, but meetings did not begin until April 4. When Federal forces threatened Jackson, the court moved to Charleston, South Carolina. In June, the court moved again, this time to Richmond, and it finally issued its findings on July 9. Those findings included some criticism of Lovell's actions but did not lay responsibility at his door. As Davis had hoped, neither did the findings assign any responsibility to him or the Navy Department. The court concluded: "General Lovell displayed great energy and an untiring industry in performing his duties. His conduct was marked by all the coolness and self-possession due to the circumstances and his position, and he evinced a high capacity for command, and the clearest foresight in many of his measures for the defense of New Orleans."[88]

Though cleared by the court, Lovell still did not receive any orders to return to active duty. Because the proceedings of the court had not officially been published and sent to Congress, Davis and the War Department could continue to keep Lovell "on the shelf." Again, Lovell appealed to his friends in Congress to request a copy of the court's report, and the House of Representatives passed a resolution to that effect on January 15, 1864. Davis did not forward the copy until June.[89]

In the meantime, Joe Johnston had assumed command of the Army of Tennessee, and he asked the War Department in January 1864 to assign Lovell to his army so he could command one of the three corps he hoped to organize. The War Department sent Johnston's request to Davis, who informed the general that he would not receive Lovell's services. Two months later, Johnston and Bragg corresponded about a new chief of artillery for Johnston's army. Bragg suggested several names, including Lovell's, and said, "Lovell was one of the best artillery officers in the old service, a good judge and fond of good horses."[90] Johnston asked for Lovell's assignment as chief of artillery, but again Davis refused the request.[91]

Lovell volunteered his services to Johnston during the Atlanta Campaign. Johnston wrote later that Lovell had done so because he was "prompted by a zeal in the cause which made him regardless of the claims of his rank."[92] Johnston assigned him to examine the crossings over the Chattahoochee River near Atlanta and prepare defenses to prevent Federal cavalry raids from surprising the city and destroying the valuable depots located there. Lovell probably worked with Gustavus Smith, now commander of the Georgia militia, in placing state troops in the defenses constructed. He traveled back and forth between Johnston's army and Atlanta during the coming weeks. In late June, Johnston sent his chief of artillery to construct redoubts along the Chattahoochee at points selected by Lovell. When Lieutenant General Alexander P. Stewart succeeded to the command of Lieutenant General Leonidas Polk's corps in late June, Johnston recommended Lovell as the new commander for Stewart's division. The War Department rejected the request. General John B. Hood replaced Johnston on July 17 and asked for Lovell's assignment to command his old corps. Richard McMurry wrote, "Hood's suggestion that Lovell be given a corps was probably based on his ability to handle troops and his familiarity with the army and its situation."[93] Hood had no better luck than had Johnston.

With this last refusal by Davis to allow him to return to field command, Lovell moved with his family to Columbia, South Carolina. Lieutenant General William J. Hardee inquired of the War Department in December 1864 whether he could assign Lovell to some command at Savannah, Georgia. As everyone before him, Hardee received a rebuff to his request. Lovell volunteered his services to the governor of South Carolina late in December. He gave as his reason his desire "to contribute to the utmost of my ability to the success of the common cause."[94] After Joe Johnston took command in the Carolinas in March 1865, he wrote to General Robert E. Lee asking him to order Lovell to report to him for duty. Lee persuaded the War Department to return Lovell to service, and, on April 7, Lovell received assignment to command of Confederate forces in South Carolina. The war was practically over, but Lovell remained on duty until late May, when he finally surrendered to Union forces.[95]

Lovell chose to remain in the South after the end of the war. He traveled to New York briefly to acquire capital that would enable him to establish and operate a rice plantation near Savannah. A tidal wave destroyed his first crop, and he was forced to return to New York City with his family. There he became a civil engineer and surveyor. Lovell served as assistant engineer under former Union general John Newton when the latter supervised the removal of obstructions in the East River at Hell's Gate. After a short illness, Lovell died June 1, 1884.[96]

It is difficult to assess Lovell's performance as a Confederate general since he held only two commands of any significance. He performed as well as anyone could have as commander at New Orleans and practically performed miracles in placing the city in a decent defensive position. He might have had more success with more support from Davis and the government in Richmond. As stated above, his quick actions in May 1862 prevented Union forces from capturing Vicksburg. Lovell participated enthusiastically in the Corinth Campaign and showed his abilities as a field commander. An examination of Lovell's war service shows him to have been a slightly above average general. If he had been given the opportunity, he would have made a good division commander in the Army of Tennessee. Despite all of the abuse heaped upon him by governmental officials, soldiers, and civilians, he remained loyal to the Confederacy to the bitter end. Dan Sutherland put it well when he wrote, "Wherever he served, Lovell won the praise of subordinates and superior alike for his courage, energy, and devotion."[97]

Notes

"Mansfield Lovell" was originally published in *Leadership During the Civil War: The 1989 Deep Delta Civil War Symposium; Themes in Honor of T. Harry Williams,* ed. Roman J. Heleniak and Lawrence L. Hewitt, 85–110 (Shippensburg, PA, 1992), and is reprinted with permission of White Mane Publishing Company.

1. U.S. War Department, *The War of the Rebellion: A Compilation of the Official Records of the Union and Confederate Armies,* 128 vols. (Washington, DC, 1880–1901), series 1, vol. 5:797. Hereinafter cited as *OR.* All references are to series 1 unless otherwise indicated).
2. Ibid., vol. 6:751.
3. Ibid., vol. 32, pt. 3:592.
4. Gustavus W. Smith, "Mansfield Lovell," in *Fifteenth Annual Reunion of the Association of the Graduates of the U. S. Military Academy* (East Saginaw, MI, 1884), 117.
5. Daniel E. Sutherland, "Mansfield Lovell's Quest for Justice: Another Look at the Fall of New Orleans," *Louisiana History* 24 (1983): 239.
6. Charles L. Dufour, "The Night the War Was Lost. The Fall of New Orleans: Causes, Consequences, Culpability," *Louisiana History* 2 (1961): 174. See also Charles L. Dufour, *The Night the War Was Lost* (New York, 1960).
7. Sutherland, "Lovell's Quest for Justice," 258–59.
8. Ezra J. Warner, *Generals in Gray: Lives of the Confederate Commanders* (Baton Rouge, LA, 1959), 194; Smith, "Mansfield Lovell," 113, 117; Mary Gillett, "Joseph Lovell," *Dictionary of American Military Biography,* 3 vols., ed. Roger J. Spiller (Westport, CT, 1984), 2:662.
9. Smith, "Mansfield Lovell," 113.
10. Ibid.; Warner, *Generals in Gray,* 194.
11. Smith, "Mansfield Lovell," 113–14; Robert E. May, *John A. Quitman: Old South Crusader* (Baton Rouge, LA, 1985), 164.
12. May, *John A. Quitman,* 164.
13. Quoted in ibid., 449n40; ibid., 164, 352, 359.
14. Ibid., 189, 194; John F. H. Claiborne, *Life and Correspondence of John A. Quitman,* 2 vols. (New York, 1860), 2:308–310; Pierre G. T. Beauregard, *With Beauregard in Mexico . . . ,* ed. T. Harry Williams (Baton Rouge, LA, 1956), 85, 97–98; Francis B. Heitman, *Historical Register of the United States Army, from Its Organization, September 29, 1789 to September 29, 1889* (Washington, DC, 1890), 644; Warner, *Generals in Gray,* 194.
15. Smith, "Mansfield Lovell," 114–15; Heitman, *Historical Register of the United States Army,* 644; Warner, *Generals in Gray,* 194; Donovan Yeuell, "Mansfield Lovell," *Dictionary of American Biography,* 20 vols., ed. Dumas Malone (New York, 1928–44), 11:441.

16. May, *John A. Quitman,* 236, 238.

17. Ibid., 238.

18. Ibid., 274, 277, 290.

19. Ibid., 295.

20. Smith, "Mansfield Lovell," 117–18.

21. Ibid., xv.

22. Jefferson Davis, *The Papers of Jefferson Davis,* vol. 2, *June 1841–July 1846,* ed. James T. McIntosh (Baton Rouge, LA, 1974), 107n104; May, *John A. Quitman,* 244, 263, 271.

23. Quoted in Stephen W. Sears, *George B. McClellan: The Young Napoleon* (New York, 1988), 52–53; Beauregard, *With Beauregard in Mexico,* 106–8.

24. Sears, *McClellan,* 56–57.

25. Quoted in Gilbert E. Govan and James W. Livingood, *A Different Valor: The Story of General Joseph E. Johnston, C.S.A.* (Indianapolis, 1956), 23–24, 26.

26. Yeuell, "Mansfield Lovell," 441; Smith, "Mansfield Lovell," 115; Gustavus W. Smith, *Confederate War Papers* (New York, 1884), 360–61.

27. Smith, "Mansfield Lovell," 115–16.

28. Alfred Roman, *The Military Operations of General Beauregard in the War between the States, 1861 to 1865,* 2 vols. (New York, 1884), 1:15.

29. Quoted in Grady McWhiney, *Braxton Bragg and Confederate Defeat,* vol. 1, *Field Command* (New York, 1969), 153.

30. Jefferson Davis, *Jefferson Davis, Constitutionalist: His Letters, Papers and Speeches,* 10 vols., ed. Dunbar Rowland (Jackson, MS, 1923), 5:55.

31. *OR,* vol. 53:127, 129–30; Roman, *Beauregard* 1:20.

32. Smith, "Mansfield Lovell," 118.

33. Quoted in Sears, *McClellan,* 66.

34. Smith, "Mansfield Lovell," 116; Smith, *Confederate War Papers,* 363–64.

35. *OR,* vol. 5:797–98; Govan and Livingood, *A Different Valor,* 73.

36. *Richmond Whig,* Sept. 13, 1861; Warner, *Generals in Gray,* 281.

37. *OR,* vol. 53:739.

38. Ibid., vol. 6:738, 744, 745, 746, 751; ibid., vol. 53:743.

39. Ibid., vol. 6:643, 751; ibid., vol. 51, pt. 2:344; ibid., vol. 53:742, 748, 748–49; Roman, *Beauregard* 1:154.

40. *OR,* vol. 5:883.

41. Ibid., vol. 6:751.

42. Ibid., vol. 6:786; John B. Jones, *A Rebel War Clerk's Diary at the Confederate States Capital,* 2 vols., ed. Howard Swiggett (New York, 1935), 1:89; entry of Jan. 6, 1862,

Thomas Bragg Diary, Thomas Bragg Papers, 1861–1862, Southern Historical Collection, Univ. of North Carolina at Chapel Hill.

43. Major General Braxton Bragg to "My dear Doctor," n.d., Braxton Bragg Papers, William P. Palmer Collection, Western Reserve Historical Society, Cleveland, Ohio (hereinafter cited as Bragg Papers, Palmer Collection).

44. Bragg to Moore, Oct. 31, 1861, Thomas O. Moore Papers, Louisiana and Lower Mississippi Valley Collection, Hill Memorial Library, Louisiana State Univ., Baton Rouge (hereinafter cited as Moore Papers).

45. Bragg to "My dear Doctor," n.d., Bragg Papers, Palmer Collection; Bragg to Moore, Nov. 14, 1861, Moore Papers; *OR,* vol. 6:759.

46. Bragg to Moore, Nov. 14, 1861, Moore Papers; Bragg to "My dear Doctor," n.d., Bragg Papers, Palmer Collection; *OR,* vol. 6:759.

47. *OR,* vol. 6:751.

48. T. Harry Williams, *P. G. T. Beauregard: Napoleon in Gray* (Baton Rouge, LA, 1955), 101.

49. *New Orleans Daily Picayune,* Oct. 17, 19, 1861.

50. Dufour, *The Night the War Was Lost,* 93.

51. *OR,* vol. 6:645–46.

52. Ibid., 798.

53. Dufour, *The Night the War Was Lost,* 340; John T. Scharf, *History of the Confederate States Navy* (1887; reprint, New York, 1977), 301.

54. Smith, "Mansfield Lovell," 119; *OR,* vol. 6:754, 758, 760, 776.

55. Roman, *Beauregard,* 154; Smith, "Mansfield Lovell," 118–19; Sutherland, "Lovell's Quest for Justice," 236; *OR,* vol. 6:646–47, 881.

56. *OR,* vol. 6:561, 823, 847; Smith, "Mansfield Lovell," 120–21; Smith, *Confederate War Papers,* 71–75.

57. *OR,* vol. 6:865–66.

58. Ibid., 646–47, 878; Dufour, *The Night the War Was Lost,* 200–202.

59. *OR,* vol. 6:510–11, 514–15, 883; Dufour, *The Night the War Was Lost,* 223–301.

60. *OR,* vol. 6:517.

61. Dufour, *The Night the War Was Lost,* 340, 342.

62. Ibid., 174.

63. Sutherland, "Lovell's Quest for Justice," 254.

64. *OR,* vol. 6:567, 651, 783, 784, 877, 884, 885.

65. Ibid., vol. 6:567, 585; ibid., vol. 15:6–8; ibid., vol. 52, pt. 2:316; U.S. Naval War Records Office, *Official Records of the Union and Confederate Navies in the War of the Rebellion,* 31 vols. (Washington, DC, 1894–1927), series 1, vol. 18:491–93, 507, 508.

66. *OR,* vol. 15:739–40, 741–42, 746; ibid., vol. 52, pt. 2:318–19. The War Department order was dated May 26, 1862, several days after the Lovell-Ruggles confrontation. The author believes that Davis directed the War Department to remove Vicksburg from Lovell's department after he learned that the general was involved in the town's defense.

67. *OR,* vol. 15:758; ibid., vol. 17, pt. 2:591, 599, 613; Robert G. Hartje, *Van Dorn: The Life and Times of a Confederate General* (Nashville, 1967), 182–83.

68. Mansfield Lovell to Emily Lovell, Sept. 20, 1862, Mansfield Lovell Papers, Keeler Collection, Henry E. Huntington Library, San Marino, CA; quoted in Clement Eaton, *Jefferson Davis* (New York, 1977), 154.

69. Quoted in Smith, *Confederate War Papers,* 96–97.

70. *OR,* vol. 52, pt. 2:325.

71. Lovell to Emily Lovell, June 26, 1862, Lovell Papers.

72. *OR,* vol. 15:15, 769; Hartje, *Van Dorn,* 183.

73. *OR,* vol. 16, pt. 2:995; ibid., vol. 17, pt. 2:697; Lovell to Emily Lovell, Sept. 11, 19, 1862, Lovell Papers; Archer Jones, *Confederate Strategy from Shiloh to Vicksburg* (Baton Rouge, LA, 1961), 76–77.

74. Lovell to Emily Lovell, Sept. 20, 1862, Lovell Papers.

75. Lovell to Emily Lovell, Sept. 25, 28, 1862, Lovell Papers; *OR,* vol. 17, pt. 2:711.

76. Lovell to Emily Lovell, Sept. 28, 1862, Lovell Papers; Hartje, *Van Dorn,* 214–15; Albert Castel, *General Sterling Price and the Civil War in the West* (Baton Rouge, LA, 1968), 106.

77. Lovell to Emily Lovell, Oct. 1, 1862, Lovell Papers.

78. *OR,* vol. 17, pt. 1:417.

79. Lovell to "My dear G. W.," Oct. 26, 1862, Lovell Papers; *OR,* vol. 17, pt. 1:404–405; Castel, *General Sterling Price,* 123–24.

80. *OR,* vol. 17, pt. 1:379–81, 405–406; Hartje, *Van Dorn,* 228–32.

81. Lovell to Emily Lovell, Oct. 13, 1862, Lovell Papers.

82. R. S. Bevier, *History of the First and Second Missouri Confederate Brigades, 1861–1865* (St. Louis, 1879), 154.

83. Castel, *General Sterling Price,* p. 124 and note 40.

84. Quoted in Hartje, *Van Dorn,* 233.

85. *OR,* vol. 52, pt. 2:381.

86. Ibid., vol. 17, pt. 2:733, 745, 779, 786, 787; ibid., vol. 52, pt. 2:381, 729; Lovell to Emily Lovell, Nov. 4, 27, 28, 30, Dec. 12, 1862, Lovell Papers.

87. Quoted in Smith, *Confederate War Papers,* 98.

88. *OR,* vol. 6:555–56, 600, 606, 639–42; quoted in Dufour, *The Night the War Was Lost,* 349; Smith, *Confederate War Papers,* 98–108; Sutherland,"Lovell's Quest for Justice,"247–53.

89. *OR,* vol. 6:554.

90. Ibid., vol. 32, pt. 2:563–64, 592.

91. Ibid., 636.

92. Joseph E. Johnston, *A Narrative of Military Operations During the Late War Between the States* (1874; reprint, New York, 1969), 332.

93. Richard M. McMurry, *John Bell Hood and the War for Southern Independence* (Lexington, KY, 1982), 137; *OR,* vol. 38, pt. 4:749, 797; *OR,* pt. 5:892; ibid., vol. 52, pt. 2:681; Johnston, *Narrative of Military Operations,* 332, 345, 572–73; Govan and Livingood, *A Different Valor,* 429n59.

94. Quoted in Smith, *Confederate War Papers,* 115; Mary Boykin Chestnut, *Mary Chestnut's Civil War,* ed. C. Vann Woodward (New Haven, 1981), 626, 630, 671, 691, 692.

95. *OR,* vol. 46, pt. 3:1339; ibid., vol. 47, pt. 2:1454; ibid., pt. 3:565, 688, 765, 817, 861, 862, 866, 872, 873.

96. Warner, *Generals in Gray,* 195; Smith,"Mansfield Lovell,"126; Yeuell,"Mansfield Lovell,"442.

97. Sutherland,"Lovell's Quest for Justice,"247.

General Braxton Bragg. Library of Congress.

Braxton Bragg and the Confederate Invasion of Kentucky in 1862

Lawrence Lee Hewitt

HISTORIANS HAVE TARNISHED THE CHRIST-LIKE IMAGE OF ROBERT E. LEE IN RECENT years while the portrait of a contentious Braxton Bragg, Lee's antithesis in the Confederacy, remains unrevised. Bragg's denigration continues with his most recent besmircher describing him as being a "disgusting" individual.[1] I ask you to rid your mind of such preconceptions and to consider a passage from Shakespeare: "The evil that men do lives after them; The good is oft interred with their bones."[2] And so it is with Bragg and his accomplishments as commander of the Western Department.

On June 17, 1862, an ailing General P. G. T Beauregard relinquished his command of the Western Department to General Braxton Bragg, and three days later President Jefferson Davis confirmed Bragg as permanent commander of that department.[3] Bragg faced a situation that had rapidly deteriorated since January when the Confederates had held all of Tennessee and a large part of Kentucky and had controlled the Mississippi River from Columbus, Kentucky, to the delta. During the interim, a succession of Confederate defeats left both states virtually under the control of Union forces. Federal flotillas now steamed unmolested both above and below Vicksburg. The Union right flank extended into Mississippi, the vanguard occupied north Alabama, and the left flank threatened Cumberland Gap. Could Bragg devise some plan to halt this momentum of the Union forces?

Bragg's immediate command consisted of the demoralized Army of the Mississippi, an army whose failure to destroy the enemy at Shiloh and its subsequent withdrawal from Corinth to Tupelo had caused its morale to plummet and its desertions to escalate. Shortages of food, of transportation, and of competent junior officers intensified Bragg's concerns as a commander.[4]

Uncertain of the limits of his command, Bragg requested clarification. He was notified that the Western Department embraced "that part of Louisiana east of the Mississippi, the entire States of Mississippi and Alabama, and that portion of Georgia and Florida west of the Chattahoochee and Apalachicola Rivers."[5] Tennessee was not mentioned. Because of this omission, Bragg believed that his command encompassed the whole of Tennessee, because the Western Department had earlier included that entire state.[6] This, Bragg's erroneous assumption concerning the extent of his command, became the primary determinate in bringing about the invasion of Kentucky.

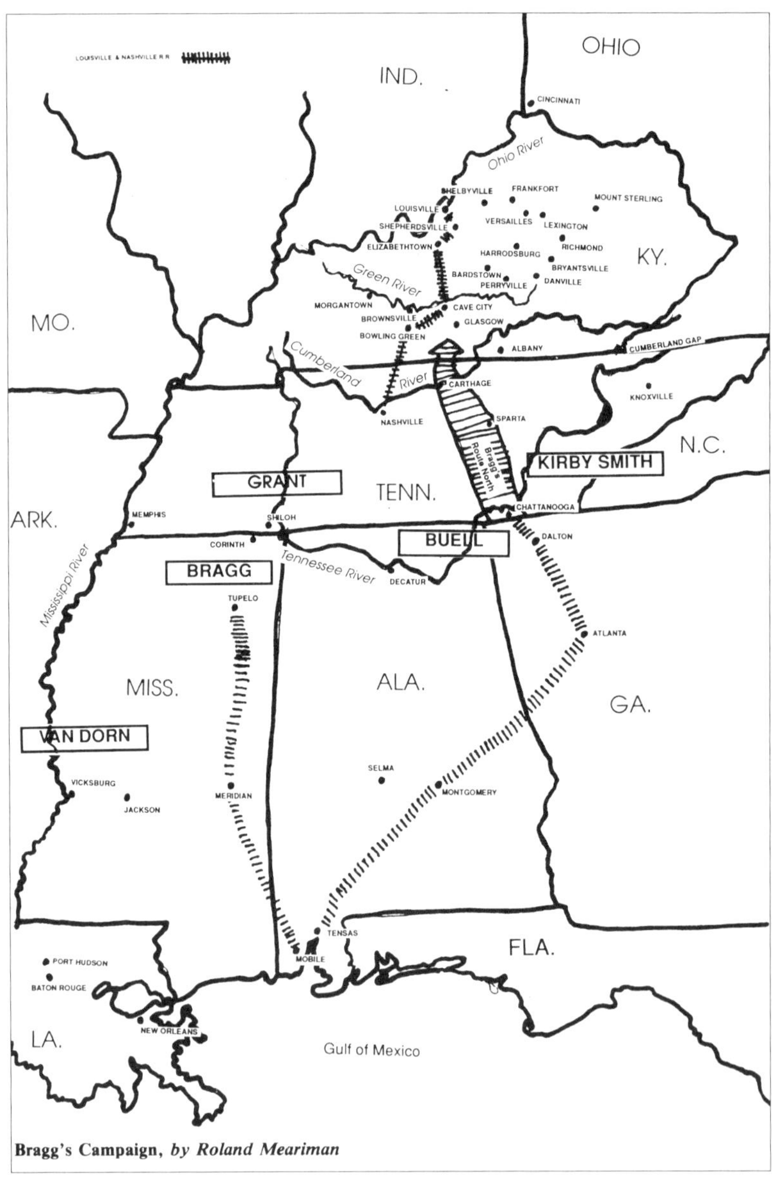

Bragg's Campaign, ***by Roland Meariman***

The situation that Bragg faced clearly called for a "defensive-offensive" policy, that is, he would assume a defensive posture and only attack should an advantageous opportunity present itself. Union Major General Henry W. Halleck had a combined force of 104,000 troops in and around Corinth. To oppose this horde, Bragg had only 42,000 men. He could augment this field force with 11,000 other Confederates who were scattered from northwestern Mississippi to southeastern Louisiana. Another 4,000 Confederates were already engaged in defending his left flank at Vicksburg, while 7,000 more guarded Mobile.[7]

Bragg considered his right flank to be East Tennessee, where Major General Edmund Kirby Smith had 11,000 men dispersed between Chattanooga and Cumberland Gap. The Federals had 8,000 soldiers in eastern Kentucky threatening Cumberland Gap, 7,000 in northern Alabama threatening Chattanooga, and an additional 7,000 occupying Middle Tennessee.[8] Although the Federals outnumbered the Confederates in East Tennessee, the threats to that area were less ominous than were those to Vicksburg and northern Mississippi.

But Bragg soon learned that fortune had smiled upon him. In early June Union General Halleck had divided his massive force into three armies; one army had the task of securing West Tennessee; another army advanced toward Tupelo, and the remaining army, under Major General Don Carlos Buell, moved eastward through northern Alabama to link up with the forces already poised for an attack upon Chattanooga.[9]

Bragg saw in the breakup of Halleck's forces an opportune moment for the launching of an offensive; however, he felt that, because of their geographically important positions, Vicksburg and East Tennessee would have to be made safe before he could justify any offensive campaign. Once he had secured Vicksburg and Chattanooga and had acquired the necessary transportation for his army, Bragg would strike at Corinth.

Vicksburg required a competent commander. President Davis's first choice had been Bragg, but when Bragg assumed Beauregard's command, Davis chose Major General Earl Van Dorn. Bragg regretted having to detach Van Dorn from Tupelo,[10] but more importantly, Bragg had to furnish him with the troops necessary to defend Vicksburg. Time prohibits a discussion of the ingenuity demonstrated by Bragg in the transfer of these troops to Vicksburg, but his strategy is noteworthy. His yielding of territory of little strategic value in northwestern Mississippi enabled him to reinforce Vicksburg as well as to retain at Tupelo the maximum number of troops for an offensive strike on Corinth.

Since Bragg could now relax his concern for Vicksburg, he focused his attention on East Tennessee. Beauregard, aware that East Tennessee had been established as a separate department in March, had disregarded Kirby Smith's requests for aid. When Adjutant and Inspector General Samuel Cooper had ordered him to reinforce East Tennessee, Beauregard had replied: "It would be fatal to detach any troops from this army at this moment, when I expect daily to meet such superior forces."[11] Mistakenly believing that his own responsibilities included East Tennessee, Bragg reacted differently than had Beauregard.

Before receiving any communication from Kirby Smith, Bragg made the necessary preparations for sending reinforcements to East Tennessee. On June 20, Kirby Smith informed Bragg that he had abandoned Cumberland Gap to avoid being invested and that at least two brigades were necessary to insure the holding of the railroad which constituted the most direct route from Virginia to the western Confederacy.[12]

Bragg, busy readying his forces at Tupelo for an assault on Corinth, withheld the reinforcements and suggested to Cooper on June 22 that East Tennessee should be reinforced from the Atlantic coast. On the following day, Secretary of War George Randolph telegraphed Bragg that reinforcements had already been sent from the Atlantic coast to Chattanooga but that more were needed. Although he urged Bragg to send the additional reinforcements,[13] Bragg took no action, still hoping that troops from elsewhere would be sent to bolster the forces in East Tennessee.

The fact that no further requests to aid East Tennessee were made during the next three days must have eased Bragg's concern for his right flank. When he arose on the morning of the twenty-seventh, he had reason to be assured that two of his three prerequisites for an offensive had been met—a firm hold on the Mississippi at Vicksburg and a stabilized and seemingly secure front in East Tennessee. Bragg issued a proclamation that day; he announced his formal assumption of permanent command of the Western Department, and he concluded: "A few more days of needful preparation and organization and I shall give your banners to the breeze."[14] Bragg merely awaited the completion of the final prerequisite—that of molding the Army of the Mississippi into a potent, offensive force—before launching an attack upon Corinth.

Later in the day a telegram from Kirby Smith shattered Bragg's illusion of East Tennessee's security: "Buell is reported crossing the river at Decatur and daily sending a regiment by rail toward Chattanooga. I have no force to repel such an attack."[15] Bragg acted promptly; he dispatched Major General John P. McCown's 3,000-man division from Tupelo to Chattanooga that same day. Bragg's movement against Corinth would have to be postponed, pending further developments at Chattanooga.[16] Bragg's commitment to securing East Tennessee before attacking Corinth allowed Kirby Smith to dictate Bragg's offensive strategy.

While Bragg waited, he continued to improve his command at Tupelo. His transfer of McCown to East Tennessee cleared the way for Major General Sterling Price to assume command of the Army of the West. This promotion convinced Price, a Missourian, to remain east of the Mississippi, and he raised the spirits of his men by promising to lead them back to their native state by way of Kentucky.[17]

Bragg, in an effort to increase the efficiency of the Army of the Mississippi, promoted Major General Leonidas Polk to second-in-command of the Western Department. Polk's elevation allowed Bragg to resign from immediate command of the Army of the Mississippi and to appoint Major General William J. Hardee as his successor. Bragg felt that Polk made a better bishop than he did a general. In Bragg's opinion, Hardee would provide better training for the troops in Polk's corps. By virtue of his prewar service, Hardee qualified as the best man available to train the Army of the Mississippi.[18] Bragg's methods, despite some discontent and criticism from the ranks, undoubtedly enhanced the condition of his forces.

Bragg also attempted other improvements within the Western Department. With Corinth garrisoned by Union forces, any strategic movement of troops by Bragg depended upon adequate rail connections through the central Confederacy. Consequently, he repeatedly urged the completion of the railroad from Meridian, Mississippi, to Selma, Alabama.[19]

While Bragg strove to increase the likelihood of his success, events were developing rapidly in East Tennessee, or so it seemed to Kirby Smith. During the first week of July, both Kirby Smith and Bragg received inaccurate reports that Union forces in northern Alabama were moving toward Nashville and even beyond.[20] Both generals chose to believe that the enemy was withdrawing, because that is what they wished to believe. In Bragg's mind, the enemy's retrograde movement would provide him with an opportunity to attack Corinth. For Kirby Smith, the withdrawal of the enemy would relieve the pressure on Chattanooga which, in turn, would enable him to operate against Cumberland Gap.

Kirby Smith did not perceive Buell's true designs until July 10, when he telegraphed Bragg that Buell continued to move on Chattanooga with nearly 30,000 men. Kirby Smith's inability to cope with this situation caused him to magnify the problem in a letter to President Davis on July 14. "I see by the Northern papers that three divisions of Grant's army are to operate against East Tennessee in connection with Buell's corps. This brings an overwhelming force, that cannot be resisted except by Bragg's cooperation."[21] However, Kirby Smith wanted more than just Bragg's cooperation—he wanted Bragg to assume command at Chattanooga personally.

Kirby Smith's letter indicates his unwillingness to assume responsibility for Chattanooga, because he did not ask for reinforcements from Bragg; instead, he wanted Bragg to move his army into Middle Tennessee. The letter also provides an insight into Kirby Smith's future plans: "The disorders in that State [Kentucky] are extremely propitious for his [John Hunt Morgan's] operations." Kirby Smith reasoned that, because Kentucky had warmly welcomed raiding Confederate cavalry, it would do the same for his entire command.[22] Thus, by July 14, Kirby Smith had determined to invade Kentucky and Bragg would find himself and his army *dragged* along.

Kirby Smith began transferring troops from Chattanooga toward Cumberland Gap on July 17, knowing full well that the threat to that city had not diminished.[23] He was shrewd enough to know that, in order to justify an advance into Kentucky, he would first have to make Chattanooga secure *or* at least, to shift the responsibility to someone else—namely, Bragg.

On July 18, the Department of East Tennessee was extended to include "that part of the State of Georgia which is north of the railroad leading from Augusta via Atlanta to West Point, and so much of North Carolina as is west of the Blue Ridge Mountains in that State."[24] In as much as this order removed that portion of Georgia which had been part of the Western Department, Bragg must have been informed of this action. It is not known when he learned of this departmental adjustment, but on the following day he received a telegram from Kirby Smith urging him to move into Middle Tennessee and promising him his cooperation.[25]

This message caused Bragg to reconsider the scope of his command. Apparently, his responsibilities did not include the defense of East Tennessee, after all. On the other hand, the messages from Kirby Smith were becoming more urgent, and if Chattanooga was his responsibility, Bragg would have to act promptly. He wrote to Kirby Smith on July 20:[26]

> I am left in doubt, from what I can learn, whether yours is a separate command, or still, as formerly a part of General [Albert Sidney] Johnston's old department and hence embraced within my command. Can you enlighten me by copies of any orders or instructions you may have? My only desire is to know the precise limits of my responsibilities, not to interfere in the least with your operations and command, as you must know best when and how to act, and have my fullest confidence.

Bragg's uncertainty, as evidenced by the above, offers a rationale for his actions up to this point, and finally, it justifies his most crucial decision, the decision to transfer the Army of the Mississippi to Chattanooga.

Bragg had always considered East Tennessee to be of more importance than northern Mississippi and Alabama. Even before he could establish the limits of his responsibilities, the threat to East Tennessee compelled Bragg to act. Leaving Price and the Army of the West to operate in northern Mississippi and northwestern Alabama, Bragg ordered the Army of the Mississippi to Chattanooga. By moving the infantry stationed at Mobile in advance, reinforcements reached Chattanooga by July 27. Despite some evidence to the contrary, Bragg still had no intention of invading Kentucky.[27] Many have disagreed with this view in the past, because to accept it is to acknowledge that Bragg had the ability to recognize the threats throughout the Western Theater and enough intelligence to determine the strategic value of each endangered region.

Still, Bragg does not deserve all the credit for moving to Chattanooga, because his failure to learn conclusively that his department did not include Chattanooga greatly influenced his decisions.[28] Had he been informed of the limits of his department prior to July 27 or 28, he undoubtedly would have attacked Corinth, believing that a successful attack there would force the recall of Buell's army.

The odds in northern Mississippi had greatly changed since June 17, when Bragg had succeeded Beauregard. The breaking up of the Union forces at Corinth, begun in early June, continued throughout July. Bragg could now throw his concentrated force against relatively isolated detachments of the enemy forces.[29] By July 27, however, Bragg was committed to transferring his army to Chattanooga.

Bragg met with Kirby Smith in Chattanooga on July 30. Two days later, in a letter to Cooper, Bragg outlined a plan of action that encompassed the entire Western Theater. Realizing that his own army would not be able to move for at least ten days, he reinforced Kirby Smith with two brigades for a movement against Cumberland Gap. Should this effort prove to be successful, Bragg would be able to withdraw a greater number of troops from the Knoxville region for his own advance against Buell in Middle Tennessee. Bragg realized that Buell would take advantage of any delay to consolidate his position in Middle Tennessee and probably to draw reinforcements from Corinth. Should such a situation develop, Price and Van Dorn could then unite their forces and launch an attack against West Tennessee.[30]

While Bragg awaited the arrival of his wagon train, which was moving overland, Kirby Smith continued to shift his own troops, as well as those on loan from Bragg, to Knoxville in preparation for his advance against Cumberland Gap. On August 9 he notified Bragg that he had a report that the Federals had a month's supply of provisions, adding that, if the information proved to be correct, it would take more time to reduce Cumberland Gap than Bragg indicated he was willing to take. Further, Kirby Smith requested that he be allowed to move on Lexington, Kentucky, thereby isolating Cumberland Gap.[31]

By this time, Bragg had also begun to think beyond Middle Tennessee. On the tenth of August, he authorized Kirby Smith to move into Kentucky, but urged him to cooperate with a Confederate column moving westward from Virginia, and warned him not to stray too far into Kentucky before Bragg had engaged Buell.[32]

Bragg's reply to Kirby Smith marks a turning point in his planning. He had realized that Buell had had ample time to concentrate his forces in the defenses around Nashville, making an attack too costly even for the combined force of both his and Kirby Smith's armies. He also knew that Buell would be strongly reinforced from West Tennessee, which would enable Price and Van Dorn to drive the Federals from that region. All of these factors prompted Bragg to formulate a new plan of operations—he would attempt, by *maneuvering* and not by fighting, to regain control of the territory in the Western Theater that had been lost by the Confederacy. In accordance with this new strategy, Bragg authorized Kirby Smith to enter Kentucky prior to the defeat of Buell and the reduction of Cumberland Gap. Likewise, when Bragg learned that Kirby Smith had determined to advance on Lexington, he fully concurred.[33]

Between August 10, when Bragg agreed to let Kirby Smith move into Kentucky, and August 24, when Bragg endorsed Kirby Smith's advance on Lexington, their correspondence and Kirby Smith's actions clearly illuminate the command relationship between the two men; Kirby Smith followed Bragg's instructions to the letter. The leading authorities on this subject contend that Bragg and Kirby Smith operated independently of each other, and that Kirby Smith, because he was a department commander, did not have to obey the orders of Bragg, at least, not while the troops of the two commanders were separated. The major weakness of Bragg's offensive, according to these historians, was that he had no control over Kirby Smith's movements and only nominal control over those of Van Dorn and Price; moreover, as each army advanced from its base, communication between Bragg and the army commanders became increasingly more difficult. What has not been pointed out is that, in view of their prior relationship, Bragg had no reason to suspect that Kirby Smith would fail to follow orders after he reached Kentucky.[34]

While Kirby Smith moved on Cumberland Gap and into Kentucky, Bragg hastily prepared his own army for an offensive campaign. In an effort to increase the efficiency of the command, he urged President Davis to relieve several high-ranking generals and to replace them by promoting junior officers, men whom Bragg considered to be more competent. Davis declined, realizing that such action would bring about a barrage of criticism. Moreover, such promotions were not his to make, according to the Confederate Constitution. Criticism voiced against Bragg's proposal resulted in a rift in the army's command system that would ultimately lead to his removal at the end of 1863.[35]

Bragg also continued to press for a concerted movement by Price and Van Dorn against West Tennessee. Such action would either prevent the transfer of Union troops from the Corinth area to reinforce Buell or result in an easy route into Middle or West Tennessee after the withdrawal of those troops. Unfortunately, Bragg failed to inform the War Department of his instructions to Price and Van Dorn.[36]

On August 28, Bragg moved out of Chattanooga—initiating a campaign that most scholars believe should have resulted in a significant Confederate victory. Historians have long speculated about Bragg's motives when he left Chattanooga. The most prominent among them intimate that Bragg wanted to maneuver Buell out of Middle Tennessee without fighting him. Moreover, they describe Bragg as being indecisive throughout the campaign.[37]

Was Bragg indecisive? Although Kirby Smith was pushing him to move into Kentucky, he also advocated that Bragg first defeat Buell. Davis recommended the same primary action to Bragg—Buell should be defeated first. Tennessee Governor Isham Harris also pressured Bragg to defeat Buell and to retake Nashville before any advance be made into Kentucky.[38] Did Bragg ignore their mutual recommendation because he was indecisive? Hardly. Bragg made his own decisions and remained committed to them until circumstances dictated otherwise. Bragg kept his own council and arrived at the final decision to invade Kentucky; likewise, he alone made the decision to bypass Nashville. He planned to conduct a campaign of maneuver and to engage in combat only when it was to his advantage to do so.

Bragg assumed immediate command of the army at Chattanooga and divided it into two corps, one commanded by Polk and the other by Hardee. Crossing the Tennessee River at Chattanooga, the Confederates took up the march toward Middle Tennessee on August 28. By September 3, the advance units had reached Sparta.[39]

Prior to his arrival at Sparta, Bragg had probably intended to move directly toward Lexington via Albany. However, upon his arrival at Sparta on September 4, he was informed that forage was scarce along his projected line of march, that Kirby Smith was in no immediate danger from Federal forces, and that Buell was still at Nashville. These three factors encouraged Bragg to attempt to cut Buell's communications with Louisville.[40]

Bragg captured Munfordville on September 17, thus effectively blocking Buell's direct line of retreat to Louisville. Although Buell had advanced to Bowling Green, Bragg chose not to attack him because of Buell's superior troop strength. Bragg was also unable to lure Buell into launching an assault. A shortage of supplies finally compelled Bragg to move, northeastward to Bardstown. If Kirby Smith's forces had been concentrated at nearby Shelbyville, as Bragg expected them to be, the two Confederate armies would then have been in position for a simultaneous advance on Louisville. But, contrary to Bragg's instructions, Kirby Smith had dallied in eastern Kentucky, and Bragg was compelled to allow Buell to reach Louisville without a fight.

Bragg has been severely criticized for not attacking Buell before he reached Louisville, but without support from Kirby Smith, Bragg could only have attacked Buell with 28,000 men at any given time. Buell's force must have numbered more than 39,000 men during the period in which Bragg might have risked battle. Prior to Bragg's withdrawal to Bardstown, Union ranks had swollen to 52,000 troops. Furthermore, Bragg was no Robert E. Lee, and none of his subordinates ever came close to emulating "Stonewall" Jackson.

A smashing victory could not have been achieved by such an inferior force, and a defeat, with Confederate units scattered throughout eastern Kentucky, might well have been disastrous to the Confederate cause. The wiser choice was for Bragg to postpone battle until he could unite with Kirby Smith and the Confederate column from Virginia. In an effort to achieve such a concentration of forces, Bragg left Polk in command at Bardstown on September 28 and went to Lexington to confer with the dilatory Kirby Smith.[41]

Bragg was already aware that the invasion had fallen apart. Kirby Smith's faulty deployment prohibited Bragg from uniting more than fifty percent of the Confederates in Kentucky against Buell. The Virginia column had failed to prevent the escape of the Union garrison at Cumberland Gap. Van Dorn and Price had failed to prevent Buell's reinforcement from the Corinth area. Operations at Baton Rouge and Port Hudson had delayed the arrival of reinforcements from Van Dorn. And finally, Kentuckians failed to rally to the Confederate standard.[42]

On October 2, while at Lexington, Bragg learned that Kirby Smith's detachment between Louisville and Frankfort had been driven back by the enemy. Believing that Buell was moving on Frankfort, Bragg planned a Napoleonic concentration of his forces on the battlefield. Kirby Smith would attack Buell's vanguard and Polk would advance directly from the south against the Union right flank. The two Confederate forces would then unite somewhere in the vicinity of Frankfort. Fortunately, Polk discovered that Buell was moving on Bardstown instead of Frankfort, and he acted accordingly. Bragg responded by planning an even larger concentration at Harrodsburg, to the southeast; Polk and Kirby Smith would be joined by other Confederate forces scattered about eastern Kentucky, by the reinforcements from Van Dorn that had finally reached Knoxville, and by a few more fresh recruits. Bragg wired Polk on October 4: "Keep the men in heart by assuring them it is not a retreat, but a concentration for a fight. We can and must defeat them."[43]

About noon on October 7, Bragg learned that Buell was advancing by several routes and that Polk was being pressed near Perryville. Bragg promptly decided to destroy Buell in detail, beginning with the Union column near Perryville. Conflicting orders issued by subordinates and Kirby Smith's failure to ascertain that the Union advance on Frankfort was merely a diversion resulted in Bragg's fighting the Battle of Perryville with only three divisions. Bragg, too, never realized that Buell was concentrating his forces near Perryville.[44] And Buell had mustered quite a force.

On September 29 the last of Buell's forces from Tennessee marched into Louisville where his Army of the Ohio was being reorganized to include reinforcements from West Tennessee and those newly mustered into service. Marching along four routes, Buell led this horde of 87,000 men against Bragg on October 1.[45]

From the outset there was continual skirmishing as the vanguard of Buell's army overtook Polk's rear guard. The pursuit of Polk to Perryville proceeded at a slower pace than did the race for Louisville. The weather was unusually dry and the shortage of water became an acute problem. Although many of the Union soldiers must have dropped out, a forced march the previous evening allowed Buell to concentrate virtually his entire command just west of Perryville by the morning of October 8.[46]

A number of bizarre factors enabled Bragg's vastly outnumbered force to win a tactical victory that day. However, before darkness ended the fighting, Bragg realized that he confronted the preponderance of Buell's army; consequently, he wisely considered discretion to be the better part of valor and made the decision to withdraw. Still hoping that he could achieve a concentration, Bragg issued orders immediately, and early on the ninth of October, the victorious but battered Confederate army moved northeastward toward Harrodsburg. The town was a death trap, because it placed Buell on an almost clear route to Bragg's base of supplies at Bryantsville and his subsequent line of retreat. For Bragg to have done otherwise would have enabled Buell to cut off Kirby Smith. The threat to his rear induced Bragg to abandon all hope of a concentration; he ordered Polk to begin moving south, even before Kirby Smith had reached Harrodsburg.[47]

Buell's slowness to advance, following the Battle of Perryville, allowed Bragg to regroup the majority of the dispersed Confederate forces in Kentucky at Bryantsville. Upon his arrival, Bragg found that very little food had been transferred from Lexington to Bryantsville. The lack of comestibles in the area necessitated a further move. Should Bragg move against Buell, or should he retreat into East Tennessee?[48] Van Dorn's rumored defeat at Corinth, the absence of support from Tennessee, the failure of reinforcements to arrive from Van Dorn, and especially the indifference of the Kentuckians—all indicated that a withdrawal was Bragg's only viable option. Although all of his chief subordinates concurred, it was Bragg who would be held responsible, because the final decision was his alone. When asked about the best test of greatness in a general, the Duke of Wellington responded: "To know when to retreat and to dare to do it."[49] It was fortunate for the Confederacy that Bragg had the courage to order the retreat.

Bragg's army left Bryantsville for Cumberland Gap on October 13, even before all of Kirby Smith's command had reached the former place. At no time during the campaign was the entire available Confederate force united on one field. The only Confederate multi-army offensive had been a failure.[50] Or had it? Just what had Bragg accomplished since succeeding Beauregard barely three months earlier?

The Confederates had invaded Federal territory, whether Kentucky was a border state or not. During the course of the campaign, they had recaptured the strong defensive position at Cumberland Gap; they had virtually annihilated the Federal force sent to Richmond, Kentucky, to stop the movement before it was well underway, and they had captured a garrison of over 4,000 men at Munfordville. In addition, Bragg had won a tactical victory at Perryville against odds of almost four to one. Following this engagement, the Confederates were allowed to make an almost unharassed retreat, undoubtedly giving the impression to many Northerners, Southerners, and Europeans alike that the Confederates could come and go almost as they pleased, regardless of the odds.

Bragg had so maneuvered his forces that northern Alabama, Middle Tennessee below Nashville, and East Tennessee were cleared of Federal troops. Bragg's strategy had forced Major General Ulysses S. Grant to detach a considerable force from West Tennessee to aid Buell, and in so doing, Grant was held in check until late in the year by the small force that Bragg had left behind in Mississippi under Van Dorn and Price; a portion of Van Dorn's command had also established a second bastion on the Mississippi River at Port Hudson, thereby giving the Confederates control of a substantial stretch of that waterway as well as the Red River. Moreover, the Federal threat to Chattanooga, an important base for offensive operations for either side, was abated for an entire year. Can it be denied that Bragg's invasion of Kentucky prolonged the war?

Bragg's campaigning by maneuver should have become the primary Confederate strategy in the Western Theater. Even if the invasions were to fail, the maneuvers would upset Federal plans and would delay their penetration of the Confederate interior. More importantly, Bragg had demonstrated what a single individual could accomplish when charged with the defense of the entire Western Theater. Between June and October of 1862, Bragg had displayed greater ability at grand strategy than did any other Confederate commander during the entire war; he had accepted responsibility for key positions hundreds of miles from his headquarters and had successfully held them for the Confederacy. The departmentalization of the Western Theater, a process which had begun even before Bragg withdrew from Kentucky, coupled with the assignment of the ineffective Joseph E. Johnston to coordinate activities in the region made a repetition of Bragg's strategy of 1862 a virtual impossibility for the future.

With the advantage of hindsight, it is certain that the rift within the high command of the Army of the Mississippi required major adjustment. It was unfortunate for the Confederacy that Davis selected Joseph E. Johnston instead of Bragg to command the Department of the West, for Bragg had already demonstrated his willingness to assume responsibility, a characteristic sadly lacking in Johnston. Bragg's elevation would have removed him from his disgruntled subordinates in the Army of the Mississippi and provided Davis with the opportunity to select a more congenial but probably less competent commander for that army.

However, Bragg was destined for lesser things. Shortly after his withdrawal from Kentucky, a resident of Middle Tennessee wrote in her diary: "When the History of this war is impartially written, it is my deliberate opinion, that to *Bragg* will be awarded the praise of *having done more with his men and means, than any other Gen. of the War, with equal resources.*"[51] Will the "disgusting" Bragg ever receive an impartial treatment? Possibly, but not before every Southerner has been convinced that Robert E. Lee was only *human,* a mere *mortal* who failed Wellington's test of greatness and senselessly risked the destruction of his army at Sharpsburg.

Notes

"Braxton Bragg and the Confederate Invasion of Kentucky in 1862" was published in *Leadership During the Civil War: The 1989 Deep Delta Civil War Symposium: Themes in Honor of T. Harry Williams,* ed. Roman J. Heleniak and Lawrence L. Hewitt, 55–72 (Shippensburg, PA, 1992), and is reprinted with permission of White Mane Publishing Company.

1. Richard M. McMurry, *Two Great Rebel Armies: An Essay in Confederate Military History* (Chapel Hill, NC, 1989), 8.
2. William Shakespeare, *The Tragedy of Julius Caesar* (New York, 1966), 56.
3. U.S. War Department, *The War of the Rebellion: A Compilation of the Official Records of the Union and Confederate Armies* (Washington, DC, 1880–1901), series 1, vol. 17, pt. 2:606, 614. Hereinafter cited as *OR.* All references are to series 1 unless otherwise indicated).
4. Ibid., 627–28.
5. Ibid., 619, 627. In addition to Samuel Cooper, copies were sent President Davis, Secretary of War George Randolph, and General Robert E. Lee.
6. Grady McWhiney, *Braxton Bragg and Confederate Defeat,* vol. 1, *Field Command* (New York, 1969), 43; U.S. War Department, *The Official Atlas of the Civil War* (1892; reprint, New York, 1958), plates CLXIII, CLXIV.
7. *OR,* vol. 10, pt. 2:235; ibid., vol. 17, pt. 2:635, 661; ibid., vol. 15:770. Unless otherwise stated, figures represent effective strength.
8. *OR,* vol. 10, pt. 2:235, 573.
9. Ibid., vol. 17, pt. 2:3.
10. Ibid., 599, 613, 627–28.
11. Ibid., 630, 644–46, 681; ibid., vol. 10, pt. 2:307–8; ibid., vol. 16, pt. 2:679–81, 683–85.
12. *OR,* vol. 17, pt. 2:618; ibid., vol. 16, pt. 2:695.

13. *OR*, vol. 16, pt. 2:701–702.

14. Ibid., vol. 17, pt. 2:622–26; ibid., vol. 16, pt. 2:704–9.

15. *OR*, vol. 16, pt. 2:709.

16. Ibid., vol. 16, pt. 2:710, Ibid., vol. 17, pt. 2:651–52.

17. *OR*, vol. 17, pt. 2:114, 616, 636, 645–646. Davis had approved the return of Price's three Missouri brigades to the Trans-Mississippi, but such a move was virtually impossible, and following Price's speech on July 17, his soldiers agreed not to make the attempt.

18. Ibid., 636; Marcus J. Wright, *General Officers of the Confederate Army . . .* (New York, 1911), 21–23; Ezra J. Warner, *Generals in Gray: Lives of the Confederate Commanders* (Baton Rouge, LA,1959), 124, 242. Hardee had served as commandant of cadets at West Point and wrote the standard textbook, *Rifle and Light Infantry Tactics* (1853–55).

19. *OR*, vol. 17, pt. 2:612, 624; Robert C. Black III, *The Railroads of the Confederacy* (Chapel Hill, NC, 1952), 156–57; McWhiney, *Braxton Bragg*, 262.

20. *OR*, vol. 16, pt. 2:719–23; Bragg's return endorsement to Kirby Smith's July 4, 1862, telegram, Braxton Bragg Papers, William P. Palmer Collection, Western Reserve Historical Society, Cleveland, OH (hereinafter cited as Bragg Papers, Palmer Collection).

21. Ibid., 725–27.

22. Ibid., 726–27.

23. Ibid., 728.

24. Ibid., 729.

25. Kirby Smith to Bragg, July 19, 1862, Bragg Papers, Palmer Collection.

26. *OR*, vol. 17, pt. 2:651–52.

27. Ibid., vol. 16, pt. 2:731, 738–39; ibid., vol. 17, pt. 2:638, 651–52. Although many historians contend that Bragg intended to invade Kentucky before his departure from Mississippi, ample evidence exists to disprove this hypothesis, including Bragg's letter to Kirby Smith on July 20 and the movements of Withers's division between June 27 and July 21.

28. *OR*, vol. 17, pt. 2:627; ibid., vol. 16, pt. 2:734–35, 745–46.

29. *OR*, vol. 17, pt. 2:90, 99–102, 114, 143–44, 638, 648. By July 21, Bragg could field a force nearly equal to the 55,000 effective Federal troops that remained in the vicinity. Of these, 13,000 were stationed at Memphis, which had no direct rail connections to Corinth and was farther away from that city than Tupelo; 8,000 men were stationed at Jackson, Tennessee, which was in direct rail communications with both Memphis and Corinth; 5,000 troops were encamped about ten miles due north of Corinth on Clear Creek; 5,000 were located in extreme north-

western Alabama; 14,000 were located in Corinth; and 9,000 were located in an arc around that city to the south at a distance of from ten to fifteen miles.

30. Ibid., 655–56, 662; ibid., vol. 16, pt. 2:741.
31. *OR,* vol. 16, pt. 2:742–45, 748.
32. Ibid., 748–49, 995.
33. Ibid., 766–67, 775.
34. McWhiney, *Braxton Bragg,* 273, 281–82; Thomas L. Connelly, *Army of the Heartland: The Army of Tennessee, 1861–1862* (Baton Rouge, LA,1967), 204, 206–9; Stanley F. Horn,"Perryville," *Civil War Times Illustrated* 4, no. 10 (Feb. 1966): 6; *OR,* vol. 16, pt. 2:734–35, 748, 751, 755.
35. *OR,* vol. 17, pt. 2:627–28, 647, 654, 658, 667–68, 671–72; Thomas L. Connelly and Archer Jones, *The Politics of Command: Factions and Ideas in Confederate Strategy* (Baton Rouge, LA, 1973), 50. Although the correspondence cited were written prior to Bragg's arrival at Chattanooga, he continued to push for the replacement of high-ranking officers.
36. *OR,* vol. 17, pt. 2:666, 675–76, 685, 688, 690, 705–706; McWhiney, *Braxton Bragg,* 238–39.
37. McWhiney, *Braxton Bragg,* 272–92, 294, 297–98; Connelly, *Army of the Heartland,* 206–9, 217, 221–22, 225, 227–28, 230–32, 234–36, 241, 244–46, 262. Connelly wrote:"The wisdom of this decision was debatable, for it remained to be seen which would secure Kentucky: an early defeat of Buell or an immediate concentration of Confederate strength in the Bluegrass." McWhiney titled the chapter dealing with July 29–September 22, 1862,"By Marching, Not By Fighting."
38. *OR,* vol. 16, pt. 2:739–40. 766–67; Jefferson Davis, *Jefferson Davis, Constitutionalist: His Letters, Papers and Speeches,* 10 vols., ed. Dunbar Rowland (Jackson, MS, 1923), 5:313; McWhiney, *Braxton Bragg,* 283; Connelly, *Army of the Heartland,* 221.
39. Organization formed at Chattanooga during mid-August 1862, Bragg Papers, Palmer Collection; *OR,* vol. 16, pt. 2:759, 772, 782, 784; ibid., vol. 17, pt. 2:618–19; ibid., vol. 16, pt. 1:1089; Connelly, *Army of the Heartland,* 224–26.
40. Connelly, *Army of the Heartland,* 224–26.
41. McWhiney, *Braxton Bragg,* 286–292, 289n34; Stanley F. Horn, *The Army of Tennessee: A Military History* (1941; reprint, Norman, OK, 1953), 170–72; Joseph H. Parks, *General Edmund Kirby Smith, C.S.A.* (Baton Rouge, LA, 1954), 229; Joseph H. Parks, *General Leonidas Polk, C.S.A.: The Fighting Bishop* (Baton Rouge, LA, 1962), 259–62; Thomas Robson Hay,"Braxton Bragg and the Southern Confederacy," *Georgia Historical Quarterly* 9 (1925): 270; Robert U. Johnson and Clarence C. Buel, eds., *Battles and Leaders of the Civil War,* 4 vols., (1887; reprint, New York, 1956), 3:8, 10; Connelly, *Army of the Heartland,* 233–36; T. Harry Williams,"The Military Leadership of North and South," in *Why the North Won the Civil War,* ed. David

Donald (New York, 1960), 42; *OR,* vol. 16, pt. 2:246–47, 461, 493, 511–14, 516, 784, 988; ibid., vol. 17, pt. 2:144, 174; ibid., vol. 16, pt. 1:6–726, 1090–91. Because Bragg failed to give battle, Williams, cited above, labeled him "the general of the lost opportunity." The "Abstract from Field Return of the Army of the Mississippi, August 27, 1862" indicates 27,816 men present for duty, including 1,700 men of Preston Smith's brigade, which had previously been transferred to Kirby Smith's command. As Buell's combined force numbered 64,000 men, and 25,000 were detached in Middle Tennessee, he must have marched into Kentucky with 39,000 troops and received an additional 8,000 men when Thomas's division arrived on September 20.

42. *OR,* vol. 16, pt. 2:876; Connelly, *Army of the Heartland,* 236–39, 241–42; McWhiney, *Braxton Bragg,* 295–300.

43. *OR,* vol. 16, pt. 2:896–98, 903–905.

44. Ibid., pt. 1:1091–92; Johnson and Buel, *Battles and Leaders* 3:47. Hardee was the immediate commander of the Confederates being pressured near Perryville.

45. *OR,* vol. 16, pt. 2:558; Frederick H. Dyer, *A Compendium of the War of the Rebellion,* 3 vols. (1908; reprint, New York, 1959), 1:432–437; Charles K. Messmer, "City in Conflict: A History of Louisville, Kentucky, 1860–1865" (master's thesis, Univ. of Kentucky, 1953), 183–84. The figure of 87,000 effective troops was arrived at after a careful examination of the returns and organization tables appearing in the *Official Records.* It was not based upon the estimates given by the various officers after the battle, especially those estimates given to the "Buell" commission. Also, the figures do not take into account any stragglers.

46. Messmer, "City in Conflict," 184–86.

47. *OR,* vol. 16, pt. 1:1121; Connelly, *Army of the Heartland,* 267–68.

48. McWhiney, *Braxton Bragg,* 321.

49. Burton Stevenson, ed., *The Home Book of Quotations: Classical and Modern,* 10th ed. (New York, 1967), 1867; McWhiney, *Braxton Bragg,* 321; Connelly, *Army of the Heartland,* 270–76.

50. Connelly, *Army of the Heartland,* 279–80.

51. Bettie Ridley Blackmore, "Behind the Lines in Middle Tennessee, 1863–1865: The Journal of Bettie Ridley Blackmore," ed. Sarah Ridley Trimble, *Tennessee Historical Quarterly* 12 (1953): 54, cited in McWhiney, *Braxton Bragg,* 383.

General Joseph Eggleston Johnston. Library of Congress.

Tennessee and Mississippi, Joe Johnston's Strategic Problem

Archer Jones

From the fall of 1862 until the summer of 1863 General Joseph E. Johnston was in command of the Confederate forces between the Appalachian Mountains and the Mississippi River. A super department had been created under his command in November, 1862, embracing the armies of John C. Pemberton in Mississippi, Braxton Bragg and Edmund Kirby Smith in Tennessee and the Confederate troops stationed in Alabama. Johnston's effective command of this area, the Department of the West, lasted until May, 1863, when he took active charge of the forces attempting to relieve Vicksburg. While there has been much consideration of the merits of the establishment of this department, there has been little attention given to Johnston's work as commander. Proponents of the controversial Johnston hold, as he did himself, that he was given an unworkable assignment, while his critics said that he wasted opportunities, was inert, or at best failed to use his power, brooding "for months over the proper exercise of his command, bemoaning the fact that he had so much authority that in reality he had none."[1] Both sides are agreed, however, that he did nothing while in purely departmental command. His biographers consequently concern themselves with his relations with the Confederate Government and his problems in Bragg's army.[2]

This curious failure to scrutinize Johnston's operations in this critical period stems in large measure from Johnston himself. His memoirs, offered as his "contribution of materials for the use of the future historian of the War between the States,"[3] contributed very little beyond a brief for his increasingly bitter feud with Jefferson Davis. Concerned with proving his case and showing that he was right in the various controversies of the time, he neglected his own activities during this western period of his service. Historians can hardly be blamed for assigning no higher value to him than he did to himself, and so neglecting significant activity on his part. Actually, however, Johnston seems to have been quite active in making strategic plans for his department, though the course of military events never brought them into play effectively.

Johnston's operations and his planning were complicated by the character of his principal subordinates, Bragg and Pemberton. In January, 1863, General Bragg's perennial difficulties with his subordinates began, and Johnston was called upon to investigate them and report. Distrust of Bragg in Richmond led Johnston, either by accident or design, to be under the impression that he must remain with Bragg's army from the latter part of March until ordered to Mississippi in May.[4] His relations with Pemberton were made difficult by differences over the proper strategy for defending Vicksburg. Johnston favored relying on a mobile army, while Pemberton counted on fortified points, President Davis agreeing.[5] Though Pemberton's successful defense during the winter of 1863 may have induced some complacency about the defense of the Mississippi on Johnston's part, he was dissatisfied with Pemberton and his arrangements. He received little intelligence from Pemberton, whom he felt "ignores his authority, is mortified at his command over him, and receives his suggestions with coldness or opposition."[6] But Johnston felt that he was unable to direct Pemberton's operations or dispositions from his distant post in Tennessee and so Pemberton did things in his own way.[7]

Johnston had assumed command in December, 1862, believing that co-operation with the Confederate forces of Lieutenant General T. H. Holmes in the Trans-Mississippi would be the best means of defending the Mississippi and would even yield an opportunity to concentrate sufficient forces for an offensive against Grant.[8] But Holmes succeeded in persuading the War Department that he was unable to send reinforcements and, by January, Johnston's hopes, "tall castles in the air," had been "blown away."[9] But even before his last hopes of co-operation from the Trans-Mississippi had faded, Johnston had developed new plans, utilizing the cavalry in his department. Originally hoping that the activities of N. B. Forrest's cavalry in West Tennessee, sent from Bragg's army, might "delay" Grant's December advance on Vicksburg,[10] Johnston came to see an increased role for cavalry, after Grant's retreat had been forced by Forrest's raids and by the destruction during December of Grant's base at Holly Springs, Mississippi, by Pemberton's subordinate, Major General Earl Van Dorn.

Fearful of a renewed advance by Grant, Johnston seized on a large cavalry force as the means to prevent another approach to Vicksburg. Obviously a Federal base would not again be such an easy mark as Holly Springs. Furthermore, if the result was to be commensurate with the means, a larger force was indicated. Here he saw an opportunity to secure the co-operation of the Tennessee and Mississippi forces of his department through the mobility of the mounted branch of the service. Forrest's cavalry was at hand in western Tennessee and more could be concentrated near the scene of any projected operations. So from his headquarters in Jackson, Mississippi, Johnston, in late December, informed Bragg of his intention to "unite Forrest and [Colonel Philip D.] Roddy with Van Dorn for future operations," and asked that they be told and that he be informed of their whereabouts.[11]

A large cavalry force was thus to be assembled, utilizing the resources of the two subordinate departments, and was to be employed, not to delay, but, significantly, to "prevent Grant's advance."[12] Bragg's defeat by Rosecrans' Federal Army at Murfreesboro, Tennessee, on December 31–January 3, 1863, changed the picture somewhat. Though it was soon clear that Rosecrans was not immediately following up his success at Murfreesboro, Bragg's safety caused Johnston considerable concern. An attempt was made to make General Bragg's heavy losses good from whatever sources were available—newly conscripted men or reinforcements from the east. But Johnston's main reliance was still to be on cavalry. The same interdepartmental cavalry force under Van Dorn was to be employed, but this time to aid the forces in Tennessee as well as those on the Mississippi. The objective of the "cavalry expedition in two departments" was reported to Pemberton as intended "to interrupt any movement into Mississippi or Middle Tennessee," to Bragg as covering his "left by preventing Federal troops from going from West to Middle Tennessee," and to President Davis that the 6,000 cavalry under Van Dorn were being sent "to Bragg's aid to operate on the enemy's communications."[13]

The complete objective in organizing and dispatching this force is further revealed in its three-fold mission with Bragg's army. Its primary purpose was to aid Bragg by taking "part in a battle, in the event of the advance of the Federal army. . . . " Until the battle took place the second part of its mission, attacking the enemy's communications, might be developed. This role was to be expanded if there seemed to be no prospects of a battle, by moving "into Kentucky or farther."[14]

An effort to reinforce Pemberton in December with infantry from Bragg's army had taken three weeks. This experience had confirmed Johnston's pessimistic view of the possibilities of co-operation between the two major armies of his department.[15] However, he was now planning to exploit on a grand scale the potentialities of the employment of cavalry against the communications of the Union armies, while, at the same time, obtaining co-ordination among the departments by using the greater mobility of cavalry. It was this large cavalry force, organized as a corps under Van Dorn, to which Johnston attached "great importance" and on which he relied to aid Bragg and supply the only "pressure" which he regarded "possible by the troops in Mississippi."[16]

The scope of this interdepartmental effort was further enlarged. Additional cavalry was obtained from Major General S. B. Buckner, now commanding the District of the Gulf, embracing most of Alabama, with its headquarters at Mobile.[17] Since the cavalry in Mississippi was "almost unorganized," General Van Dorn would require considerable time to concentrate and organize the various forces which were to make up his corps. Some of the troops from Alabama were even without arms.[18]

One of the objectives of Van Dorn's corps, operating on Rosecrans' communications, failed to develop as Johnston planned. In spite of the fact that the cavalry was "so poorly equipped, and the difficulties of supplying deficiencies so great," that the troops were unable to leave northern Mississippi until February 7, they made the overland march to Bragg's army in two weeks and Van Dorn was able to report his command for duty on February 22.[19] In response to Van Dorn's query as to what his position would he, Johnston, now himself in Tennessee, enlightened him by explaining that he was intended to aid Bragg in the event of a battle or to raid Rosecrans' communications if a battle was not imminent. He was instructed further that "movements in General Bragg's theatre of operations will be, necessarily, under his control; those from beyond it" Johnston would "at least inaugurate."[20]

Though he agreed with Van Dorn's desire to "cross the Cumberland and operate on the north bank of the Cumberland and on the banks of the Ohio," what seemed to be an impending advance by Rosecrans detained not only Van Dorn's cavalry corps but that of Major General Joseph Wheeler as well.[21] Guarding the flanks of the army and engaging in skirmishing and minor actions with the enemy was hardly the duty for which Van Dorn or Johnston had envisaged the enormous cavalry force (nearly one-third of Bragg's 48,000 men) which had been assembled in Tennessee.

It was not entirely fear of a Federal advance which held the cavalry with the army. Van Dorn's corps accomplished another very important, if perhaps not anticipated, service. The stationing of his cavalry on the left flank of the army near Columbia, Tennessee enabled Bragg "to command all the resources of the Duck River Valley and the country southward," making it possible for him to feed his army and so remain in Middle Tennessee.[22] Johnston found the question of food supplies, subsistence, "the vital one," as the army was "living from hand to mouth."[23]

Bragg's army had been subsisting in the same area for some months and not only had the territory been combed by the army's agents but by those of the Commissary General of Subsistence at Richmond as well. After a long and sometimes severe correspondence in which Johnston was unable to obtain any appreciable quantity of subsistence from the Commissary General's depot at Atlanta, he was obliged to establish what was, in effect, his own commissary bureau for his department. He sent out his own agents to obtain supplies for Bragg's army throughout the Department of the West.[24] While a similar dearth of subsistence afflicted the Department of East Tennessee, which Johnston sought to remedy by foraging raids into Kentucky, the situation was critical only in Bragg's army. In Mississippi and the District of the Gulf supplies were plentiful.[25] The problems of subsistence for Bragg's army absorbed more and more of Johnston's efforts and attention as time wore on, and the situation remained critical and hand-to-mouth.[26] So, while the duty of Van Dorn's cavalry was more prosaic and much less spectacular than that of raids on the enemy's communications, it was no less important for holding the Confederate position in Middle Tennessee.

Johnston's attention was concentrated upon General Bragg's army, posted near Tullahoma, Tennessee, northwest of Chattanooga. He was fairly confident that Bragg could maintain his position if Rosecrans attacked with his present strength. The Confederates were disposed more south than southwest of the Union position at Murfreesboro, in order to cover the area from which they drew their supplies. While this position made their left flank quite secure, it kept them off the direct line between the Union army and its objective, Chattanooga. For the same reason they were vulnerable to being outflanked on their right. Should the enemy pursue this most likely line of action, Johnston and Bragg thought that if only a part of Rosecrans' army attempted it, either part of the divided forces might be successfully attacked, while the flank of the whole Union army would be vulnerable on a flank march. Even if Rosecrans should succeed in a flanking movement with his whole army, Bragg could "exchange bases with him very advantageously," as his supplies were drawn from south of Murfreesboro already.[27]

If Rosecrans' army were reinforced, however, Bragg's would be forced to yield to an advance, as the Federal army would be strong enough to threaten the Confederate communications, while maintaining in front a force too strong to be attacked. Should the Confederates be thus compelled to abandon Middle Tennessee, unpleasant alternatives were foreseen by Johnston. If the army should take up a position at Chattanooga, the lack of forage in East Tennessee and northern Georgia would not only make it necessary for the cavalry to be separated from the army but it was even "doubtful if forage for a reasonable baggage train" would be available. It would be unlikely, therefore, that the army, weakened by the loss of its cavalry, would be able to hold an area so weak defensively as Johnston regarded East Tennessee. The indicated course would then be to withdraw south of the Tennessee River and defend East Tennessee and Georgia from there.

The alternative to this line would be to take the army into northern Mississippi, possibly also occupying part of western Tennessee. The movement there "would be attended with great risk" due to the lack of supplies en route through a country now held by the enemy as well as the difficulty of crossing the unbridged Tennessee River on the way. While supply difficulties were contemplated there also, and the presence of strong Federal forces at Corinth, Mississippi complicated matters, there were compensating advantages. Not only would the Army of Tennessee be able to co-operate with Pemberton's army in Mississippi and draw reinforcements from it, but, in that region of more plentiful forage, the cavalry could be kept with the army. From this position Johnston thought that the army might be able to recross the

Tennessee River in order to prevent any invasion of East Tennessee or to defend northern Georgia. Its position would, of course, be most advantageous if Rosecrans' army should, instead of marching toward Atlanta, come to Grant's assistance.[28]

Johnston thought the retreat of the army into northern Mississippi was the preferable alternative, especially as he believed that the communications of any Federal advance on Atlanta would be very vulnerable. He made his arrangements accordingly. He had already set on foot efforts to obtain pontoons and now instructed General Pemberton to have his supply organization provide depots of supplies on or near the railway running north from Meridian to Corinth. While Pemberton said that these preparations were under way, Johnston sent agents of his own departmental supply organization into the area to establish depots, so, presumably, as not to have to rely in an emergency on Pemberton's supposedly deficient subsistence organization.[29]

While concerned with the contingency of defeat, Johnston expanded his defensive preparations. Concerned about the Department of East Tennessee, "a mere line"[30] running from northeast to southwest along the railroad from Knoxville to Chattanooga, Johnston noted the badly scattered condition of the troops there and determined to remedy this and turn the department into an asset for the defense of Middle Tennessee. He rigorously applied the principle of concentration, ordering the new commander, Brigadier General Daniel S. Donelson, to "form a strong reserve" from "his infantry not employed in guarding bridges or keeping the disloyal in subjection." Regarding the railway bridge defenses as "too extensive," he had them reconstructed in order to require only the minimum garrisons. The troops thus concentrated were to be kept "at two or three points near the railroad" which were "selected with reference to movement into Middle Tennessee or toward the gaps in the Cumberland Mountains." Nor was there any difficulty in having his instructions carried out, for the department commander reported in early April that, with a few exceptions, "the entire infantry force of this department can be thrown to the railroad in three hours time. . . . "[31] Having had the infantry in the department concentrated and organized into brigades, he had the cavalry sent into Kentucky in order that they might supply themselves. The cavalry force beyond the mountains in Kentucky had the additional purpose of collecting "cattle for the army and be in position to learn the enemy's intentions and report them. . . . "[32]

Having placed the Department of East Tennessee in the best posture for defense and so arranged the forces that they might immediately reinforce General Bragg, Johnston looked as well to providing support for East Tennessee itself, should it be menaced. Apparently believing that Bragg could spare no infantry, Johnston projected for East Tennessee the same system of cavalry assistance which he contemplated for Pemberton. He thought that the difficulty of supplying an army invading East Tennessee would force it to bring its own supplies, thus making its supply trains an attractive and vulnerable target for cavalry. Even if the enemy occupied East Tennessee, cavalry, breaking his communications, might well force his withdrawal. He intended that part of Bragg's cavalry should be used to help East Tennessee in this way, with Brigadier General John Hunt Morgan's large brigade particularly in mind.[33]

Just as Johnston still looked for the forces of the Trans-Mississippi to co-operate and assist in the defense of the Mississippi,[34] so he hoped to secure aid for East Tennessee from southwestern Virginia. Though President Davis agreed that assistance should be provided from Major General Samuel Jones and the other forces in southwestern Virginia, Johnston had made constant appeals for reinforcements from that quarter and both he and the Richmond authorities had organized diversionary raids into Kentucky with the assistance of General Jones, there had been no positive general instructions to provide assistance if need arose.[35] However, despite frequent changes in the leadership in East Tennessee, the commanders of the adjacent departments got together on a program of reinforcement. With the assistance of urgings from the Secretary of War, fellow Virginians Samuel Jones in southwestern Virginia and Dabney H. Maury, currently commanding in East Tennessee, reached an agreement for assistance to Maury, if he were threatened.[36]

In spite of the difficulty of keeping the cavalry force in position in Kentucky, the capacity of East Tennessee to resist attack had been enhanced, both by providing for assistance from southwestern Virginia and the central cavalry force and by the new dispositions which were to give warning of an attack and permit the concentration of 5,000 infantry at the threatened point. Perhaps more important than the increased defensive capabilities was the fact that the infantry was now in a position rapidly to come to the aid of Bragg's army in Middle Tennessee, while, even in the absence of the infantry, warning of an enemy advance could be given by the cavalry and assistance would be at hand from General Jones.[37]

Provision for the rapid infantry reinforcement of General Bragg from the Department of East Tennessee was paralleled by similar arrangements in the case of Mississippi. In order to overcome the already demonstrated obstacle of the great time consumed in transferring troops between Mississippi and Tennessee, Johnston intended that part of Pemberton's reinforcements should be at Jackson and Meridian, Mississippi in order to be partially along the railroad to Tennessee, yet within easy reach of the Vicksburg area. This policy was to be extended to include the infantry stationed at Mobile. In the event that rapid reinforcement would be necessary, the troops at Mobile were to start to Tennessee, to be replaced, if necessary, by the last infantry to leave Mississippi. Thus Johnston hoped to expedite the transfer of forces from Mississippi to Middle Tennessee by having the "pipeline" half full before the movement began.[38]

As a further measure of protection for Bragg, in order to "halt or delay any proposed attack on Bragg," misleading reports of Bragg's strength were placed not only in the press but into Federal intelligence channels through intercepted messages.[39] One letter, intercepted by the Federals, exclaimed over Bragg's cavalry and described "a grand review of the army" at which the troops were reviewed by General Johnston and 60,000 infantry—nearly double Bragg's actual strength in infantry—"marched in the grandest order before that old chieftain."[40]

Thus, in summary, Johnston's various arrangements for defense envisaged co-operation from the adjacent departments for his central and, as he felt, most menaced army in Middle Tennessee. He had arranged for assistance to come to Bragg's department from the northeast and southwest and had so organized things that this aid could come more rapidly and in greater strength than heretofore. While no central reserve had been provided, the half filled pipeline from Mississippi to Tennessee partially filled this role. In a sense also a pipeline was filled from southwestern Virginia, in that, if troops from East Tennessee should aid Bragg, their places might be taken by contingents from Virginia. On the other hand, the defense of the peripheral departments was to be aided by co-operation with their adjacent departments in Arkansas and Virginia and by the great concentration of cavalry at the center with Bragg's army, which could strike in either direction against an advancing enemy and his communications.

Not only was Johnston naturally secretive about his plans, but he had stressed to the President and the Secretary of War the difficulty, if not the impossibility, of co-ordinating the forces in Mississippi and Tennessee. These factors together would probably have prevented him from revealing to Richmond more than parts of these strategic arrangements, but Johnston undoubtedly did not realize the extent to which this collection of isolated measures had made a comprehensive whole. He remained unconscious, therefore, that what he had done in this planning phase, nevertheless, was what he had been sent west to accomplish and had himself said could not be done.

Johnston's arrangements received two actual trials in action against the Federals. The first was precipitated by an apparent Federal concentration against Bragg. Since the beginning of March, 1863, there had been Union activity in northern Mississippi and on the Tennessee River which had caused anxiety as to whether it was intended as a flanking movement against Bragg or as reinforcement for Rosecrans.[41] Though Secretary James A. Seddon was quite alarmed in early March, Johnston was not particularly concerned at that time, for he thought that any movement was going to be directed against Bragg, and that it would not be begun until the country in front of Rosecrans' army was suitable for a co-operating advance by his army.[42]

At the beginning of April, however, Johnston became more and more concerned. A report came in which indicated that Grant's troops might be in the process of being withdrawn from before Vicksburg to advance up the Tennessee River to get in the rear of Bragg.[43] Johnston instructed Pemberton that, if he thought that Rosecrans was being reinforced, he was immediately to dispatch Stevenson's division or an equal number of men to Tullahoma. Indications of withdrawal from the Mississippi and movement toward Bragg became stronger as the month wore on.[44] Finally, as "Intelligence from Louisville, Nashville, and Memphis" indicated that "Grant's army may join that of Rosecrans," Johnston notified Buckner in Mobile of the plan, explaining that if Pemberton should decide that Grant was actually reinforcing Rosecrans, he would be notified by Pemberton, who would be sending troops to Bragg. When he heard from Pemberton, Buckner was to send his infantry "as expeditiously as possible" to Tullahoma. Since they would start before Pemberton's reinforcements, they "could easily keep the lead." To Pemberton, Johnston sent co-ordinating instructions, and suggested "placing a brigade at Jackson and another at Meridian."[45]

Hardly had these arrangements been made, when a message was received from Pemberton with a scout's report even more conclusive that Grant was going to Rosecrans' assistance. Pemberton announced that he was collecting troops at Jackson and could "send 4,000 at once, if absolutely necessary." Johnston responded, ordering the 4,000 men "immediately" and directing Pemberton "to prepare more troops for movement." Then a telegram arrived from Pemberton announcing that he was "satisfied Rosecrans will be re-enforced from Grant's army" and was forwarding "about 8,000 men" as "fast as transportation can be furnished."[46] But the troops which began the journey from Mississippi were not followed by more as Pemberton began to have doubts four days later, on April 16, and thought that "no large part of Grant's army will be sent away." By the eighteenth the men in transit, some of whom had gotten beyond Atlanta, were stopped by General Johnston and were almost immediately on their way back to Mississippi. The men from Buckner's department did, however, reach Tullahoma.[47] While Johnston was uncertain for a day or two longer, it soon became apparent that all the Federal activity was intended to confuse the Confederates as to their real objectives and to cover cavalry raids into Mississippi and northern Alabama.[48]

This movement by Grant, though a deception, had been apparently the type of Federal operation with which Johnston's plans were designed to deal. The plans had worked smoothly, co-ordination with Buckner and Pemberton had been excellent and the troops had moved with great celerity. Reinforcements had not been sent to Bragg from East Tennessee also due to fear of a Federal advance there.[49] To meet a crisis in Pemberton's command, one which was precipitated by Grant in early May, Johnston's arrangements were not so complete and they failed not only to be fully applicable to Grant's surprise offensive, but an unexpected obstacle impeded their limited workings.

Johnston had intended to break the communications of any advance by Grant against Vicksburg by utilizing the large interdepartmental cavalry force in Middle Tennessee. Grant, having learned the same lesson as had Johnston the preceding December, not only advanced on the west side of the Mississippi River, but soon dispensed with communications altogether. Nevertheless, 3,000 cavalry were dispatched from Middle Tennessee in response to Pemberton's urgent requests. No longer fighting in forts and bayous, Pemberton desperately needed more cavalry. But the Federal forces which had advanced eastward from Memphis and Corinth interposed, and "employed" the Confederate cavalry, preventing it from reaching Pemberton.[50]

With his cavalry blocked, there was little that Johnston could do for Pemberton. He hoped that the new commander of the Trans-Mississippi, an officer of proven vigor and capacity, Edmund Kirby Smith, would actively intervene. Johnston wrote Pemberton, inquiring about Kirby Smith, saying: "Now is his time to co-operate."[51] He and Pemberton were to be disappointed by Kirby Smith, however. Johnston remained convinced that nothing beyond the cavalry could be sent from Bragg "without giving up Tennessee."[52]

By the third week in May, 1863, Grant had bottled Pemberton up in Vicksburg and Johnston was commanding the army unsuccessfully attempting to relieve the city. The failure of his defensive arrangements adequately to meet Grant's masterful campaign should not obscure the fact that, without realizing it, Johnston had comprehensively organized the resources of his department and planned for the united efforts of his forces against many contingent operations of his powerful and skillful Union opponents.

Notes

"Tennessee and Mississippi, Joe Johnston's Strategic Problem" first appeared in the *Tennessee Historical Quarterly* 18 (June 1959): 134–47, and is reprinted with permission of the Tennessee Historical Society.

1. Frank E. Vandiver, *Rebel Brass: The Confederate Command System* (Baton Rouge, 1956), 58.
2. Robert M. Hughes, *General Johnston* (New York, 1893); Gilbert E. Govan and James W. Livingood, *A Different Valor: The Story of General Joseph E. Johnston C.S.A.* (Indianapolis, 1956).
3. Joseph E. Johnston, *Narrative of Military Operations During the Late War Between the States* (New York, 1874), dedication.
4. Govan and Livingood, *A Different Valor*, 175–86.
5. Thomas Robson Hay, "Confederate Leadership at Vicksburg," *Mississippi Valley Historical Review* 11 (Mar. 1925): 558–59; Govan and Livingood, *A Different Valor*, 195–96; Sarah A. Dorsey, *Recollections of Henry Watkins Allen . . .* (New York, 1866), 173.

6. U.S. War Department, *The War of the Rebellion: A Compilation of the Official Records of the Union and Confederate Armies,* 128 vols. (Washington, DC, 1880–1901), series 1, vol. 23, pt. 2:761. Hereinafter cited as *OR.* All references are to Series 1 unless otherwise indicated.
7. Ibid.; Joseph E. Johnston, "Some War Letters of General Joseph E. Johnston," ed. R. M. Hughes, *Journal of the Military Service Institution of the United States* 50 (May–June 1912): 319–20; J. E. Johnston to Louis T. Wigfall, Mar. 4, 1863, Wigfall Papers, Library of Congress.
8. Johnston, *Narrative of Military Operations,* 155.
9. Mrs. D. Giraud Wright, *A Southern Girl in '61* (New York, 1905), 104–105.
10. *OR,* vol. 20, pt. 2:441.
11. Ibid., vol. 17, pt. 2:811; ibid., vol. 20, pt. 2:469.
12. *OR,* vol. 17, pt. 2:813.
13. Ibid., vol. 20, pt. 2:487–88, 489, 492–93, 498, 499; ibid., vol. 17, pt. 2:832, 838; ibid., vol. 52, pt. 2:404, 410.
14. *OR,* vol. 23, pt. 2:646; ibid., vol. 52, pt. 2:425.
15. Govan and Livingood, *A Different Valor,* 169, 173.
16. *OR,* vol. 17, pt. 2:832, 838, 846–47.
17. Ibid., 832, 833.
18. Johnston, *Narrative of Military Operations,* 160; *OR,* vol. 17, pt. 2:833.
19. Johnston, *Narrative of Military Operations,* 161; *OR,* vol. 24, pt. 3:618; ibid., vol. 23, pt. 2:638; ibid., vol. 52, pt. 2:425.
20. *OR,* vol. 23, pt. 2:646.
21. Ibid., 673, 688, 701; ibid., vol. 52, pt. 2:432, 433.
22. William Roscoe Livermore, *The Story of the Civil War* (New York, 1913), pt. 2:367; Johnston, *Narrative of Military Operations,* 161; *OR,* vol. 24, pt. 3:686.
23. *OR,* vol. 23, pt. 2:700, 759.
24. Ibid., 618–19, 625–26, 630, 647, 657–58, 661, 663, 674–75, 680, 688–89, 816–17; ibid., vol. 24, pt. 3:786–87; Donald Bridgman Sanger, "Some Problems Facing Joseph E. Johnston in the Spring of 1863," in *Essays in Honor of William E. Dodd,* ed. Avery Craven (Chicago, 1935), 268–69.
25. Sanger, "Some Problems Facing Joseph E. Johnston," 269–71; *OR,* vol. 23, pt. 2:816–17.
26. *OR,* vol. 23, pt. 2:724, 757–73, 775–76.
27. Ibid., 659–60, 741, 745–46, 799.
28. Ibid.

29. Ibid., vol. 23, pt. 2:730, 786–87; ibid., vol. 24, pt. 3:734, 739, 768.

30. *OR,* vol. 23, pt. 2:745.

31. Ibid., 712, 727, 742, 743.

32. Ibid., 727.

33. Ibid., 824.

34. *OR,* vol. 24, pt.3:670.

35. *OR,* vol. 23, pt. 2:660, 705–706, 712, 713, 726, 751–52, 753; ibid., vol. 52, pt. 2:435, 457.

36. *OR,* vol. 23, pt. 2:800, 803, 807, 809, 813–14, 823.

37. Ibid., 627, 746–47, 798, 818–19, 820.

38. Ibid., vol. 24, pt. 3:597; ibid., vol. 23, pt. 2:750.

39. Sanger, "Some Problems Facing Joseph E. Johnston," 282–83.

40. *OR,* vol. 23, pt. 2:750–51.

41. Ibid., vol. 24, pt. 3:652, 655, 658, 671, 673, 688–89.

42. Ibid., vol. 23, pt. 2:674, 685.

43. Ibid., vol. 24, pt. 3:712.

44. Ibid., 714, 719, 730, 731.

45. Ibid., vol. 23, pt. 2:752; ibid., vol. 24, pt. 3:734.

46. *OR,* vol. 24, pt. 3:734, 738.

47. Ibid., 747, 751, 751–52, 753, 760, 761.

48. Ibid., 766.

49. Among other communications: *OR,* vol. 23, pt. 2:799; ibid., vol. 24, pt. 3:769.

50. *OR,* vol. 24, pt. 3:808.

51. Ibid., 808, 844, 846.

52. Ibid., pt. 1:214, 239.

Lieutenant General John Clifford Pemberton. Courtesy of the Museum of the Confederacy, Richmond, Virginia.

Misused Merit: The Tragedy of John C. Pemberton

Michael B. Ballard

His heart was never in the Civil War, and his level of talent was not sufficient to meet his superiors' expectations of him. Such was the dilemma that confronted and eventually overwhelmed John Clifford Pemberton after his country divided and went to war in 1861.

Born in 1814 to a Philadelphia, Pennsylvania, Quaker family, he had walked away from the nonviolent tenets of the Friends' faith.[1] Influenced by his father's friendship with military hero Andrew Jackson, Pemberton, through Jackson, received an appointment to the U.S. Military Academy in 1833.[2]

Pemberton's years at West Point demonstrated that he had neither the soul nor the mind of a warrior. His best marks were in drawing, and his best times were partying, with both female guests and fellow cadets. He did excel at currying favor with the academy commandant, and that no doubt worked in his favor when he was nearly expelled during his senior year for dereliction of duty.[3] Pemberton finished twenty-seventh out of fifty in the class of 1837; he had accumulated 163 demerits during his four years at West Point.[4]

Following graduation, Pemberton was commissioned second lieutenant in the Fourth U.S. Artillery. He also dallied with the idea of becoming a husband. His family protested, and ultimately he ended the relationship, enlisting his father's help to break the news to the young lady in question. Lieutenant Pemberton thus displayed marks of immaturity and an unwillingness to take charge of his own responsibilities.[5]

Pemberton's first combat experience came in the Second Seminole War in Florida. Actually he participated in little fighting, for he soon used the stroking abilities he had honed at West Point to lobby, albeit unsuccessfully, for an adjutant position under a colonel. Later he did become head of an ordnance depot and landed an administrative post at a fort. Though he may have cozied up to superior officers in order to get jobs, Pemberton excelled at such duties. By the time he left Florida, he had earned a reputation as a solid staff officer.[6]

The young Pennsylvanian continued to gain staff experience in the milieu of service at various army posts in the South, Midwest, and Northeast until the outbreak of the Mexican War in 1846.[7] By that time he had met and fallen in love with Martha "Pattie" Thompson, a Norfolk, Virginia, girl whom he would marry after his return from Mexico.[8] His union with her would cause him great anguish about choosing sides in the Civil War.

Aside from his fateful love affair, Pemberton had undergone other experiences that did not portend well for his future as a general in the Confederate army. Service at isolated forts had hardened him. He was no longer the carefree cadet of West Point days. He occasionally clashed with recruits and fellow officers, frequently over mundane matters, setting a pattern that would continue throughout the remainder of his military career.[9]

The Mexican War was a proving ground of sorts for young West Point alumni who would experience their first combat in a conventional conflict. John Pemberton's experience was very positive and indicated an excellent future. He was on the field at Palo Alto, but his first real fight occurred at Resaca de la Palma during Zachary Taylor's campaign. He performed ably in a hotly contested part of the battlefield and experienced a rush of adrenaline that convinced him that he had made the right career decision.[10]

His enthusiasm waned as the war dragged on. As Taylor led his army further into Mexico, Pemberton followed his usual pattern of meeting the right person and landing a better job. He became aide-de-camp to Gen. William Jenkins Worth, a quality combat veteran who, despite being rash, abrupt, inflexible, and self-centered, was in other ways a worthy role model. Unfortunately, some of Worth's negative qualities rubbed off on, and at times dominated, the personality of John Pemberton.[11]

Worth cited Pemberton for "alacrity, zeal and gallantry" at the battle of Monterrey. Later, after Worth's division had been reassigned to Winfield Scott's expedition against Mexico City, Pemberton received a promotion to brevet captain for "gallant and meritorious conduct" at the battle of Molino del Rey.[12] By the time his service in Mexico ended, Pemberton had established himself as a solid soldier. However, he still had not proved himself as a combat leader. His preferences and strengths in Mexico had been in the administrative area. He was, and always would be, a staff officer at heart. The inclination had been evident in Florida and had solidified in Mexico.

After the war, Pemberton married Pattie, and they began their family. Pemberton drew duties in several areas of the country, returning for a time to Florida and serving at posts near New Orleans, in New York, and out west. In 1850 he was promoted to the regular rank of captain. Although he occasionally went out on patrols, especially during Indian trouble in the West,

Pemberton's duties were mostly administrative, and though he was disappointed that he had not been able to get an assignment in Europe, he was relatively content. But civil war loomed on the horizon, and when it came, it would turn John Pemberton's world upside down. He and his family were at the remote outpost of Fort Ridgely, Minnesota, when he was recalled to Washington in early 1861.[13]

As war became a matter of when rather than if, Pemberton agonized over what he would do. Initially, he decided to remain with the North unless Virginia, Pattie's home state and a state he had grown fond of during a brief antebellum tour of duty there, left the Union. His Philadelphia family breathed a sigh of relief, but their anxiety soon returned.[14]

As the boom of guns at Fort Sumter echoed across the South, Virginia left the Union on April 17. Pemberton's older brother Israel hurried to Washington to dissuade his sibling from resigning from the U.S. Army. John wavered and did not follow through on his threat to head South once Virginia seceded. He remained in the army, obeyed orders, and listened to Israel's exhortations. The decisive words, though, came from Pattie, who was in Norfolk with the children. "Why do you stay?" she inquired, assuring him that Confederate President Jefferson Davis would have a place for him in the Southern army. Davis was in Montgomery putting together a new government, and whether he had made any promise regarding John Pemberton is not known. Whatever the case, the pressure from his beloved wife led to a decision to resign on April 24. The Philadelphia Pembertons were disappointed but not surprised and had no option but to accept John's action and hope for the best.[15]

Pemberton's decision to turn his back on the army that had been his career for almost three decades was the toughest of his life. Despite claims of his apologists to the contrary, he was not a fanatical supporter of the Southern cause. Had he been, he would not have hesitated to go South once war came and Virginia joined the Confederacy. John Pemberton was no fire-eater, no states' rights dogmatist. In his mind, he had only two choices: he could sit out the war with its possibilities of rapid promotion and battlefield glory, hardly a pleasing option for a career officer, or he could fight for Pattie's homeland. His strong relationship with his wife settled the issue. At the same time, there seems little doubt that Pemberton's heart was not in this conflict that threatened to destroy the Union. When he turned his back on the land of his nativity, he became a traitor in the North and a suspect Yankee in the South. He had placed himself in a most uncomfortable position, and his low-key service in Virginia during the early months of the war indicates that he recognized the benefits of anonymity.[16]

Originally joining the provisional Army of Virginia, Pemberton later went into the regular Confederate army. By June 17, 1861, he held the rank of brigadier general in the artillery corps. This rapid rise in rank could be attributed to several factors. Pemberton habitually practiced the blandishment of superior officers, and Pattie's family had some influence. At what point Pemberton may have met Jefferson Davis is not known, but Davis's future actions indicated that he liked this Pennsylvanian. Also, Pemberton had originally reported for duty to Gen. Joseph E. Johnston, whom he had probably met in Mexico. At this point in the war, Johnston apparently had a good relationship with Pemberton.[17]

Despite his high rank, Pemberton remained unobtrusive. Through the summer and fall of 1861, he was in Norfolk, first supervising an artillery instructional camp and then directing his artillery brigade in operations around the Smithfield-Suffolk region west of Norfolk. When the first great engagement of the war took place at Manassas, Pemberton was supervising the placement of guns along the James River.[18] He may well have been content to serve out the war in this area close to Pattie's home. There is no surviving correspondence to indicate that he sought other duty. But other duty soon sought him. He was about to begin his journey into Civil Was infamy.

The first phase of that journey began soon after Robert E. Lee assumed command of the defenses along the South Carolina and Georgia coasts. Lee needed brigadiers, and the War Department ordered John Pemberton to South Carolina, where Lee had set up headquarters in the Charleston area.[19] He was probably chosen because of his experience with coastal defenses in Virginia. Pemberton obeyed the orders, perhaps with some trepidation. A Pennsylvania-born general could not have been anxious to assume a post in the state that had been the cradle of secession.

As long as Lee remained in South Carolina, Pemberton continued to stay in the background as he had in Virginia. He no doubt observed Lee; his later actions indicated that he had been influenced by Lee's strategic ideas. For example, Lee concentrated on key areas of defense rather than scattering his limited resources along outlying areas of the coast. Lee also constructed an interior line of defense far inland out of range of Federal gunboats, and he used trains to move troops where needed. Lee, the man and the soldier, impressed the irascible Governor Francis Pickens of South Carolina. Pickens found Lee to be a quiet, dignified man and a "thorough and scientific officer."[20] As he later demonstrated in his relationship with Jefferson Davis, Lee knew how to handle people, especially civilian officials. Whoever followed Lee in South Carolina would face a formidable task.

When Lee was recalled to Richmond, to advise Davis during the peninsula campaign in early 1862, fate, and the Confederate War Department, looked to John Pemberton to take on that task. Why Pemberton was chosen is a difficult question to answer. Why did Jefferson Davis, no matter how much he may have liked Pemberton, think that a Northern-born general could handle a commander's position in rabidly Confederate South Carolina? A possible explanation is that Davis and his advisers did not think that the situation in South Carolina and Georgia was urgent, and at this point, it was not. Therefore, it may have seemed a safe risk to assign command to an unproven commander. It is possible, too, that Davis thought that placing a Pennsylvanian in command in South Carolina might be a good propaganda move to embarrass the North.

Pemberton had done nothing spectacular since coming to the state, but he must have impressed Lee. In February 1862 Pemberton was promoted to major general. By date of appointment, he outranked the other brigadiers present, so his promotion was in a sense logical. But there was more to it than that. It seemed apparent that Pemberton was being groomed to take Lee's place. On March 4 Pemberton assumed temporary command, and ten days later it became permanent.[21]

In many respects, Pemberton proved to be a good man for the job. The tasks of the command primarily required bureaucratic skills, which he possessed and had honed through his many years in the army. Lee, too, had been a staff officer, and perhaps it was this aspect of Pemberton's work that led Lee to agree to leave his post in command of his senior lieutenant. In many ways, Pemberton's first taste of command was a staff officer's dream—and nightmare. One of his first acts was to reorganize the districts within what had become known as the Department of South Carolina and Georgia (for a brief time, parts of Florida would be added to the department).[22] There was no apparent need for the reorganization; it was probably just too much of a temptation for a general with a staff officer mentality.

Other more immediate issues inundated Pemberton, and he and his staff handled most of them quite deftly. These included securing construction material for boats and railroads, getting funds from Richmond to pay his soldiers, obtaining ordnance and food, keeping his forces at an adequate strength in the face of frequent demands for reinforcements elsewhere in the Confederacy, building adequate fortifications when neither he nor any of his staff had engineering experience or skills, making politically sensitive decisions on such issues as martial law and the use of slave labor, and devising strategies such as developing his interior rail lines, blocking waterways, and restricting his defensive lines to defend his department against an ever-increasing presence of the Federal navy.[23]

Although Pemberton exhibited quality administrative leadership in handling many of these tasks, it soon became apparent that he lacked Lee's adroitness in handling government officials. Governor Pickens was the titular head of state government and had considerable influence, but an organization called the South Carolina Executive Council, in which Pickens served, had been set up as a war measure and had a heavy hand in directing the state's affairs. The council consisted of the governor, lieutenant governor, and three men elected by South Carolina's secession convention.[24]

Pemberton never understood, as did Lee, that powerful political leaders must be cultivated and kept informed about military decisions. He made his decisions and rarely volunteered explanations, often not commenting at all until pressed by the council. A case in point was Pemberton's decision to abandon several outlying areas along the coast. Rather than scatter his forces, he sought, as Lee had done, to concentrate them for more effect. Pickens, no doubt acting with the approval of the rest of the council, complained to Lee in Richmond that the inner defenses were too thin for Pemberton to pull the troops in from the outer defensive works. Also, when the troops left, people living in those areas would be at the mercy of the Yankees. Pemberton should have at least held off to give those who wanted to evacuate time to do so.[25]

Lee had one of his aides pen a message of fatherly advice to his successor at Charleston. "It is respectfully submitted to your judgment whether, in order to preserve harmony between the State and Confederate authorities, it would not be better to notify the governor whenever you determine to abandon any position of your defenses, in order that he may give due notice to the inhabitants to look out for their own security."[26] Pemberton further undermined his credibility when he deceived himself into believing that Fort Pulaski near Savannah, Georgia, could hold out against Federal attacks. Lee had suspected that the fort's demise was inevitable and had concentrated on strengthening the interior lines around Savannah. Pemberton visited the fort on April 10, gave assurances that it would hold, and was subjected to a great deal of embarrassment when it fell the next day.[27] Misreading military situations would haunt Pemberton even more in his next command.

For the present, he continued under a cloud of suspicion that he was not up to meeting the demands of his job. He finally decided to move his headquarters to Charleston from the small town of Pocotaligo. In Charleston he could more closely supervise the constriction of his defensive lines. Shortly before he moved, he suffered yet another blotch on his public image. When a Confederate steamer's officers and men all went ashore in violation of regulations, the ship's slave crew absconded with the steamer and its five cannon and escaped into Union lines.[28]

At times, Pemberton seemed to go out of his way to feed the growing distrust of him by politicians and the general public. This was especially true in his handling of questions regarding the defense of Charleston. He proposed the dismantling of some of the forts defending the city, including Fort Sumter, the symbol of secession and the Confederate cause. Those forts might be taken by the enemy and their guns turned on the very streets they were built to defend. In suggesting the destruction of Sumter, Pemberton clearly showed that he was totally out of touch with political reality.[29]

Charleston's mayor grew concerned that abandoning the outer coastal defenses might open the way for the Federal navy to enter Charleston Harbor. He pointedly asked Pemberton if Charleston would be defended or abandoned to the enemy. Pemberton refused to commit himself based on a hypothetical situation, but he later commented that if future events should warrant the withdrawal of his troops he would withdraw them.[30]

Pickens pleaded with Pemberton to take a firm stand. Even if the fight meant reducing the city to rubble, Pickens wrote, that would be preferable to giving up Charleston without a desperate struggle. "We can afford to lose our city, but not our honor."[31] Unlike Lee, Pemberton had no understanding of or sympathy for Southern honor; he simply would do what the military situation dictated.

At this point, Lee again wired Pemberton, this time with more strongly worded advice. Pickens had kept Lee informed of the Charleston controversy, and Lee scolded Pemberton for the latter's negativism. Savannah and Charleston both had to be strongly defended; Charleston was especially important as a supply artery. Pemberton must make known his resolve to fight it out "street by street and house by house as long as we have a foot of ground to stand upon."[32] Lee no doubt intended to make Pemberton understand the importance of building morale and sustaining optimism. Future events would suggest that Pemberton took Lee's words all too literally.

Federal pressure along the coast exacerbated the tension between Pemberton and South Carolinians. Designed primarily to keep Pemberton from sending reinforcements elsewhere, Union operations kept Charleston in an almost constant state of panic. On June 16, 1862, Confederate forces met and defeated Federal troops on James Island in the battle of Secessionville. The battle proved to be another defining moment for Pemberton, who made no effort to go to the front. Ever the staff officer and administrator, he stayed at headquarters monitoring the situation. Clearly, he was still no warrior.[33]

The victory at Secessionville did not deter a growing movement, led by Governor Pickens, to get Pemberton replaced. Lee and Jefferson Davis received a steady flow of complaints about Pemberton's leadership.[34] He did "not possess the confidence of his officers, his troops, or the people of Charleston. Whether justly or unjustly, rely upon it the fact is so."[35] Famous South Carolina diarist Mary Boykin Chesnut noted that "Pemberton said to have no heart in this business, so the city cannot be defended." Another diarist observed of Pemberton's problems, "as he is a Pennsylvanian, [he] engenders suspicion. Everybody has lost confidence in Pemberton and many even suspect treachery, though it cannot be proved of course."[36]

Aside from politicians and other civilians, Pemberton had problems with a troublesome subordinate general, Roswell Ripley, and soldiers who felt that Pemberton was inviting defeat by continuing his strategy of constricting defensive lines. Like many Mexican War veteran officers, he did not think highly of volunteer soldiers, and he did little to cultivate good relations with the men in the ranks.[37]

As Pemberton's status in South Carolina deteriorated, Lee advised Davis that a new commander must be found. Confederate Adjutant and Inspector General Samuel Cooper came to the same conclusion after visiting Pemberton. Cooper thought that Pemberton had done a good job, yet "with such an opposition as constantly surrounds him it would be difficult for any commander situated as he is to effect much." Davis hesitated to act, but on August 29 he named P. G. T. Beauregard to replace Pemberton.[38] The news produced a smattering of support for him, but Pemberton was glad to be moving on to whatever the future might hold. He made it clear that he wanted to return to Virginia, perhaps to a low-profile assignment. His self-confidence had sunk to a low ebb; he wrote to Richmond that he was sure that Beauregard would prove to be "far more capable" in South Carolina.[39]

John Pemberton's time there had been illuminating. His administrative skills had served him well, but it had become obvious that he did not have Lee's talent for dealing with people. Lee had managed to gain the respect, admiration, confidence, and affection of civilians and soldiers alike, whereas Pemberton had failed in all four areas. Cadet Pemberton would probably have fared better than hardened career soldier Pemberton. Also, Lee was destined for battlefield greatness, but Pemberton had shown no affinity at all for the front. The fact seemed simple enough; he had been promoted and assigned above his capabilities in several areas. He probably realized that he had taken on too much; certainly as he entrained for Virginia he had no inkling that he would be thrust into a similar or possibly more difficult situation. Unfortunately for this forlorn Confederate Yankee, Jefferson Davis had other ideas.

On October 1, 1862, Pemberton received what he must have considered later to be an ominous order to take command of "the State of Mississippi and that part of Louisiana east of the Mississippi River." With the new job came another promotion, to lieutenant general.[40] This new assignment offered monumental challenges far greater than he had faced on the Atlantic coast.

Unlike the South Carolina command, which contained no urgent Union target, Pemberton's Department of Mississippi and East Louisiana contained a major Federal objective—the Mississippi River town of Vicksburg. Since the outbreak of hostilities, combined Union army-navy operations had been working to clear the river of Confederates. Northern control of the Mississippi would split the Confederacy and restore unfettered commercial shipping to and from midwestern states. By the time Pemberton arrived in Mississippi on October 9, the major river towns of New Orleans, Baton Rouge, and Natchez south of Vicksburg had been surrendered. North of Vicksburg, Federal forces had won a major victory at Island Number 10 near the Kentucky-Tennessee border and had forced the surrender of Memphis. That left Vicksburg as the last great obstacle. Strongly fortified, the town stood on high bluffs along a sharp hairpin turn in the river. Union President Abraham Lincoln had the strategic insight to understand that Vicksburg was the key to Union victory on the Mississippi.[41]

In light of the obvious significance of fortress Vicksburg to the Confederate cause in the western theater, the question loomed then and now: why was the command given to John Pemberton? A Mississippi acquaintance of Jefferson Davis would later claim that during a personal conversation with Davis, the latter confided that he had a high opinion of Pemberton and that both Lee and Cooper, consulted by the president separately, had named Pemberton as their choice.[42] If this account is valid, the endorsement of Pemberton by Davis, Lee, and Cooper is truly incredible. Did they think that what had happened in South Carolina was an aberration?

With no other known accounts of the decision-making process, one can only speculate on the rationales that resulted in Pemberton's assignment in Mississippi. Angered by South Carolina's criticism of one of his favorites, Davis may well have decided to give Pemberton another important command as a way of taunting critics. Those same critics had often given Davis a hard time. The president could also, with some validity, argue that Pemberton had done a good job in South Carolina. Despite the criticism, most of Pemberton's former district was still in Confederate hands. Davis characteristically focused on the positive when it came to his friends, no matter how much the negatives might taint future performances.

Lee's role may be more instructive. By the time Pemberton left South Carolina, Lee had taken command of the Army of Northern Virginia, had won the Peninsula and Second Manassas campaigns, and had escaped with a drawn battle at Sharpsburg. During these early months of command, Lee made it clear that he did not want problem generals in Virginia. Many such Confederate officers who started their service in the Virginia theater would eventually end up in other areas.[43] This may account for Lee supporting Pemberton. Lee also was Virginia-focused and resisted any proposal for sending his best generals west.

As for Cooper, after his visit to Charleston, he may have felt that Pemberton had been treated unjustly and deserved another chance. Whatever their motives, if these three men honestly thought that Pemberton could get the job done, they grossly miscalculated his ability. Moreover, they never seemed to appreciate fully, until it was too late, the significance or complexities of the challenges that awaited any commander of the Vicksburg area.

The Mississippi command would have challenged a commander of supreme ability. Myriad waterways offered the Union navy several possible approaches to Vicksburg. The Confederates had very little in the way of a navy to contest those approaches. Various roads, though admittedly not very good ones, gave Federal commanders several possible invasion routes. For Confederate forces, railroads were a blessing and a curse. Two ran north and south and one east and west, the latter connecting Meridian and points east with Vicksburg. They were vital to supplying Pemberton's army and providing flexibility for troop movements, but their length made them vulnerable to Union destruction or use as a supply line for General Grant's invading force in north Mississippi.[44]

Pemberton did not have numbers to match Grant, and with a few diversions, Grant could so scatter Pemberton's forces as to make them almost impotent.[45] Grant understood that and would put such a strategy to effective use. Obviously, Pemberton faced great challenges in protecting Vicksburg and in keeping his army supplied.

As the Vicksburg campaign progressed, the command structure in the West further compounded Pemberton's problems. Joseph E. Johnston, severely wounded during the early stages of the Peninsula Campaign, had permanently lost command of the Virginia army to Robert E. Lee. President Davis therefore appointed Johnston to command the western theater. Johnston's command included most of the territory between the Blue Ridge Mountains in North Carolina and the Mississippi River. He would be the immediate superior of both Pemberton and Braxton Bragg, who commanded the Army of Tennessee. Pemberton continued to report directly to Richmond as he had done before Johnston's appointment. Bragg also occasionally bypassed Johnston, leaving the latter angry but powerless to remedy the situation without the support of Davis and Cooper.[46]

Johnston also had no control over the Trans-Mississippi. Commanders there did not have to obey Johnston's wishes for them to cooperate with Pemberton. Davis refused to intervene, and the result was very little coordination between Confederate forces on opposite sides of the Mississippi.[47]

Not fully recovered from his severe wound, Johnston accepted the assignment with trepidation. Consequently, his heart was never in the Vicksburg campaign, at least not in it the way Richmond wanted it fought. Johnston would give up geographic points in favor of combining Pemberton's and Bragg's armies.[48] Davis would have none of it, and Pemberton would soon find himself trying to juggle his orders to appease both his immediate superior and his commander in chief. Pemberton sided with Davis; he had learned through his South Carolina experience that Richmond ultimately ruled. But the president and the War Department had created a dysfunctional command system and had placed an insecure general in the midst of it. It was surely a situation with disastrous potential.

Pemberton's actions during his initial days in Mississippi were equally foreboding. Despite the great pressure that Grant was putting on Confederate forces in north Mississippi, pushing south with the idea of taking Jackson and then moving west to Vicksburg, Pemberton preferred to spend time behind his desk in Jackson. There he sought to get his department in order through piles of paperwork, while occasionally checking on the battlefront.

Also, unlike in South Carolina, he had developed a decent working relationship with the governor of Mississippi, John J. Pettus. Certainly Pemberton was no slacker; passersby often observed him working late into the night. One of his staff noted that confusion had reigned in all departments before his boss's arrival, but that order had gradually been restored thanks to the general's leadership. But Pemberton's frequent absences from the front were another indication that he was no battlefield commander. If anyone in Richmond found this disturbing, there is no record of it.[49]

As Grant pressed on, Pemberton pulled his army slowly southward to prevent Union troops from turning vulnerable flanks. His forces had retreated all the way to Grenada before salvation came. One of Pemberton's officers suggested a raid on Grant's supply base at Holly Springs. Given his ever-lengthening supply line, the destruction of this base might force Grant to retreat into west Tennessee. That Pemberton had not thought about such an option himself says much about his tactical and strategic abilities, but at least he had the gumption to order the raid.

Led by his predecessor, Earl Van Dorn, Confederate cavalry swooped down on Holly Springs on December 20, destroyed huge piles of Union food and ordnance, and escaped. A disgusted Grant retreated as Pemberton had hoped, and the north Mississippi campaign ended.[50] But a concomitant campaign was still alive. William T. Sherman had taken a large force down the Mississippi from Memphis to attack Vicksburg. The idea was to link up with Grant, but then came Holly Springs. Sherman should have abandoned his part of the plan, but he decided to forge ahead and attack entrenched Confederates along bluffs overlooking Chickasaw Bayou just north of Vicksburg. Pemberton, who was enjoying himself reviewing troops with visitors Jefferson Davis and Joseph E. Johnston, hurried to Vicksburg after receiving news of Sherman's approach. Actually, reports of Sherman's move down the Mississippi had begun arriving shortly after Van Dorn's raid, but Pemberton had virtually ignored them. A general with a noncombative, nonaggressive nature, Pemberton had a penchant for ignoring intelligence reports that indicated that the enemy was on the offensive.

At Vicksburg, Pemberton stayed away from the front during the battle of December 27–29. Wearing the self-imposed mantle of staff officer, he ordered reinforcements from north Mississippi to Vicksburg and funneled them to the Chickasaw front, where his subordinates Martin Luther Smith and Stephen Dill Lee orchestrated Sherman's defeat. Pemberton's staff-officer role was yet another warning ignored in Richmond.[51]

Of course there was cause for celebration. Grant and Sherman had been repulsed. Pemberton thus far had excelled in reorganizing his department, and he had kept the enemy at bay. But his battlefield successes had been as much the result of Grant's failure to guard his supplies properly and Sherman's impetuosity as of anything Pemberton had done. Despite the victories, Pemberton had developed a strictly defensive state of mind. As a result, and despite his disappointment, Grant had been allowed to set the pace of the campaign. He had not so much seized the initiative from Pemberton; rather, Pemberton had handed it over. The tenacious Grant was not one to give it back, and as the new year of 1863 approached, he began a season of diversions that left his opponent looking anxiously this way and that in a state of constant reaction to Grant's moves.

Grant put his troops to work digging a canal across from Vicksburg that would have diverted the Mississippi away from the city's fortified bluffs. The canal had been tried before, and as before, the project failed, as did a similar project called the Duckport Canal. Grant also tried cutting a path through various waterways to reach the Red River below Vicksburg. The Red emptied into the Mississippi, so such a route would make Vicksburg superfluous. But this gambit, known as the Lake Providence expedition, also failed.[52]

Two other expeditions likewise failed, but they underscored two glaring Pemberton weaknesses. The Yazoo Pass operation, an attempt to reach the Yazoo River via rivers flowing south from northwest Mississippi, resulted in a repulse by Confederates at Fort Pemberton near Greenwood. During the course of this successful campaign, Pemberton agitated William Loring, an aggravating general of the Ripley stripe and another eastern-theater reject. Loring complained bitterly because Pemberton would not send him more guns and men. Pemberton had few soldiers and little ordnance to spare, and he considered Loring's request unnecessary. Loring, along with one of his lieutenants, Gen. Lloyd Tilghman, formed an anti-Pemberton clique. Tilghman, an old Pemberton friend, was still bitter about a run-in he had had with the commander in north Mississippi. For the remainder of the Vicksburg campaign, dissent would grow among Pemberton's generals, and there is no evidence that he ever did anything to address the problem. Rather than build an effective team, he simply immersed himself in paperwork, leaving his command structure at times in an almost anarchical state.[53]

That state was clearly in evidence during another Grant scheme, the Deer Creek (also known as Steele's Bayou) expedition by David Dixon Porter, commander of the Union navy at Vicksburg. With Grant's blessing, Porter hoped to bypass Confederate works on Yazoo River bluffs above Vicksburg. By negotiating a series of waterways, Porter intended to reenter the Yazoo above Confederate artillery emplacements, threaten Fort Pemberton upriver, or go downriver to shell rebel works. Instead, Porter almost got trapped in the narrow waterways and was fortunate to keep his boats from falling into Confederate hands. Pemberton, anchored to his Jackson desk, was not on hand to take charge. The capture of Porter's boats would have been a significant turning point in the Confederates' favor, but Pemberton's generals in the area were more concerned with stopping Porter than with aggressively attacking him. The mentality of the commanding general was filtering down to his subordinates, with predictable results. Pemberton had become so absorbed in work that should have been delegated to his staff that he ignored the duties of leadership, and a golden opportunity slipped by when Porter's fleet steamed safely back into the Mississippi.[54]

General Grant noted his rival's weaknesses and kept up the pressure. A Federal foray inland from Greenville convinced Pemberton that Grant had given up a direct assault on Vicksburg and was going to try north Mississippi again. But by far the most spectacular and effective plan Grant devised was a cavalry raid from northwest Mississippi through the heart of the state and on to Baton Rouge. Led by Benjamin Grierson, the raid entranced Pemberton, who focused all his attention on stopping the Yankee horsemen who destroyed much rebel property, including railroads. Grierson and his men escaped practically unscathed, and while a frustrated Pemberton fumed over the raid, Grant was making a decisive move.[55]

After his series of failed efforts to approach Vicksburg, Grant had decided to move his army south down the Louisiana side of the Mississippi and cross below Vicksburg. Porter's boats ran the gauntlet of Confederate cannon to meet the army and ferry it across. Despite numerous warnings from his scouts and generals, and despite the obvious giveaway that something was afoot when Porter's boats steamed past his batteries, Pemberton continued to track Grierson until it was too late. During Sherman's Chickasaw Bayou campaign, Pemberton had ignored warnings but salvaged a victory. Now he ignored warnings and gave Grant the edge he needed to break the Vicksburg stalemate. Temporarily thwarted by rebel batteries at Grand Gulf, Grant moved his landing point farther south to Bruinsburg, where he crossed, and on May 1 he won the battle of Port Gibson, which gave him a vital foothold.[56]

With no direct rail line from Vicksburg to Port Gibson over which to send reinforcements, Pemberton had to sit by and watch his outnumbered troops fight well but ultimately lose to Grant's superior numbers. Had Pemberton paid attention to reports coming into his office from his most competent general, John Bowen, he would have had time to concentrate forces and possibly beat Grant back into the river. Trying to recover, Pemberton revealed the degree to which he had lost control of himself and the campaign. He sent an urgent message to commanders in the Jackson-Vicksburg area to proceed at once; he forgot to tell them where to go.[57]

With Grant now threatening the south flank of Vicksburg, Pemberton pulled his men toward the city and set up a defensive perimeter, entrenching along the bluffs of the west bank of the Big Black River. This was sound strategy and might have worked had not the flawed western Confederate command system intervened.[58]

Joseph Johnston, in the dark for several days about what had happened at Port Gibson because Pemberton was reporting directly to Richmond, now began to take an active and, for the Confederacy, unfortunate role in the campaign. First he advised Pemberton to forget Vicksburg and unite his forces to defeat Grant at the Mississippi. The orders came too late, for Grant was now driving inland. Meanwhile, Davis sent Pemberton word that Vicksburg and Port Hudson (a Confederate stronghold on the Mississippi above Baton Rouge) must be held at all costs. In Johnston's mind, holding Vicksburg was not as important as uniting forces to defeat Grant; Davis disagreed, and Pemberton was caught in between.[59]

Grant's army, meanwhile, won another battle at Raymond, and Grant, hearing of Johnston's arrival at Jackson, decided to attack the Mississippi capital city before turning west toward Vicksburg. Johnston, more timid than his orders, concluded that he was too late to do anything, ordered Pemberton to march toward Jackson to catch the Federals between the two Confederate forces, then abandoned Jackson after offering only token resistance. He still wanted Pemberton to unite with him, but Johnston retreated north to Canton, leaving Grant between his small force and Pemberton's. Trying to appease both Johnston, who wanted him to move, and Davis, who wanted Vicksburg protected, Pemberton marched his army away from the Big Black bluffs to hit Grant's supply line. Such a move would keep his army between Vicksburg and Grant. The march to the Raymond road from Edwards, Mississippi, near the Big Black was a comedy of errors. Logistical support had been botched, and ineffective scouting had delayed the advance.[60]

As Pemberton carried out his plan, he received word that Johnston had evacuated Jackson and the latter's most recent message ordering a junction of Confederate troops. Grant, presented with a copy of Johnston's message by a Federal spy, intervened as Pemberton tried to carry out the order, and the resulting battle of Champion Hill proved to be the decisive conflict of the campaign.[61]

Champion Hill was the first and only battle in which John Pemberton commanded an army. All the aspects of his personality that prevented him from being a good field general, and all the seeds sown by his mistakes thus far, combined for the predictable disaster. Disgruntled generals, especially Loring and, to a lesser degree, Bowen, ignored his orders. Pemberton's staff—like their boss, inexperienced at controlling a large army in battle—did not perform well. Pemberton remained confused and uncertain the whole day, just as he had been through much of the campaign. Some of his lieutenants fought well but the lack of command coordination doomed their efforts. Pemberton could not retreat to his office and analyze the situation and write out orders. Decisions had to be made quickly, instinctively, and forcefully, and he simply was not up to the task.[62]

When it became apparent that the day was lost, Pemberton ordered a retreat back to the Big Black. Loring's division got cut off, and as Pemberton kept waiting for word from his protagonist, Grant launched an assault that sent Pemberton's army reeling across the Big Black and into Vicksburg (Loring managed to escape Grant's grasp and join Johnston at Canton).[63] As Pemberton waited with his army in Vicksburg for Grant's approach, another message from Johnston arrived ordering the evacuation of the city. But Pemberton, more attentive to Davis's orders than those of his immediate superior, decided to try to save Vicksburg. He had learned his lesson in Charleston. After two attacks on rebel works failed, Grant settled in for a siege. Johnston marked time, more interested in debating the size of his army with Richmond than in relieving Pemberton. No appreciable help came from the Trans-Mississippi, and in the end, Pemberton had no choice but to surrender the city he had tried so hard to protect.[64]

Pemberton chose, he later claimed, to give up the fight on July 4 to get better terms. This only fueled the criticism of him, which had been growing ever since Grant crossed the Mississippi and was now practically hysterical. The name Pemberton became anathema throughout the South. President Davis defended his general but was unable to find him another suitable command. Ultimately, Pemberton settled for a reduction in rank and reassignment in the eastern theater, where he served out the remainder of the war quietly. As a result of Johnston's published criticisms of him, he demanded but never received a court of inquiry. The feud lasted into the postwar years. Johnston published his memoirs, but Pemberton, though writing defenses of his actions, a part of which was published by his grandson in a 1942 biography, never got his story out to the degree that Johnston had. He died on a hot July 13 in 1881, his death barely noticed in the South for which he had sacrificed so much. His passing probably would not have gained much attention anyway but was certainly overshadowed by the ultimately successful assassination attempt on President James Garfield.[65]

The Confederacy never recovered from the loss of Vicksburg. Pemberton's army was paroled but would not fight again as an army. The Union navy had uncontested control of the Mississippi, and vital supply lines from the Trans-Mississippi were cut off from the rest of the Confederacy.

Whether the disaster to Confederate arms could have been avoided is a question that cannot be answered. But there can be no doubt that the Confederate government sent the wrong man to the Vicksburg theater. John Pemberton had much merit as a staff officer who could organize and direct noncombat affairs of a military department. He had no merit at all as a field general who was required to create public confidence and an effective command structure, motivate soldiers, and deftly handle tactical and strategic challenges. His actions in South Carolina and Mississippi clearly demonstrated these facts, yet Jefferson Davis and his cohorts blindly placed this beleaguered general in a vital command. The results were devastating, both for the Confederacy and for John Clifford Pemberton.

Notes

"Misused Merit: The Tragedy of John C. Pemberton" was first published in *Civil War Generals in Defeat*, ed. Steven E. Woodworth, 141–60 (Lawrence, KS, 1999), and is reprinted with permission of the University Press of Kansas.

1. The story of Pemberton's roots and pre–West Point years is in Michael B. Ballard, *Pemberton: A Biography* (Jackson, MS, 1991), 3–11.
2. Ibid., 6, 10–11.
3. For details of Pemberton's West Point years, see ibid., 12–25.
4. Pemberton Cadet Record, John C. Pemberton Papers, Pemberton Family Papers, Historical Society of Pennsylvania, Philadelphia.
5. Adjutant General's Office to John C. Pemberton, June 29, 1837; General Orders No. 46, Head Quarters of the Army Adjutant General's Office, Washington, DC, July 12, 1837, all in ibid.; Ballard, *Pemberton,* 22, 26–27, 30–31, 32–33.
6. Pemberton to Parents, Jan. 27, 1838, to John Pemberton, Apr. 22, 1838, to Anna Pemberton, Jan. 23, 1839, to Rebecca Pemberton, Apr. 16, 1839, and to Israel Pemberton, Feb. 13, 1839, John C. Pemberton Papers, Pemberton Family Papers; Rebecca to Israel, Nov. 16, 1838, Israel Pemberton Papers, Pemberton Family Papers, Historical Society of Pennsylvania, Philadelphia.
7. See Ballard, *Pemberton,* 35–40.
8. Ibid., 40–41; Marriage Record of John Clifford Pemberton and Martha Thompson, John C. Pemberton Papers, Southern Historical Collection, Univ. of North Carolina at Chapel Hill.
9. The changes in Pemberton's personality are discussed in Ballard, *Pemberton,* 37, 58, 77–78, 79.
10. Pemberton to Israel, May 28, 1846, to John Pemberton, May 10, June 9, 1846, John C. Pemberton Papers, Pemberton Family Papers; George Meade to Israel, June 10, 1846, Israel Pemberton Papers, Pemberton Family Papers. For Pemberton's experiences in the Mexican War, see Ballard, *Pemberton,* 43–63.
11. Pemberton to Rebecca, Aug. 5, 1846, John C. Pemberton Papers, Pemberton Family Papers; Ballard, *Pemberton,* 52.
12. *Niles' National Register* 71 (Nov. 21, 1846), 180–81; G. W. Cullum, *Biographical Register of Officers and Graduates, U.S. Military Academy* (West Point, NY, 1891), excerpt in Pemberton Papers, Southern Historical Collection.
13. Pemberton's years between the Mexican and Civil Wars are covered in Ballard, *Pemberton,* 64–82.
14. Israel Pemberton to Pemberton, Apr. 15, 1861, Israel Pemberton Papers, Pemberton Family Papers.
15. Ibid.; Rebecca Pemberton to Caroline Hollingsworth Pemberton, Apr. 23, 1861, Pemberton Papers, Southern Historical Collection; U.S. War Department, *The War*

of the Rebellion: A Compilation of the Official Records of the Union and Confederate Armies, 128 vols. (Washington, DC, 1881–1901), series 1, vol. 2:580. Hereinafter cited as *OR.* All references are to series 1 unless otherwise indicated.

16. Pemberton's decision is analyzed in Ballard, *Pemberton,* 85–86.

17. *OR,* vol. 2:856, 963; Certificates of Rank, Pemberton Papers, Southern Historical Collection; Joseph E. Johnston, *Narrative of Military Operations, Directed, during the Late War between the States* (New York, 1874), 12.

18. *OR,* vol. 2:982–83; ibid., 4:666.

19. *OR,* vol. 6:334.

20. Douglas Southall Freeman, *R. E. Lee: A Biography,* 4 vols. (New York, 1934–1935), 1: 613, 630–31; *OR,* vol. 6:366.

21. Certificates of Rank, Pemberton Papers, Southern Historical Collection; *OR,* vol. 6:402, 407.

22. *OR,* vol. 6:414, 523.

23. Pemberton's administrative challenges in South Carolina are discussed at length in Ballard, *Pemberton,* 91–106.

24. Charles E. Cauthen, *South Carolina Goes to War, 1860–1865* (Chapel Hill, 1950), 142, 144, 161. The council was generally unpopular among the South Carolina public and was eventually abolished, but not until Pemberton had left the state.

25. Pemberton to Samuel Cooper, Mar. 27, 1862, Letters Sent and Received by Gen. J. C. Pemberton, chap. 2, vol. 21, Record Group 109, War Department Collection of Confederate Records, National Archives (hereinafter cited as RG109); *OR,* vol. 14:423–24.

26. *OR,* vol. 14:424–25.

27. Freeman, *R. E. Lee,* 1:627; Pemberton to R. E. Lee, Apr. 10, 1862, Pemberton Letters Sent and Received, chap. 2, vol. 21, RG109; *OR,* vol. 6:166,

28. *OR,* vol. 14:13–15, 483; Pemberton to W. H. Taylor, May 14, 1862, and to A. L. Long, May 21, 1862, Pemberton Letters Sent and Received, chap. 2, vol. 21, RG109.

29. Pemberton to A, L. Long, May 21, 1862, ibid.

30. *OR,* vol. 14:490–91, 513–14,

31. Francis Pickens to Pemberton, May 23, 1862, Pemberton Letters Sent and Received, chap. 2, vol. 21, RG109.

32. *OR,* vol. 14: 523–24. See also F. W. Pickens to Jefferson Davis, Aug. 20, 1862, in Jefferson Davis, *Jefferson Davis, Constitutionalist: His Letters. Papers, and Speeches,* ed. Dunbar Rowland, 10 vols. (Jackson, MS, 1923), 5:326–27.

33. Pemberton to Jefferson Davis, June 2, 5, 1862, to George Randolph, June 3, 1862 (three messages that date), and to James Chesnut, June 6, 1862, all in Pemberton Letters Sent and Received, chap. 2, vol. 21, RG109; *OR,* vol. 14:89, 90, 567; Milby Burton, *The Siege of Charleston, 1861–1865* (Columbia, SC, 1970), 99, 105–9.

34. For examples, see Ballard, *Pemberton,* 107–10.

35. *OR,* vol. 14:560.

36. Mary Boykin Chesnut, *Mary Chesnut's Civil War,* ed. C. Vann Woodward (New Haven, 1981), 332; Emma Holmes, *The Diary of Miss Emma Holmes 1861–1866,* ed. John F. Marszalek (Baton Rouge, 1979), 174, 177.

37. Pemberton to A. L. Long, May 21, 1862, Pemberton Letters Sent and Received, chap. 2, vol. 21, RG109; *OR,* vol. 14:503–4, 524; Ballard, *Pemberton,* 107–8.

38. *OR,* vol. 14:560, 569–70, 601.

39. Ibid., 601. For examples of support, see *Charleston Daily Courier,* Sept. 26, 1862, and *Charleston Mercury,* Sept. 29, 1862.

40. *OR,* vol. 15:820; ibid., vol. 17, pt.2: 718; Ballard, *Pemberton,* 120.

41. See Michael B. Ballard, *The Campaign for Vicksburg* (Conshohocken, PA, 1996), 1–6, for background prior to Pemberton's arrival; David. D. Porter, *Incidents and Anecdotes of the Civil War* (New York, 1885), 95–96.

42. *New York Herald,* Aug. 17, 1881, clipping in John Clifford Pemberton Papers, Library of Congress.

43. Richard M. McMurry, *Two Great Rebel Armies: An Essay in Confederate Military History* (Chapel Hill, NC, 1989), 133–34.

44. Ballard, *Campaign for Vicksburg,* 5, 6, 13, 22; ibid., *Pemberton,* 123, 125.

45. For comparisons of troop strength, see *OR,* vol. 17, pt. 2:576–78, 661–62, 699.

46. Ibid., 757–58; Ballard, *Pemberton,* 118–19.

47. Ballard, *Pemberton,* 119–20, 144.

48. *OR,* vol. 17, pt. 2:758.

49. *Jackson Daily Mississippian* quoted in *Memphis Daily Appeal,* Nov. 28 [29], 1862; R. W. Memminger, "The Surrender of Vicksburg—A Defence of General Pemberton," *Southern Historical Society Papers* 12 (July, Aug., Sept. 1984): 353; Ballard, *Pemberton,* 121–23.

50. Robert G. Hartje, *Van Dorn: The Life and Times of a Confederate General* (Nashville, 1967), 254–69; *OR,* vol. 17, pt. 2:451; Ballard, *Pemberton,* 127.

51. Edwin Cole Bearss, *The Vicksburg Campaign,* 3 vols. (Dayton, OH, 1985–1986), 1:143–46, 154, 172, 190, 200, 208–9; R. S. Bevier, *History of the First and Second Missouri Brigades* (1879; reprint, St. Louis, 1985), 166–67; *OR,* vol. 17, pt. 2:669; Ballard, *Pemberton,* 128–29.

52. For a brief survey of Grant's diversionary activities, see Ballard, *Campaign for Vicksburg,* 18–24.

53. Bearss, *Vicksburg Campaign,* 1:475–507, 539; Ezra J. Warner, *Generals in Gray: Lives of the Confederate Commanders* (1959; reprint, Baton Rouge, 1970), 194; William A. Drermon Diary, May 30–July 4, 1862, p. 6, Mississippi Department of Archives

and History, Jackson; F. W. M.,"Career and Fate of Gen. Lloyd Tilghman," *Confederate Veteran* 1 (Sept. 1893): 274–75; Pemberton, General Orders No. 33, Feb. 1, 1863, Orders and Circulars, 1862–1865, Records of the Department of Mississippi and East Louisiana and of the Department of Alabama, Mississippi, and East Louisiana, entry 94; Lloyd Tilghman to John Waddy, Jan. 27, 1863, General John C. Pemberton Papers, 1862–1864, entry 131, all in RG109.

54. The expedition is discussed in detail in Bearss, *Vicksburg Campaign,* 1:549–91. The failure of Confederate command during this episode is discussed in Samuel W. Ferguson Memoir, pp. 38–41, Vicksburg National Military Park, Vicksburg, Miss.

55. The standard account of Grierson's raid is D. Alexander Brown, *Grierson's Raid* (Urbana, 1962). See also OR, vol. 24, pt. 1:532; ibid., pt. 3:776, 783–87, 791, 792, 794, 802–3; Pemberton to Daniel Ruggles, Apr. 20, 1863, John C. Pemberton Papers, U.S. Naval Academy Archives, Annapolis.

56. Ballard, *Campaign for Vicksburg,* 21–32.

57. *OR,* vol. 24, pt. 3:810–13. Details of the struggle at Port Gibson are in Bearss, *Vicksburg Campaign,* 2:317–407.

58. *OR,* vol. 24, pt. 3:821.

59. Ibid., 808, 815, 839, 842.

60. Bearss, *Vicksburg Campaign,* 2:480, 512–14; Johnston, *Narrative of Military Operations,* 506; *OR,* vol. 24, pt. 1:239–41, 262–63; ibid., pt. 2:74, 75; ibid., pt. 3:876, 882, 883; Ballard, *Pemberton,* 156–59.

61. *OR,* vol. 24, pt. 1:263.

62. Ballard, *Pemberton,* 160–64.

63. Ibid., 164–65.

64. *OR,* vol. 24, pt. 1:241, 272. On Pemberton and the siege, see Ballard, *Pemberton,* 167–80.

65. *OR,* vol. 24, pt. 1:285; Ballard, *Pemberton,* 153–54, 182–86, 200–201. A lengthy account of the Vicksburg campaign by Pemberton has recently been discovered and is in private hands. The owner has assured the author that the contents do not differ markedly from those of Pemberton's published writings in the *OR;* John C. Pemberton [III], *Pemberton: Defender of Vicksburg* (Chapel Hill, 1942), 281–319; Robert U. Johnson and Clarence C. Buel, eds., *Battles and Leaders of the Civil War,* 4 vols. (1887; reprint, New York, 1956), 3:543–45. *Editors' note:* The unpublished manuscript referred to above was published in 1999: John C. Pemberton, *Compelled to Appear in Print: The Vicksburg Manuscript of General John C. Pemberton* (Cincinnati, OH: Ironclad Publishing, 1999), ed. David M. Smith.

Lieutenant General Daniel Harvey Hill. Courtesy of the Museum of the Confederacy, Richmond, Virginia.

Soldier with a Blunted Sword: Braxton Bragg and His Lieutenants in the Chickamauga Campaign

Steven E. Woodworth

WELL BEFORE NATHAN BEDFORD FORREST'S GRAY-CLAD TROOPERS CLASHED WITH ELEments of George H. Thomas's Union army corps in the woods around Jay's Mill, at the northeastern end of the Chickamauga battlefield, much of the course and even the outcome of the struggle that was to rage over this ground during the next thirty-six hours had already been set up by developments within the Confederate Army of Tennessee's high command. Though the soldiers of both sides would have to fight and die in order to make reality of the situations their generals had developed, the fact remains that the Confederate soldiers, though they did not know it, were condemned to fight against the long odds of a frontal assault, contending for what at best could be little more than an indecisive victory. They were neither the first nor the last Civil War soldiers asked to face such odds in hope of such scant gains, but their lot appears the more dramatic in the light of history because, up until just a few days before, it could have been entirely different. That it was not was the fault of the Army of Tennessee's generals, and for once, their failure had been not so much the result of incompetence as of contentiousness. Even more ironic, their commander, General Braxton Bragg, was almost as much the victim of this situation as were the common soldiers themselves, about to be fed into the meat-grinder of battle. The fact was that the fires of contention and bickering within the top ranks of the Army of Tennessee, originally kindled by just a few discontented individuals, had by the time of the Chickamauga Campaign spread so far and grown so hot that the entire command structure was warped and almost useless. Despite the strategic insights of Bragg and the courage of his troops, the army had become no longer a sharp weapon that could strike fatal blows to the opposing army, but rather a blunt instrument fit only for the crudest sort of bludgeoning.

To understand the situation in the Army of Tennessee's high command at the time of the Chickamauga Campaign, it is necessary to step back some twelve months, to the Confederate invasion of Kentucky. In the late summer of 1862 the Confederacy had seemed about to reap a rich harvest of victories that might well bring Southern independence. Robert E. Lee was advancing into Maryland for the first time in the war, and in Kentucky separate Confederate forces under Bragg and Edmund Kirby Smith controlled the Bluegrass Region and seemed to threaten both Cincinnati and Louisville. After the summer of 1862, however, Confederate fortunes had steadily declined.

That this was the case in Kentucky was largely the fault of Confederate President Jefferson Davis. Davis, at the outset of the invasion, had failed to place Smith under Bragg's orders. Consequently, at crucial moments during the campaign, Smith had refused to cooperate with his senior general, making it impossible for Bragg to win the victory over Buell that his superior strategy and hard marching should have gained him. Another error of Davis's was his refusal, before Bragg's army had left its former base in Mississippi, to take the commander's advice and remove incompetent generals—Bragg called them "dead weight"—from leadership of its corps and divisions. At the brigade-command level were a number of young, energetic, and talented officers such as Patrick R. Cleburne and Alexander P. Stewart, and Bragg wanted to see men like these promoted to fill the higher spots. The president's reason for refusing to act on Bragg's recommendation was the presence among the high-ranking incompetents of his old friend and West Point crony, Leonidas Polk.[1]

Polk had been a year ahead of Davis at the academy back in the 1820s. He was highly persuasive and had a winning personality. After graduation Polk had resigned immediately and become an Episcopal clergyman—eventually a bishop—and in the thirty-four years between that time and the outbreak of the Civil War, he apparently never read a single book on military matters. In fact, Polk did little reading of any sort. His strong suit was influencing people, and that he did very well indeed. When war came, Davis had made Polk a major general. Only Davis's inordinate dependence on old friends could explain his giving that kind of rank to a man with Polk's almost complete lack of qualifications, and the new general's performance early in the war proved to be every bit as questionable—even disastrous—as his poor credentials gave cause to expect.[2]

Bragg, who was a good judge of military competence, though he tended to be rather naive about other men's personalities, had not taken long to realize that the bishop-general was a liability to the Confederate war effort. Thus when he had written to the president before the Kentucky Campaign, he had suggested that of his top officers, only William J. Hardee was worth keeping. The implication was, of course, that Polk was not. That was the truth, but Davis would hear none of it. He refused Bragg's request, and that was that. The senior generals would stay, competent or not. During the Kentucky Campaign Polk gave his commander plenty of cause to be dissatisfied. He provided Bragg confused reports of the situation in his front, delayed carrying out orders and thus endangered the army, and on one occasion simply refused to obey an order from his commanding general. Several months later, Bragg summed up a large part of the problem with Polk when he wrote in a letter to the president, "Genl. Polk by education and habit is unfitted for executing the plans of others. He will convince himself his own are better and follow them without reflecting on the consequences." The bishop-general was proud and willful and would make trouble if he did not get his way.[3]

And make trouble he did. After the army's return from Kentucky, Polk began a regular campaign to have Bragg removed from command. The reasons for this were plain: first, Polk resented anyone who gave him orders. Then, as second-ranking man in the army, Polk would—theoretically at least—succeed to army command should Bragg be removed. To accomplish his goal, Polk launched a stream of misleading letters to his old West Point friend Davis, begging for Bragg's removal. He fed false information to congressmen. He boasted in letters to friends that things would have been different if he "had been in chief command." In that case, he crowed, defeating Buell and taking Kentucky "could have been easily done." Even worse, he began to stir up discontent among the army's other officers.[4]

Being an influential man, he soon won over to his opinion his fellow corps commander William J. Hardee. Hardee was inclined to be critical of superiors in any case, though reluctant to assume responsibility himself. He had far more practical influence within the army's officer corps than Polk could ever hope to have because of his reputation as an expert on tactics. Prior to the war he had produced the army's standard manual of infantry tactics, and although the work was largely cribbed from previous French publications and was not particularly brilliant for all that, it gave Hardee a formidable reputation. Within his corps of the Army of Tennessee he held regular classes for his officers, instructing them in the basics of troop handling and command, and, as he came increasingly under the influence of Polk, also insinuating to them that Bragg was incompetent and could do nothing right. Most of the junior officers, from division commanders on down, often had no way of seeing the larger strategic and tactical picture, and so they drank in Hardee's distortions about Bragg.[5]

Another group within the army's officer corps that had a special grudge against Bragg consisted of officers from Kentucky. Such influential men as former Vice President John C. Breckinridge and Simon B. Buckner, along with several other Kentucky generals, hated Bragg because it was the most obvious alternative to facing the fact that their home state by and large had not chosen to side with the Confederacy. Bragg had pointed this out after the Kentucky Campaign, citing the obvious fact that if the people of Kentucky did not support them, Confederate troops could not possibly remain in the state for long. It was no more than the truth, but the Kentucky Confederates could not accept that and instead chose to believe, falsely, that Bragg had lost the campaign through incompetence and thus betrayed their beloved state.[6]

The discontent within the Army of Tennessee's officers increased after the Battle of Murfreesboro. At that battle Bragg had handled the Union army about as roughly as Lee and "Stonewall" Jackson were to do with Joseph Hooker's troops at Chancellorsville four months later. But Rosecrans was not Hooker, and the Army of the Cumberland was not the Army of the Potomac. The Federals did not retreat, and as they continued to bring up reinforcements, Bragg found that it was he who would have to retreat. Though the move was taken only at the intense urging of a number of his generals, including Polk, Bragg immediately came under intense criticism throughout the Confederacy and in the press for supposedly having thrown away another victory. The denunciations were almost certainly fed at least in part by Polk and possibly others within the army. Bragg, naively believing that his generals would support him since he had acted on their advice, sent each of them a note asking if they had advised retreat and stating as sort of a closing flourish that if he found that he had indeed acted without the support of his generals in this, he would resign at once. Polk, Hardee and his minions, and the Kentucky clique saw their chance. Admitting that they had counseled withdrawal after Murfreesboro, most of them went on to state bluntly that they thought the army would be better off without Bragg.

When President Davis learned of this bizarre situation, he ordered the western theater commander, Joseph E. Johnston, to go to the army's new base at Tullahoma, Tennessee, and investigate the matter. Johnston went and found the army in good shape and well led, the only problem being the attitudes of many of its top generals. He reported this to the president, but by this time, the constant bombardment of negative distortions about Bragg from Polk and whomever else Polk could recruit for the task as well as Kentucky generals, congressmen, and the like had finally persuaded Davis that the Army of Tennessee's commander must go. The president thought it would be easy enough to supersede Bragg. Since Johnston was Bragg's superior and theater commander, whenever he was present in person with the army, he would, of course, command. All that was necessary was that Johnston stay with the Army of Tennessee and actually exercise that command. But the president had not reckoned on the stubborn and contrary Joe Johnston. That Davis wanted a thing was almost reason enough for Johnston to oppose it, and besides that, the Virginian coveted glory but feared responsibility more, and though he would complain at not being given a field army to command, he would not take one when it was given. Consequently, and despite the president's wishes, Bragg remained in command of the Army of Tennessee.[7]

Eventually, Davis acquiesced, but the hostile generals within the army never did. Polk continued his campaign of trying to undermine Bragg, concentrating on those outside the army, especially President Davis, while Hardee continued to spread the poison of distrust within the army's high command. Relations between Bragg and many of his generals, particularly his two corps commanders, grew so bad that communication between them virtually stopped. Thus Bragg's subordinates had little knowledge and less understanding of their commander's operational plans and ideas and had become convinced that any movement Bragg ordered simply had to be ill conceived and potentially disastrous—merely because Bragg had ordered it. This situation, far more than any strategic mistake Bragg could possibly have made, was to have disastrous results of its own.[8]

On June 24, 1863, the series of events leading directly to the battlefield of Chickamauga began, miles to the northwest of the meandering Georgia stream of that name. Rosecrans, moving south from his base at Murfreesboro, Tennessee, advanced toward Tullahoma, in front of which Bragg had positioned the two corps of the Army of Tennessee, blocking the road to Chattanooga. Bragg hoped to be able to deal Rosecrans a severe defeat, and for that purpose he knew he would have to do more than stand passively on the defensive. Rosecrans was a resourceful officer, and given time he could be counted on to find a way to put the Confederates at a disadvantage. Bragg did not plan to give him that opportunity. As he envisioned it, Hardee's corps, stationed nearer to the road and railroad lines of supply to which Rosecrans was tethered, was to serve as a blocking force, compelling the Federals to deploy in line against it and delaying them just long enough to allow Bragg to land the haymaker of this one-two combination. Polk's corps was positioned just to the west, and Bragg envisioned it swinging to the right, striking the flank of the Federal forces deployed to face Hardee, and winning a crushing victory, much like Lee's at Second Manassas.

By the time Rosecrans began his advance, Bragg must have suspected that the plan probably was not going to work. Relations between himself and his two top generals had been so bad that they were hardly on speaking terms with each other. Consequently, it had been difficult to make sure that Hardee and Polk each understood what was expected of them. Bragg apparently thought they did. They apparently did not, and in any case they were not much inclined to cooperate in any plan of Bragg's. Hardee had almost inexplicably failed to entrench the position from which he was to stall the oncoming Union army, and Polk seemed to have no idea of cooperation. Rosecrans opened his campaign with a clever feint that seemed to fool Hardee completely and easily got past the Confederate positions. Bragg fell back on his prepared positions around Tullahoma, hoping to make a stand there. But by this time Hardee and Polk were convinced that Rosecrans could not be stopped. More to the point, they probably had convinced themselves by this time that any plan of Bragg's would fail. Their belief became, in effect, a self-fulfilling prophecy.

The two corps commanders pressed Bragg to agree to retreat farther, but Bragg resisted at first. At this point, the malcontents actually appear to have considered the possibility of mutiny—forcibly removing Bragg and seizing control of the army. That would seem to be the only reasonable interpretation of a letter Hardee sent to Polk about that time. He labeled it "confidential," and from the nature of the contents, one can easily see why. Hardee said that he "had been thinking seriously of the condition of affairs with this army." He believed that Bragg was not fit to command it. That being the case, he proceeded to ask, "What shall we do?" "What is best to be done to save this army and its honor?" Apparently, the thought of obeying orders and cooperating with the commanding general did not occur to him, for he continued, "I think we ought to counsel together." Concerned that they have the complicity of other top-ranking generals, particularly the influential and embittered Kentucky clique, Hardee went on to ask, "Where is Buckner? I would like Buckner to be present."[9] Whether Hardee and Polk would have had the nerve to go through with such a plan had it come to that, we can never know. Before matters had progressed much further, Bragg, weakened by physical illness and bombarded by his subordinates' demands for retreat, finally gave in and ordered the army to fall back.

In the next few days, Rosecrans followed one flanking movement with another, and the Army of Tennessee, with its almost dysfunctional high command, seemed less and less able to respond adequately and rapidly. The constant agitation of Bragg's generals had produced a situation in which the army's command apparatus had become virtually paralyzed and an easy prey for a resourceful enemy. The series of retreats that began in Middle Tennessee ended with Bragg sadly abandoning Chattanooga and falling back into Georgia.[10]

In Richmond, Jefferson Davis was not unmindful of the plight of Bragg's army and the Confederacy's fortunes in Tennessee. To the enormous disgust of his old friend Polk—and by now of many others as well—he did not see Bragg's removal as the solution to the problem. With Joseph E. Johnston on duty in Mississippi and unavailable for command of the Army of Tennessee, Davis knew of no one else whom he preferred to Bragg. While Bragg had never been a personal friend of the president, Davis did respect Bragg's ability and his loyalty to the Confederacy. To Polk, this was nothing but willful stubbornness. In a remarkable letter written about this time, the bishop-general told a friend, "The truth is, I am somewhat afraid of Davis. He has so much at stake on this issue [Bragg] that I do not find myself willing to risk his judgment. . . . He is proud, self-reliant, and I fear stubborn." As far as Polk was concerned, Davis should "lean a little less on his own understanding and realize that there were some minds in the land from whom he might obtain counsel worth having." Davis's friendship toward Polk was the only thing that had raised the pompous and self-important bishop as high as he was in the Confederate army and also the only thing that had kept him there. Now Polk was becoming bitter and contemptuous of Davis because presidential favoritism was not forthcoming to raise him the rest of the way to an independent army command. The letter reveals a great deal about Polk's personality.[11]

Rather than remove Bragg, Davis's solution to the plight of the Army of Tennessee was to reinforce it from Lee's army in Virginia. The decision to take this step was made before Bragg was forced out of Chattanooga, and was motivated primarily by the desire to prevent the cutting of the railroad line that ran through Chattanooga, Knoxville, and southwestern Virginia to Richmond. Davis ordered Lieutenant General James Longstreet and two divisions, about ten thousand men from Lee's army, to be detached and sent to Bragg. Substantial reinforcements were also ordered to Bragg from Mississippi. Once all these forces arrived in northwestern Georgia, Bragg would enjoy a luxury rare for Confederate commanders—he would actually outnumber Rosecrans by a small margin.

While the forces were on the way, however, opportunity presented itself to Bragg in a form that was rare for any commander on any side in any war. Rosecrans, elated at the bloodless success he had achieved simply through maneuvering, now seemed to forget that the Army of Tennessee was not a disorganized foe fleeing from a crushing defeat, but was still intact, a dangerous army spoiling for a fight, if its generals could ever give it a fair chance. Bragg encouraged Rosecrans in the delusion of Confederate disintegration by sending fake deserters into Federal lines with made-up stories of demoralization, desertion, and headlong flight among the Rebels. Under this false impression, the Federal commander divided the three corps of his army and sent them on widely divergent paths through the mountain passes in hopes of catching Bragg or at least keeping him on the run. The result was Rosecrans's presenting Bragg with an unbelievable opportunity to destroy the Federal army piece by piece, before the widely separated Union columns could come within supporting distance of each other.[12]

Bragg was not slow in responding to this invitation. On the afternoon of September 9, 1863, he set troops in motion to crush a substantial segment of Rosecrans's center column, the army corps of Major General George H. Thomas. By early morning of the tenth, all was to be in readiness. Confederate forces with more than a twofold superiority in numbers would strike both flanks of the foremost Union division. Calling on the troops closest to hand, lest the opportunity slip away while others were coming up, Bragg ordered the division of Thomas Hindman to strike one side and that of Patrick Cleburne, a part of Daniel Harvey Hill's corps, to hit the other. Then things began to go wrong. A good bit of what went wrong at this point had to do with Hill himself.[13]

Hill had served with the Army of Northern Virginia earlier in the war and had earned a well-deserved reputation as a ferocious fighter. Yet although he was one of that hard-fighting army's best combat leaders, he had proved to be a detriment to it nevertheless. His problem was his personality. He was bitter, sour, critical, and contentious, ever ready to pick nits and find fault. Lee also believed that Hill's organizational abilities were not all that might be desired. At any rate, probably both for that reason and because of his habit of carping criticism, Lee had suggested to Davis that Hill be transferred to other service. As with most of Lee's suggestions to the president, this one had met with swift compliance, and Hill had spent the year of the war leading up to Chickamauga in relative backwater commands. Sent to the Army of Tennessee to replace Hardee, who was needed at the time in Mississippi, Hill filled his predecessor's place in more ways than one, and proved to be almost the worst possible addition to the Army of Tennessee. The sour, critical Hill quickly picked up the bad attitude that infected its high command, and he was soon parroting Polk's denunciations of its commander.[14]

Thus he, like many of the other officers of the army, was conditioned to believe that an order from Bragg could not possibly be wise. When, on the night of September 9, he received Bragg's order to send Cleburne's division against the exposed portion of Thomas's corps, he began thinking of obstacles to action and excuses for inaction. He claimed the roads were obstructed but took no steps to clear them, and he complained that Cleburne was ill but made no move to substitute another commander or another division. Indeed, evidence suggests that Cleburne may not have been sick at all. The plain fact was that Hill did not trust Bragg and therefore feared to make the movement. Hindman, for his part, was little better. An officer of uncertain capability, Hindman had a service record that could, at best, be described as uneven. Now, as senior officer of the attack Bragg had ordered to beat the Army of the Cumberland in detail, Hindman lost his nerve and ordered the movement halted.[15]

Still, an opportunity remained for September 11. The Union officers did not yet realize their predicament and the exposed force remained in position. Since Hill seemed certain that Cleburne's division would not be available, Bragg selected the next nearest troops, Buckner's, to make the attack, sending an order for the Kentuckian to send two divisions to reinforce Hindman. Buckner, too, was little inclined to trust Bragg. As soon as he could meet with

Hindman, the two of them discussed Bragg's order and decided it had better not be carried out. They so informed their commander. Bragg was not to be deterred from such a golden opportunity and insisted that they go ahead. At that, Buckner and Hindman simply refused to obey Bragg's order and pulled their troops back into defensive positions.[16]

No third chance to destroy Thomas's force presented itself, as the Union general became aware of his danger and pulled back out of reach, but again on September 13 an opportunity almost as inviting beckoned to the Confederates when Rosecrans's left wing was found to be isolated from the rest of the Federal army. Again Bragg set up his forces to crush the exposed Union corps. This time the task of making the attack fell to the troops of Polk. The bishop-general, however, believed he was facing an overwhelming Union force, when in fact matters were quite the other way around. He informed Bragg that he would not attack but instead would take up a defensive position. Again, Bragg insisted that the chance be seized and the attack made. "We must force him to fight," he wrote, "at the earliest moment and before his combinations can be carried out." He assured Polk that his information about Federal vulnerability was solid and promised to increase the already decisive superiority Polk enjoyed by even further weakening other parts of the army to reinforce him, and he stressed that the attack must be made soon, "lest another golden opportunity . . . be lost by the withdrawal of our game." Of course, Polk disobeyed his orders and remained on the defensive, and the Union force escaped.[17]

Six days later Bragg launched the Battle of Chickamauga. His tactics were simple, a frontal attack en echelon intended to hammer the enemy back and to one side, push him into a mountain cul-de-sac called McLemmore's Cove, and destroy him. Over the years Bragg has been criticized for the use of such a straightforward and unimaginative plan. If he had been a competent general, his critics argue, he would have attempted some sort of turning or flanking maneuver or have endeavored to catch and defeat the enemy in detail. They overlook the fact that Bragg had tried and succeeded in doing just that, not once but three times, during the fortnight prior to the battle. That no such relatively crushing and inexpensive victory was won was not the fault of Bragg so much as of the endemic dissension within the army's high command and of the officers—particularly Polk—who had fostered that dissension. Perhaps, if Bragg had been among the handful of truly great commanders in the history of warfare, if he had been another Lee or another Jackson, he might have succeeded in winning a decisive victory anyway. Yet even such an extraordinary commander might well have failed under these circumstances. As it was, Bragg had done all that should have been necessary to secure success, but he fought with a disadvantage he could not overcome. He was like a warrior with a dull and heavy sword. The instrument in his hands was simply not suited for quick, skillful, and dexterous work. Thus if he fell to hacking and bludgeoning his foe in a rather unsophisticated manner, we must look for the reason not merely by analyzing his own skills as a tactician, but also by considering the nature of the army—and particularly the generals—he commanded.

Notes

"Soldier with a Blunted Sword: Braxton Bragg and His Lieutenants in the Chickamauga Campaign" previously appeared in Steven E. Woodworth's *No Band of Brothers: Problems of the Rebel High Command* (Columbia, MO, 1999), 70–80, and is reprinted with permission of the University of Missouri Press.

1. U.S. War Department, *The War of the Rebellion: A Compilation of the Official Records of the Union and Confederate Armies,* 128 vols. (Washington, DC, 1880–1901), series 1, vol. 16, pt. 2:745–46, 748, 751–53, 766–67, 869, 866; vol.17, pt. 2:627–28, 654–55, 658, 667–68, 673 (hereinafter cited as *OR;* all references are to series 1 unless otherwise indicated); Gary Donaldson, "'Into Africa': Kirby Smith and Braxton Bragg's Invasion of Kentucky," *Filson Club Historical Quarterly* 61 (Oct. 1987): 464; Thomas Lawrence Connelly, *Army of the Heartland: The Army of Tennessee, 1861–1862* (Baton Rouge, LA, 1967), 195.
2. Woodworth, *No Band of Brothers,* 12–18.
3. Grady McWhiney, *Braxton Bragg and Confederate Defeat,* vol. 1, *Field Command,* (1969; reprint, Tuscaloosa, AL, 1991), 328–29.
4. Ibid.; Thomas Lawrence Connelly, *Autumn of Glory: The Army of Tennessee, 1862–1865* (Baton Rouge, LA, 1971), 20–21.
5. Nathaniel C. Hughes Jr., *General William J. Hardee: Old Reliable* (Baton Rouge, LA, 1965), 3–85; Connelly, *Autumn of Glory,* 20–21, 90.
6. McWhiney, *Braxton Bragg,* 329–33.
7. Steven E. Woodworth, *Jefferson Davis and His Generals: The Failure of Confederate Command in the West* (Lawrence, KS, 1990), 187–98; McWhiney, *Braxton Bragg;* Judith Lee Hallock, *Braxton Bragg and Confederate Defeat,* vol. 2 (Tuscaloosa, AL, 1991).
8. Connelly, *Autumn of Glory,* 26–28, 123–29.
9. Quoted in Joseph E. Parks, *General Leonidas Polk, C.S.A.: The Fighting Bishop* (Baton Rouge, LA,1962), 315.
10. Regarding the Tullahoma campaign, see Steven E. Woodworth, "Braxton Bragg and the Tullahoma Campaign," in *The Art of Command in the Civil War* (Lincoln, NE, 1998).
11. Parks, *Polk,* 321.

12. *OR,* vol. 30, pt. 2:22, 26–37.

13. On the Chickamauga campaign, see Steven E. Woodworth, *Six Armies in Tennessee: The Chickamauga and Chattanooga Campaigns* (Lincoln, NE, 1998).

14. Douglas Southall Freeman, *Lee's Lieutenants: A Study in Command,* 3 vols. (New York, 1942–44), 1:19–22.

15. *OR,* vol. 23, pt. 2:26–37; *OR,* vol. 23, pt. 4:634; Glen Tucker, *Chickamauga: Bloody Battle in the West* (Indianapolis, 1961), 67–69.

16. *OR,* vol. 30, pt. 4:636.

17. Ibid., 26–37, 49, 636.

Major General Patrick Ronayne Cleburne. Courtesy of the Alabama Department of Archives and History, Montgomery.

Patrick Cleburne's Defense of Tunnel Hill Revisited

Craig L. Symonds

Of all the generals who served in the Western Theater, few compiled a combat record as impressive as that of Major General Patrick Ronayne Cleburne, the Irish immigrant from County Cork by way of Helena, Arkansas. Like most immigrants, Cleburne sought to immerse himself in his adopted community. After immigrating in 1849 and settling in what was then the frontier town of Helena, Arkansas, he became first a pharmacist, then a lawyer, and eventually a man of some status in the community. Despite a personal shyness, he was politically active in the campaign against the anti-immigrant Know-Nothing movement in 1856. As the secession crisis drew near, he joined the local militia unit, the Yell Rifles, and was elected its captain. After war broke out, Cleburne was elected colonel of what became the 15th Arkansas Infantry Regiment, and he earned a rapid promotion to brigadier general. He compiled a glittering combat record at Shiloh and during the Kentucky Campaign, at Murfreesboro and at Chickamauga. By November of 1863, he was a major general in command of a division in the Army of Tennessee, the highest ranking foreign born officer in the Confederate Army.

A number of personal characteristics made Cleburne especially effective in his new role. One was that he was able to remain calm in the midst of crisis. A virtual stoic on the battlefield, he was the calm eye at the center of the storm. He never lost his nerve or his temper, or even became particularly animated. His placid demeanor and sense of calm infected everyone around him and encouraged his subordinates to remain unruffled even amidst the chaos of battle. Another key characteristic was Cleburne's effort to encourage a sense of partnership with his subordinates so that they did not fear to make suggestions or act independently to changing circumstances. Like Admiral Horatio Nelson's captains, Cleburne's brigadiers were a band of brothers. Finally, Cleburne paid attention to the hundreds of tiny details that another commander might have assumed belonged to others. He was never off duty.

An excellent example of how Pat Cleburne made use of these characteristics to win an unlikely victory is his defense of Tunnel Hill in the fighting for Chattanooga in the fall of 1863. Dating back to the stand of the handful of Spartans who defended the pass at Thermopylae, there are many examples of a relatively small unit holding off much larger attacking forces. Bunker Hill and the Alamo have become icons of American military history, and the Civil War, too, had its share of such actions from the stand of Thomas J. Jackson's Brigade at First Bull Run (Manassas) to that of George Greene's Brigade on Culp's Hill at Gettysburg. Pat Cleburne's successful defense of the northern end of Missionary Ridge on November 25, 1863, belongs in this category. Against odds of roughly seven to one, Cleburne's one reinforced division not only held its position in a series of battles that lasted virtually all day, but may in the process have saved the Army of Tennessee from destruction. Historian Wiley Sword dubbed this engagement "one of the Civil War's most remarkable encounters."[1]

The particular circumstances that led to this remarkable encounter grew out of Braxton Bragg's attempt to conduct a siege of Chattanooga in the fall of 1863. The twin blows of Gettysburg and Vicksburg that summer had been devastating to Confederate arms. Chastised by his reversal in Pennsylvania, Robert E. Lee agreed to send James Longstreet and two divisions to the Western Theater to enable Bragg to go on the offensive. With these reinforcements, Bragg won a victory of sorts at Chickamauga in September, but it had cost him dearly—more, in fact, than the South could afford (as so many Confederate victories seemed to do). And not only had Bragg failed to follow up his victory with a meaningful pursuit, thus allowing the Federal army to dig in at Chattanooga, but he and Longstreet had become engaged in a bitter and completely profitless quarrel.[2]

Of course Bragg managed to find a way to quarrel with nearly everyone—except President Jefferson Davis (he knew where his bread was buttered). But Longstreet was pretty quarrelsome in his own right, and his contrariness on this occasion was exacerbated by his disappointment at not being granted command of the army after coming west. Like Achilles, he sulked in his tent. Indeed, Longstreet's lack of energy in superintending his portion of the siege lines was one reason why the Federals were able to open a precarious supply line into Chattanooga (the "cracker line") in October. After that, relations between the two Confederate generals deteriorated further, and Old Pete insisted that he be liberated from Bragg's command. Happy to oblige, Bragg agreed to send him and his two divisions to Knoxville—a foolish decision that deprived Bragg of nearly a third of his force, and which made little strategic sense anyway since the reoccupation of Knoxville was not critical to Confederate fortunes. With Longstreet's departure, Bragg's army was outnumbered by the Union force it was supposedly besieging in Chattanooga.[3]

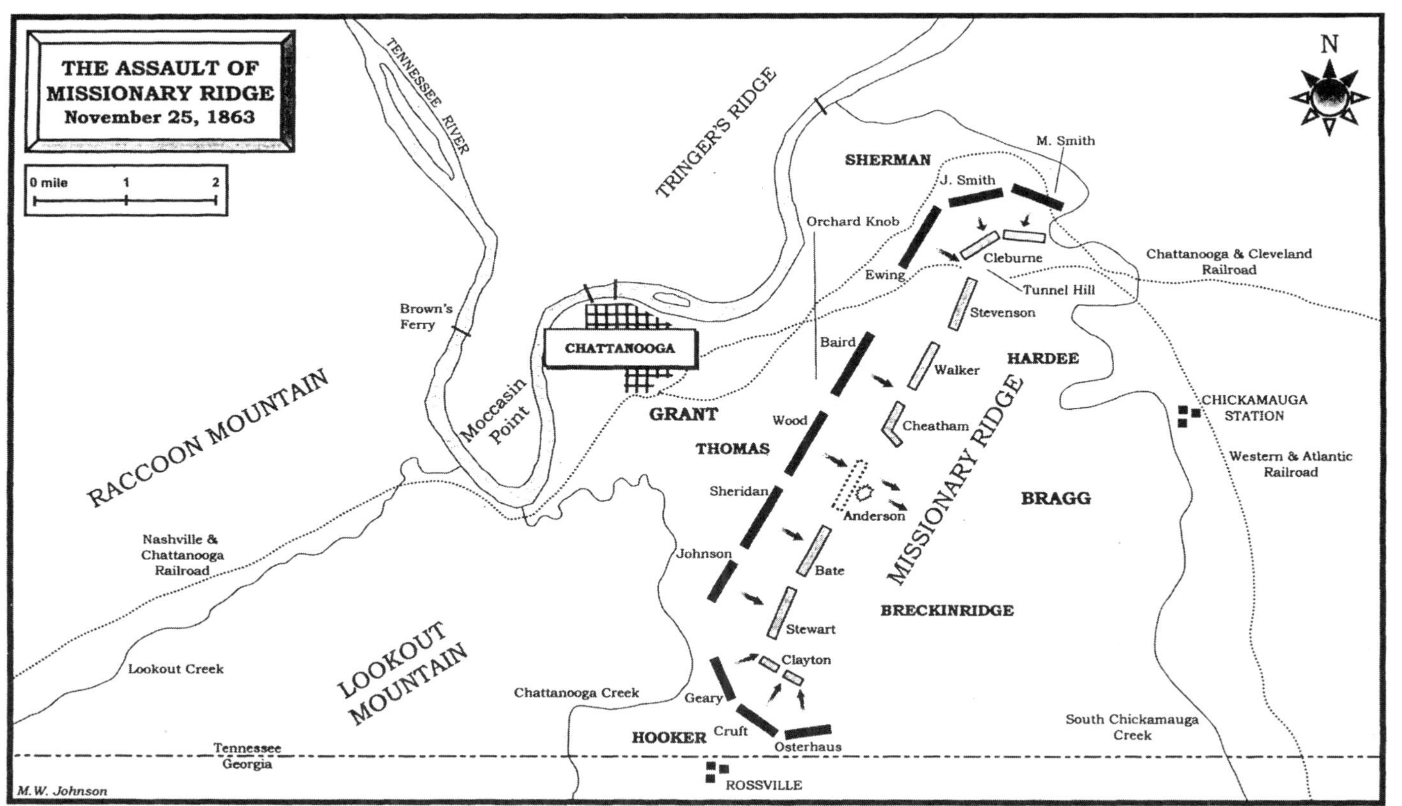

Courtesy of the Savas Publishing Company.

Of course having to make do with inferior numbers was not unusual for Confederate armies, and the fact that Bragg's army occupied the dominating heights of Missionary Ridge and Lookout Mountain seemed to compensate for the numerical difference between the armies. Bragg conducted his "siege" of Chattanooga by occupying only the territory south of the river. The difficult terrain north of the Tennessee made access to the city from that direction impractical except by mule train along narrow footpaths. Large wheeled vehicles—such as supply wagons—had to come via the road that ran parallel to the railroad on the south bank of the Tennessee River opposite Moccasin Point. As long as the Confederates held Lookout Mountain, they controlled the traffic along the river. This included not only the wagon road and the railroad, but also the river traffic: supply vessels coming upriver could not round Moccasin Point under the guns of Confederate batteries on Lookout Mountain. Therefore, even though Bragg's forces did not control any of the ground north of Chattanooga, they did control the transport in and out of the city.

Another apparent Confederate advantage was the location of Bragg's main defensive line along the top of Missionary Ridge. Unlike other battlefield "ridges"—notably Cemetery Ridge at Gettysburg—Missionary Ridge was not a gentle fold in the ground, but a genuine mountain 400 feet high with a gradient of better than 30 degrees and which, at places, seems to be virtually perpendicular. Surely, Bragg's position was all but impregnable.

A number of factors, however, made Bragg's position less secure than it seemed. First, after Longstreet's departure, Bragg's army numbered only about 35,000 men, and it occupied lines that were much too extended for an army of that size. From the nose of Lookout Mountain to the northern end of Missionary Ridge, it is seven miles as the crow flies, and better than ten miles along the winding Confederate siege lines. This resulted in an average of no more than two soldiers per yard—and that is assuming that every soldier was in the front line with no reserve. Those circumstances might not have been fatal but for the second factor: the difficult terrain that Bragg counted on to deter the enemy from breaking out of the city also prevented him from easily shuttling his own forces from one point to another along the ridge crest. As a result, each division would be pretty much on its own if it had to beat off a Union assault. Finally, no position is stronger than its flanks, and if the Federals seized either Lookout Mountain on his left, or Tunnel Hill on his right, Bragg could not hold Missionary Ridge.[4]

It should have been a fatal blow to Bragg's hopes, therefore, when, on November 24, Lookout Mountain fell to the men of Joseph Hooker's command in the so-called "Battle Above the Clouds." After that, it was clear to most knowledgeable observers that the game was up: the loss of Lookout Mountain not only opened up the roads into Chattanooga from the west and effectively ended the siege, but it also exposed Bragg's forces on Missionary Ridge to a flanking maneuver through Chattanooga Valley. Given those circumstances, Ulysses S. Grant expected that Bragg would try to retreat under the cover of darkness and, in fact, that is exactly what the Confederate commander planned to do.[5]

When Lookout Mountain came under attack on November 24, Patrick Cleburne was not on Missionary Ridge. He was five miles away at Chickamauga Station on the Western & Atlantic Railroad superintending the embarkation of two divisions—his own and Simon Buckner's—which Bragg had ordered to Knoxville to support Longstreet. Apparently Bragg hoped that with this reinforcement Longstreet could complete his conquest of that city quickly and return with his entire force to help turn back the expected Federal sortie from Chattanooga. Buckner's men had already departed, and Cleburne's men were preparing to leave, when a courier galloped up to hand Cleburne new orders to suspend the transfer until further orders. Within minutes, subsequent messages ordered him to recall the departed troops and return with them to Missionary Ridge: the enemy was coming out! Cleburne telegraphed ahead to order Buckner's men to turn around, then he set his own division in motion marching back up the reverse slope of the ridge, and he rode ahead of the column to Bragg's headquarters. From there he watched the shell bursts amidst the clouds and fog as Hooker's men pried the Confederates out of their positions on Lookout Mountain. Bad as this was, worse news soon arrived. Even as Hooker's corps assailed Lookout Mountain, several more Federal divisions emerged from behind the screening bulk of Tringer's Ridge north of the Tennessee River and crossed to the south side of the river beyond the Confederate right. To counter this threat, Bragg ordered Cleburne to take his division to Tunnel Hill at the north end of the extended Confederate line, and to hold the position "at all hazards."[6]

At the northern end of the extended (one might say overextended) rebel line, Missionary Ridge disintegrated into a number of detached hills separated by gullies, disappearing altogether along the banks of South Chickamauga Creek. The most prominent of the small hills at the north end of the ridge line was Billy Goat Hill. About a mile south of it, the Chattanooga & Cleveland Railroad passed through a tunnel under the ridge before linking up with the Western & Atlantic. The knob-like eminence above the railroad tunnel was called Tunnel Hill, and it was as critical to the security of the Confederate right as Lookout Mountain was to the Confederate left. If the Federal army succeeded in gaining a purchase on Tunnel Hill, it could sweep southward along the ridge line, in effect rolling up Bragg's army like a hallway carpet.

To hold this critical position, Cleburne had only three brigades. His fourth brigade—that of Lucius Polk—had been detached by Bragg to guard the bridges over South Chickamauga Creek. That left Cleburne with James A. Smith's Texas brigade, Daniel Govan's Arkansas brigade, and Mark P. Lowrey's brigade of Alabamians and Mississippians, a total of just under 4,000 men. He also had three artillery batteries—Semple's, Swett's, and Key's—with a total of twelve guns.

Arrayed against him was literally an entire army. The Union troops that were pouring across the Tennessee River belonged to the Army of the Tennessee, commanded by Grant's favorite subordinate, William Tecumseh Sherman. Grant's plan to pry Bragg from Missionary Ridge was to keep the main Confederate line occupied with a credible feint against the center conducted by George Thomas's Army of the Cumberland, while Sherman seized control of the north end of the ridge with the Army of the Tennessee. To accomplish his task, Sherman had a total of four divisions—those of Peter Osterhaus, Morgan Smith, John E. Smith, and Hugh Ewing (Sherman's foster brother)—and twelve batteries of artillery. Eventually Grant would add Oliver O. Howard's XI Corps to Sherman's command, giving him a grand total of six divisions and 30,000 men. In the forthcoming engagement Sherman would have more *divisions* than Cleburne had *brigades,* more batteries than Cleburne had guns.[7]

Sherman not only had the numbers, he had a head start. Uncle Billy started his move across the Tennessee near the mouth of South Chickamauga Creek at dawn. Bragg did not dispatch Cleburne to Tunnel Hill until 2:00 P.M. By the time Cleburne arrived at Tunnel Hill ahead of his troops at about 2:30 P.M., two of Sherman's divisions were already over the river, and more were streaming across the new pontoon bridge that Sherman's engineers had laid. Informed by a signalman that Federals were even then advancing up the north slope of Billy Goat Hill, Cleburne sent Smith's Texas brigade—marching in the van of his column—with orders to try to beat the Federals to the crest. It was a close race, but the Federals won it. Smith's men arrived near the top of the hill to find it already occupied, and they were driven back by superior numbers.

That disappointment turned out to be much less devastating than it seemed at the time. Unfamiliar with the terrain, Sherman believed that Billy Goat Hill, which was somewhat higher in elevation than Tunnel Hill, was continuous with Missionary Ridge and that in seizing it, he had already captured his principal objective. He therefore ordered his subordinates to halt there and entrench, anticipating an easy victory the next day—if, that is, the Confederates stayed through the night to contest him, which he probably thought unlikely.[8]

Cleburne thought it unlikely, too. He deployed Smith's Brigade around the nose of Tunnel Hill. To guard the army's open right flank, Cleburne pushed Govan and Lowrey out to Smith's right, but he sent most of his artillery (all but two guns) to the rear in the full expectation that with the fall of Lookout Mountain, Bragg would now be compelled to order a general retreat that night.

And, indeed, Bragg thought about it. But as dogmatic as he was when it came to issues of little consequence, Bragg could be criminally indecisive about more important issues. He ought to have retreated—and at once. He had been effectively outflanked at both ends of his line. Even if he could repel the expected Federal attack on Missionary Ridge, he should have suspected that Hooker would be working his way around to his rear through Chattanooga Valley to cut him off from his only remaining line of supply—or means of escape. It was John C. Breckinridge, the former vice president who now commanded one of Bragg's two infantry corps, who convinced him to stay and fight. Breckinridge emphasized that Missionary Ridge was a position of great natural strength. If the Army of Tennessee could not fight there, he declared, it could not fight anywhere. In effect, Breckinridge shamed Bragg into staying, perhaps the worst possible reason for making a command decision. Moreover, the strategy conference at Bragg's headquarters consumed several valuable hours. Consequently, it was not until after midnight when Cleburne's young aide returned from Bragg's headquarters with the news that the army was not going to retreat that night after all, and that Cleburne would have to defend Tunnel Hill the next morning against Sherman's gathering horde.

Immediately, Cleburne recalled his artillery and set his men to digging defensive lines. At least it was bright enough to see, for until midnight there was a full moon. But then, as if it were a portent of things to come, a dark shadow began to creep across the moon, which went into a full eclipse.[9]

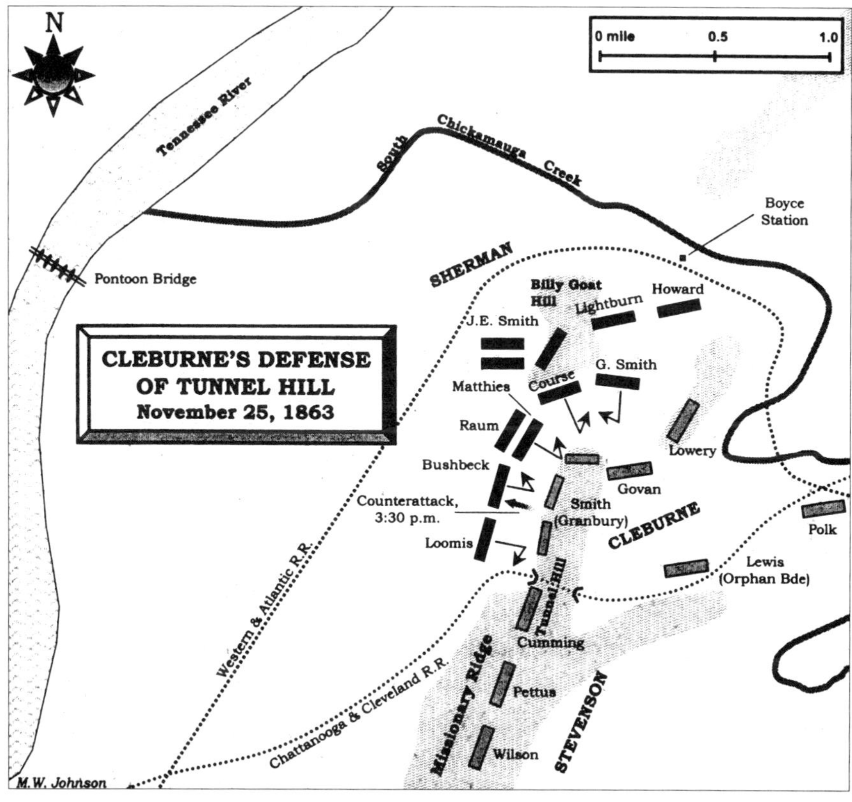

Courtesy of the Savas Publishing Company.

The twenty-fifth of November dawned clear but cold with wisps of ground fog in the gullies and ravines north of Tunnel Hill. Remarkably, Sherman showed no evidence of any eagerness to initiate the fight. Though Grant had ordered him to attack "at dawn of day," Sherman spent most of the morning reorganizing his forces. His behavior suggested that he believed it was at least possible that the Confederates were going to attack *him.* Because he assumed that his troops already occupied the critical hill at the north end of Missionary Ridge, he suspected the rebels would try to take it back. Not until mid-morning did he discover that Billy Goat Hill was separated from the main ridge line by a deep ravine, and that instead of defending his hilltop, he would have to attack the rebels on their hilltop to secure his foothold on the north end of Missionary Ridge. Even then he did not accelerate his timetable. He assigned three brigades (the equivalent of Cleburne's entire command) to "hold our hill [that is, Billy Goat Hill] as the key point," sending only one brigade to attack Cleburne. Sherman's excessive caution, despite his overwhelming numerical and materiel superiority, would prove to be a critical factor in Cleburne's eventual success.[10]

Cleburne effectively used the respite that Sherman's caution provided by supervising continued work on his defensive lines. He took personal charge of sighting the artillery batteries, placing one (Semple's) in support of Govan on the right, and the other two (Key's and Swett's) atop Tunnel Hill where they could be moved across the chord of the defensive arc to support either the western or northern approaches. In addition, Cleburne received valuable reinforcements when Brigadier General Joseph Lewis and his Kentucky (Orphan) brigade reported to him. Cleburne posted these men in reserve behind his main line.

Not until 11:00 A.M. did Sherman launch the first of what would be a series of offensives against Cleburne's position by sending a single brigade (that of John M. Corse) against the nose of Tunnel Hill where Smith's Texans were entrenched. Corse's men fought bravely, mounting three consecutive assaults and getting within fifty yards of the Confederate line before finally falling back. When they did so, Smith sought permission from Cleburne to launch a counterattack, which was immediately approved. Smith personally led his men down the slope and drove Corse's brigade back to its starting position at the foot of the hill. Smith, however, was wounded in the charge, and had to be replaced in command of the Texas brigade by Colonel Hiram Granbury.[11]

An hour later, Sherman tried again, this time using elements of two divisions (those of Hugh Ewing and John E. Smith) to conduct a coordinated attack from both the north and west. Cleburne coolly shifted both troops and guns to strengthen whichever segment of his line was assailed, and again his defenses held. This time, however, the Federals did not fall back. Instead they took cover in the thick trees and boulders only a few score yards from Cleburne's line and kept up a steady fire fight. Cleburne later wrote that the rifle fire was so heavy it was "like one continuous sheet of hissing, flying lead." The slope in front of Cleburne's position was so steep that the defenders had to rise up out of their entrenchments to shoot downhill, which made them easy targets. To dislodge the bluecoats, Cleburne's men took to rolling large boulders down the slope; when the Federals jumped up to get out of the way, the Confederates shot them down.[12]

The fighting was furious on both sides and it did not abate. The Federals took heavy casualties, but even so, the tide of battle seemed to swing in their favor. A battle of attrition necessarily favors the larger force, and things became critical when the heavy Union fire drove the exposed Confederate gunners from their batteries. Without artillery support, Cleburne's outnumbered division might easily be overwhelmed. Knowing that he had to bring the artillery back into the fight, and that to do so he needed to drive away the line of skirmishers in front of his lines, Cleburne organized another counterattack. He called upon Brigadier General Alfred Cumming's Georgia brigade from Carter Stevenson's division for support, and at about 3:30 in the afternoon, he sent four regiments charging downhill. Their onslaught was so unexpected, and caused such confusion, that some Federal officers later reported incorrectly that the rebels had charged out of the mouth of the railroad tunnel.

The Federal soldiers who were the target of this onslaught had been having a difficult time of it, too. They had been clinging to the steep slope of Tunnel Hill for several hours, loading prone and firing blind, and they were both tired and low on ammunition. When the Georgia and Texas troops charged downhill, the Federals either scrambled quickly for safety or threw up their hands. The Confederates pursued their enemy all the way to the bottom of the ridge, then slowly retreated back to the crest carrying with them eight captured battle flags and herding more than 500 prisoners.[13]

Cumming's downhill attack proved to be the decisive moment in the battle for Tunnel Hill. Sherman's will to continue the effort was broken. He pulled his men back so far that they again came under fire from Key's guns atop Tunnel Hill. Amazingly, Sherman's reports to Grant were so pessimistic that Grant began to fear the Confederates on Tunnel Hill might go over onto the offensive. To reduce the pressure on Sherman, he ordered Thomas to make a "demonstration" in front of Bragg's main line. Gamely, the men of Thomas J. Wood's and Philip H. Sheridan's Federal divisions assailed the Confederate rifle pits at the base of the Missionary Ridge. Unwilling to remain there as passive targets in the face of plunging fire from atop the ridge, they began to ascend it. An astonished Grant demanded to know "Who ordered those men up the ridge?" Defensively, Thomas denied that he had done so. In fact, the troops were doing it largely on their own. Their spontaneous ascent proved to be the decisive attack of the day as they gained the crest and put the Confederate defenders to flight.[14]

At Tunnel Hill, meanwhile, Cleburne was entirely unaware of the debacle underway three miles to the south; he was occupied in securing his position. At 5:00 P.M., already growing dark in late November, Cleburne ordered another counterattack—his third of the day—to clear his front. In the twilight darkness his men again leapt from their entrenchments and charged downhill. The enemy was gone.

In his official report, written three weeks later, Sherman asserted that his attack on Cleburne's position was a success. "The movement, seen from Chattanooga, 5 miles off, gave rise to the report . . . that we were repulsed," he wrote defensively to Grant. "Not so!" he insisted. "The real attacking columns of General [John M.] Corse, Colonel [John M.] Loomis, and General [John E.] Smith were not repulsed. They engaged in a close struggle all day, persistently, stubbornly, and well." That they did. But they also failed to take Tunnel Hill, which was their objective, and were driven from it by Cleburne's afternoon counterattack. Though Grant surely knew better, he sustained Sherman in the fiction that his attack had achieved its objective. "Sherman's assaulting column advanced to the very rifle-pits of the enemy," Grant wrote in his report to Henry Halleck, "and held their positions firmly and without wavering." Indeed, in his otherwise admirable *Personal Memoirs,* written twenty years later, Grant began Chapter 43 with the heading: "Sherman Carries Missionary Ridge," something which Sherman did not do until after the Confederate evacuation. Indeed, Grant and Sherman cooperated in a kind of unspoken conspiracy after the battle, the purpose of which was to suggest that Grant's plan all along had been for Sherman to feint against northern Missionary Ridge while Thomas's men assaulted the center. That is surely what happened, but it is not what Grant had envisioned. The scenario Grant had envisioned had been smashed beyond recovery by Patrick Cleburne and his division on Tunnel Hill.[15]

The official version of the fighting on Missionary Ridge influenced the popular histories of the immediate postwar years. Perhaps this explains why the National Park Service property atop Tunnel Hill is called the "Sherman Reservation" after the man who unsuccessfully attacked it rather than the "Cleburne Reservation" after the man who successfully defended it. Ironically, the other major NPS holding atop the ridge line is called the "Bragg Reservation." Thus the only two significant NPS reservations on Missionary Ridge are named for the battle's principal losers.

Casualties in the day-long fight for Tunnel Hill are hard to determine with precision. Cleburne reported 62 killed, 367 wounded, and 12 missing, for a total of 441, or just over eleven percent. But those figures include his losses in the sustained fighting at Ringgold Gap two days later as well as on Tunnel Hill. The four Federal divisions who participated in the fighting on Tunnel Hill reported losses of 222 killed, 1,263 wounded, and 210 missing, for a total of 1,695, or about seventeen percent of the approximately 10,000 Federals that were actively engaged. These Federal numbers are probably closer to the actual totals, for although they include the fighting at Ringgold Gap, none of these divisions were engaged at Ringgold Gap. A conservative estimate of the comparative losses at Tunnel Hill suggests that Cleburne's Division inflicted approximately five times the casualties on their assailants as they suffered themselves.[16]

For more than six hours, against odds of nearly seven to one, Cleburne's one reinforced division successfully defended Tunnel Hill against four Federal divisions and effectively stymied the major offensive effort of the Federal army. To be sure, Sherman contributed to Cleburne's success not only by his excessive caution, but by failing to take full advantage of his numerical superiority—he actively employed only about one-third of the 30,000 men available to him. But Cleburne's performance was nevertheless remarkable. He demonstrated an ability to make effective use of advantageous terrain, to exercise personal oversight in the coordination of different units under difficult circumstances, to use his artillery effectively, and above all to maintain his coolness in a crisis.

A good example of this last characteristic is captured in an anecdote told by Cleburne's assistant adjutant general, Irving Buck. According to Buck, a young staff officer who was returning from other duty just after the decisive counterattack at 3:30 A.M., noted a large number of blue-uniformed soldiers tramping across the rear of Cleburne's position. Alarmed at what he believed was an assault from the rear, he hurried up to Cleburne in great anxiety to report impending disaster. Cleburne, who always kept the entire battlefield clear in his mind, laughed and reassured the young officer that it was, in fact, a column of enemy prisoners being herded to the rear.[17]

At around 6:00 P.M., as Cleburne rode slowly southward along the crest of the ridge line to report the successful defense of Tunnel Hill to William J. Hardee and Bragg, he met a courier en route to find him who brought the news that the enemy had broken through the center and that the army was in full retreat. The courier also brought new orders. Cleburne was to take charge of the rear guard and establish a new defensive line perpendicular to the ridge and hold it long enough to allow the main body to escape. Cleburne rotated his defense 180 degrees to defend against an assault from along the ridge from the south, and assumed effective command of the right wing of the army as it evacuated Missionary Ridge.

Two days later, Cleburne's Division would save the army yet again by successfully defending the narrow defile at Ringgold Gap against Joe Hooker's entire Federal corps. Amidst the gloom of the astonishing Confederate defeat on Missionary Ridge and the subsequent retreat into Georgia, the performance of Cleburne and his division was the one bright spot for Confederate arms. His performance in this campaign earned him the praise of the commanding general and the thanks of the Confederate Congress. Indeed, his role in the campaign gave him a new nickname. In receipt of the reports from the Western Theater, Jefferson Davis noted that Cleburne's men followed him with the same implicit confidence that Stonewall Jackson's men had given their commander. Cleburne's many admirers began to refer to him as the "Stonewall of the West."[18]

There were other battles and other victories for Pat Cleburne. At Pickett's Mill, his division once again saved the Army of Tennessee from potential disaster by turning back a thrust by Howard's XI Corps; at the Battle of Atlanta, his men drove XVII Corps from its position and achieved the closest thing General John Bell Hood ever got to a tactical victory; at Jonesboro, south of Atlanta, Cleburne commanded a corps in action for the only time in the war. And at Franklin, exactly one year after his remarkable defense of Tunnel Hill, he led his division in the grandest and most tragic charge Confederate soldiers ever executed, and was shot dead, sword in hand, fifty yards from the Federal breastworks.

Cleburne's battlefield success has led more than one student of the war to wonder if he could have been equally successful in higher command responsibility. He was passed over for corps command after Hood took command of the Army of Tennessee in July of 1864, and he was never really in position to contend for command of the army itself. Some have attributed this to his astonishing proposal in December of 1863 to free all the slaves and invite the males to serve in the Confederate army. His proposal unleashed a storm of protest from the other generals, and was quietly squashed by President Davis. Even without that, however, it is unlikely that Pat Cleburne would even have been a serious contender for corps command, much less army command. He was a foreigner, he was not West Point trained, and his skills lent themselves best to the role of a hand's on commander. He was always more a leader than a chess master. In the end, then, Pat Cleburne was probably exactly where his unquestioned gifts should have placed him: in command of the most successful fighting division in the Army of Tennessee, and one of the best in the Civil War.

Notes

The foundational version of"Patrick Cleburne's Defense of Tunnel Hill Revisited" appeared as"Stonewall of the West: Pat Cleburne and the Defense of Tunnel Hill," in *Civil War Regiments: A Journal of the American Civil War* 7, no. 1 (2000): 75–90, and is reprinted with permission of Savas Publishing Company, Mason City, IA.

1. Wiley Sword, *Mountains Touched with Fire: Chattanooga Besieged, 1863* (New York, 1995), 241.
2. The most recent assessments of the origins of this campaign are Sword, *Mountains Touched with Fire,* and Steven Woodworth, *Six Armies in Tennessee: The Chickamauga and Chattanooga Campaigns* (Lincoln, NE, 1998).

3. The suggestion to send Longstreet to Knoxville apparently originated with Jefferson Davis, but the Confederate president was motivated at least as much by a desire to restore harmony to the high command as by grand strategy. See U.S. War Department, *The War of the Rebellion: A Compilation of the Official Records of the Union and Confederate Armies,* 128 vols. (Washington, DC, 1880–1901), series 1, vol. 51, pt. 2:557. Hereinafter cited as *OR.* All references are to series 1 unless otherwise indicated.
4. Craig L. Symonds, *Stonewall of the West: Patrick Cleburne and the American Civil War* (Lawrence, KS, 1997), 161.
5. Sword, *Mountains Touched with Fire,* 195–221; Woodworth, *Six Armies in Tennessee,* 185–89.
6. *OR,* vol. 32, pt. 2:745–46.
7. The order of battle for both sides is printed in *OR,* vol. 31, pt. 3:564–71 (Federal), and vol. 31, pt. 2:657–64 (Confederate). An order of battle is also available in Sword, *Mountains Touched with Fire,* 360–69.
8. OR, vol. 31, pt. 2:573.
9. Irving A. Buck,"Cleburne and his Division at Missionary Ridge and Ringgold Gap," *Southern Historical Society Papers* 8 (1880): 466.
10. *OR,* vol. 31, pt. 2:574. In this report, Sherman tried to disguise both his excessive caution and his confusion about the local geography. He asserted that"the sun had hardly risen"when he launched his initial attack, even though no serious attack commenced before 11:00 A.M.
11. The most vivid account of this assault is in Samuel T. Foster, *One of Cleburne's Command: The Civil War Reminiscences and Diary of Capt. Samuel T. Foster, Texas Brigade, C.S.A.,* ed. Norman D. Brown (Austin, TX, 1980), 60ff. See also Sword, *Mountains Touched with Fire,* 244–46.
12. *OR,* vol. 31, pt. 2:750.
13. Sword, *Mountains Touched with Fire,* 255–58; Symonds, *Stonewall of the West,* 168–69.
14. Robert U. Johnson and Clarence C. Buel, eds., *Battles and Leaders of the Civil War,* 4 vols. (New York, 1884–88), 3:725.
15. *OR,* vol. 31, pt. 2:34, 575; Ulysses S. Grant, *Personal Memoirs of U. S. Grant,* 2 vols. (New York, 1886), 2:61.
16. The figures given here are presented on the National Park Service signage on Tunnel Hill.
17. Johnson and Buel, *Battles and Leaders* 3:468.
18. Davis did not write these words until after the war in his *Rise and Fall of the Confederate Government,* 2 vols. (New York, 1881), 2:557. But the nickname"Stonewall of the West"became current in the Army of Tennessee during the 1864 campaign.

Lieutenant General Nathan Bedford Forrest. Courtesy of the U.S. Army Military History Institute, Carlisle, Pennsylvania.

Bedford Forrest and His "Critter" Cavalry at Brice's Cross Roads

Edwin C. Bearss

The spring and summer of 1864 found the attention of the people of the North and of the South focused on the fighting in Virginia and Georgia. In these states, mighty armies fought battles that were to decide whether the United States was to be one nation or two. Interwoven with and having important repercussions on the fighting in Georgia were military operations in northeast Mississippi designed to prevent a Confederate cavalry corps under Nathan Bedford Forrest from striking into Middle Tennessee and destroying the single-track railroads over which Major General William T. Sherman's armies drew their supplies. The Battle of Brice's Cross Roads was fought to protect these railroads.

By the spring of 1864 almost three years of bloodshed and heartbreak had passed since the firing on Fort Sumter signaled the beginning of the Civil War. In the West, an army led by Major General Ulysses S. Grant, supported by the U.S. Navy, had won a series of victories and had forced the surrender of Vicksburg in July of 1863. The fall of Port Hudson a few days later gave Union forces control of the Mississippi River and divided the Confederacy. At Missionary Ridge in the fourth week of November, 1863, armies under Grant had driven the Confederates from the approaches to Chattanooga and recovered the initiative that had belonged to the South in that region since the Battle of Chickamauga in September. In the east, General Robert E. Lee's Army of Northern Virginia, despite its costly defeat at Gettysburg, remained a powerful fighting machine and guarded the approaches to Richmond.

Because of Grant's successes in the west, President Abraham Lincoln brought him east and in early March of 1864, promoted him to lieutenant general, and gave him command of all United States armies. Vowing to defeat the Confederacy, Grant proposed to employ the North's superior resources to grind it down in a war of attrition. Costs would be high, but the North could replace its losses while the South could not. In his planning, there was one factor that Grant could not overlook; if the major Confederate armies were still in the field in November, the electorate might send the Lincoln administration down to defeat at the polls. It was crucial that Northern armies either defeat the South or score sweeping successes by November. A stalemate would be as bad a blow as a defeat.

Grant proposed to concentrate his efforts on the destruction of the two major Confederate armies and thus end the long, drawn-out war. He would personally oversee the movements of the forces whose goal was the defeat of Lee's Army of Northern Virginia, by maintaining his headquarters with the Army of the Potomac. In the west, Sherman, who had succeeded Grant as commander of the Military Division of the Mississippi, was to destroy General Joseph E. Johnston's Army of Tennessee.

On May 7, 1864, coordinating his movements with Grant's, Sherman put his armies in motion through the maze of timber-clad rocky ridges of northwestern Georgia, skillfully employing his superior numbers to outflank successive Confederate positions and compelling Johnston to fall back again and again. But, by May 25, the Federal advance had been checked, for the time being, in front of New Hope Church. Although he had thrust deeply into Confederate territory, Sherman had failed to defeat Johnston, as the Southern leader yielded ground to gain time. As Sherman's troops battled their way forward, their supply lines lengthened and became increasingly vulnerable to Confederate cavalry raids.

There was only one cavalry leader, in the North or the South, whom Sherman respected and that was forty-two-year-old Nathan Bedford Forrest. A self-made man with only six months of schooling, Forrest had entered Confederate service as a private, and by repeated demonstrations of personal bravery, leadership, and audacity, he had risen to the rank of major general.

Holding no respect for soldiers who fought by the book, Forrest attributed his many successes to the simple fact that he "got there first with the most men." Standing six feet two inches tall and being of powerful build, he was always ready to engage the foe personally or to thrash any of his own men guilty of malingering. Wounded four times, no other American general has killed as many enemy soldiers with his own hands or has had as many horses—twenty-nine—shot from under him. His words of command as he led a charge were "Forward, men, and mix with 'em!"[1]

Forrest led a cavalry corps based in northeast Mississippi. His corps was effective, because he used it as mounted infantry. The men rode horses and mules to the scene of action, but Forrest habitually made them fight on foot. Unlike most cavalry units, his men worked hard and could wreck a railroad as efficiently as Sherman's infantry. As Sherman's supply line lengthened, the Federal commander feared that "that devil Forrest" would get into Middle Tennessee and break the railroads behind him.

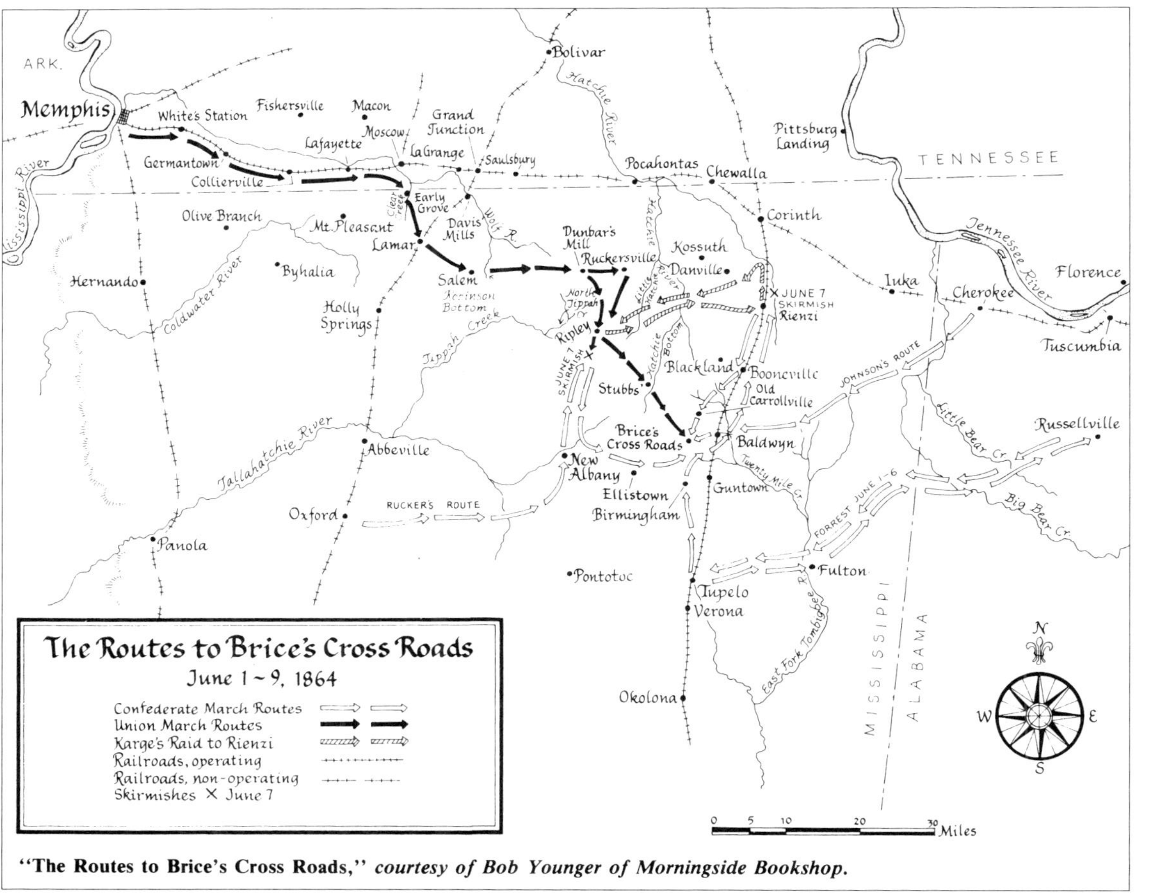

"The Routes to Brice's Cross Roads," *courtesy of Bob Younger of Morningside Bookshop.*

When Forrest was in West Tennessee and western Kentucky in March and April, Sherman telegraphed the commander in Memphis, Major General Cadwalader C. Washburn, not to disturb the Confederate cavalryman, because Forrest could do much more harm by cavorting all over the country as he could there in West Tennessee. Grant, however, saw that Forrest was recruiting his command while harassing and destroying isolated Union garrisons, and he directed Sherman to send enough troops to Memphis to chase Forrest back into Mississippi.

Sherman's first task was to find an officer equal to the challenge. Brigadier General Samuel D. Sturgis, an 1846 graduate of the U.S. Military Academy and an "old army" man, was chosen. A veteran of Wilson's Creek, where he won promotion for "gallant and meritorious conduct," Sturgis had also served in the Army of the Potomac. In the winter of 1863–64, he had been commended for his cavalry leadership in East Tennessee. He had a chance to test Forrest in the first week of May while the Federals were chasing the Confederates out of West Tennessee. His columns, however, moved too slowly and the Southerners outdistanced them. Reporting to his superiors on May 7, Sturgis wrote that "it was with greatest reluctance and after mature deliberation with myself and my principal officers that I resolved to abandon the chase as hopeless. . . . Though we could not catch the scoundrel we are at least rid of him, and that is something."[2]

As Sherman's armies pressed deeper into Georgia, Johnston knew that the only way to stop their advance was to destroy the Federal supply line—the railroads from Nashville and Chattanooga. Accordingly, he appealed to Major General Stephen D. Lee for help in breaking these rail lines.

Lee, a graduate of the U.S. Military Academy, commanded the Department of Alabama, Mississippi and East Louisiana. Destined at thirty to become the Confederacy's youngest lieutenant general, his dark hair, beard, and eyes gave him a cavalier look. He had been present when the first shot was fired at Fort Sumter and had served as an artillerist in the Army of Northern Virginia until after the Battle of Antietam in September of 1862. He had then been transferred to the west and led an infantry brigade in Lieutenant General John C. Pemberton's Army of Vicksburg. After Pemberton's surrender in July 1863, Lee was exchanged and placed in command of all the cavalry in Mississippi. He had assumed responsibility for the department in May, 1864, after the previous commander, Lieutenant General Leonidas Polk, and most of Polk's soldiers, joined Johnston for the Georgia campaign.

Responding to Johnston's plea, Lee ordered Forrest and his cavalry to advance into Middle Tennessee and wreck the Nashville & Chattanooga Railroad. Forrest moved promptly. On June 1, he rode out of Tupelo, Mississippi, with two thousand hell-for-leather horsemen and a battery of artillery. Three days later, at Russellville in north Alabama, Forrest was overtaken by a courier with a message from Lee reporting that a powerful Union column had left Memphis to invade Mississippi. Forrest was told to forget the Middle Tennessee raid and to return to Tupelo.

In response to a call from Sherman, to send a formidable expedition toward Tupelo or in whatever direction Forrest happened to be, Washburn ordered Sturgis to advance from Memphis. Sturgis's force, which moved out on June 1, mustered forty-eight hundred infantry, thirty-three hundred cavalry, four hundred artillerists with twenty-two cannon, and a large supply train. The little army was organized into two divisions—one of infantry and the other of cavalry. Brigadier General Benjamin H. Grierson, a former music teacher from Jacksonville, Illinois, and now one of the North's best cavalrymen, led Sturgis's horse-soldiers, one regiment of whom were armed with seven-shot Spencer carbines. Colonel William McMillen, who had served with the Russian army as a surgeon in the Crimean War and had led the 95th Ohio Infantry during the Vicksburg campaign, commanded the infantry. McMillen was hot tempered and fond of whiskey.

Sturgis was to strike the Mobile & Ohio Railroad at or near Corinth and to destroy any force that might be posted there. He then was to proceed down the railroad, at least as far as Okolona, wreck it and to return to Memphis by way of Grenada. Forrest's Corps, during this sweep, would be dispersed and destroyed and the countryside devastated.

At Lamar, Sturgis's scouts told him that the Confederates had evacuated Corinth and retired down the railroad. Sturgis changed his line of march to intersect the railroad farther south. On the fifth, Grierson, who had the lead, sent four hundred of his horse-soldiers racing eastward to Rienzi. The Federals reached the railroad and did some damage and then rode northward until they found their way barred by rain-swollen Tuscumbia Creek.

Heavy rains pelted Sturgis's column as it bore southeastward through an area that had been ravaged by two years of raids and counterraids. It was June 7 before Sturgis's infantry reached Ripley, seventy-five miles from Memphis. During the day, one of Grierson's brigades reconnoitered the New Albany Road and encountered a roadblock manned by two regiments sent out by Forrest to feel for the Federals. In the meantime, Sturgis advanced down Guntown Road. By nightfall on the ninth his command was concentrated and camped on Stubbs's farm, nine miles northeast of Brice's Cross Roads.

Forrest had returned to Tupelo from the aborted Middle Tennessee raid on June 5. Told about the general direction of Sturgis's march and that the Federals had broken the railroad at Rienzi, Forrest ordered two brigades to that point. Colonel Tyree Bell's Tennessee brigade stopped at Rienzi, and Forrest, with his artillery and escort and Colonel Hylan B. Lyon's brigade, took position at Booneville, where he was joined by Colonel Edward Rucker's brigade on the evening of June 9. Rucker's people had clashed with Sturgis's vanguard near New Albany, forty-eight hours earlier. Forrest also had one brigade—Colonel William A. Johnson's—at Baldwyn. Lee reached Booneville on the ninth and went immediately to Forrest's headquarters for a briefing.

After examining Forrest's troop returns and learning that his force numbered about forty-nine hundred cavalry and twelve cannon, Lee suggested that Forrest retire toward Okolona and let Sturgis push deeper into Mississippi and farther from his base before giving him battle. Forrest was to move out on June 10 in the direction of Brice's Cross Roads and then to continue on toward Okolona. Lee, accompanied by two batteries of artillery, then boarded a southbound train for Okolona.

That evening, June 9, Forrest called a number of his officers together. He told them that his spies had reported the Federals encamped at Stubbs's farm and that, while he would prefer to get them into the open country where he could "get a good look at them" as Lee desired, the Confederates might be drawn into a battle before that could be realized. Orders were issued for the brigade and artillery commanders to have their men ready to ride before daylight and to push forward as rapidly as possible toward Brice's Cross Roads.[3]

Torrential rains, which did not cease until after midnight, turned the roads into ribbons of mud. When the sun rose, many knew that it would be one of those hot humid days which saps a man's energy. As Forrest traveled with Rucker's brigade, he told Rucker that he intended to attack Sturgis at Brice's Cross Roads and outlined his battle plan:

> I know they greatly outnumber the troops I have at hand but the road along which they will march is narrow and muddy; they will make slow progress. The country is densely wooded and the undergrowth so heavy that when we strike them they will not know how few men we have. Their cavalry will move out ahead of the infantry, and should reach the crossroads three hours in advance. We can whip their cavalry in that time. As soon as the fight opens they will send back to have infantry hurried up. It is going to be as hot as hell, and coming on a run for five or six miles over such roads, their infantry will be so tired out we will ride over them.[4]

Forrest then pushed ahead to join his advance brigade of Kentuckians commanded by Lyon, an officer described by a Federal general as "overbearing, and . . . a very rude character." Unlike the Confederates, the Federals did not make an early start on June 10. It was 5:30 A.M. before Grierson's horse-soldiers swung into their saddles and started down Guntown Road; it was seven o'clock before the infantry marched.[5]

Grierson's vanguard encountered, charged, and scattered a scouting party, chasing the Rebels across the narrow bridge spanning Tishomingo

Creek. By 9:45 the Union cavalry held Brice's Cross Roads, and one brigade had followed the retreating Confederates for a mile along Baldwyn Road. At the edge of a field, the Federals reined up their horses when they sighted Lyon's vanguard on the opposite side of the field, four hundred yards away. While two of his companies charged the surprised Yanks, Lyon dismounted and deployed his eight hundred-man brigade. Grierson did likewise, forming his thirty-two hundred horse soldiers to the left and right of the road. Like the Confederates, most of the Federals fought on foot.

Forrest knew that he would have to gain and hold the initiative, or the "bulge," as he called it. Lyon's Kentuckians were advanced. For almost an hour, the Confederates drove forward, then retired, and advanced again. A great quantity of powder was burned and a few men killed or wounded, but Grierson, unfortunately for the Union, had allowed himself to be bluffed by a smaller force.

Rucker's brigade now came up on a trot and formed on Lyon's left. Once again, Forrest waved his men forward. There was the sharp crack of small arms and the roar of artillery as the Confederates moved forward. On Forrest's left, one of his battalions advanced too far and was sent reeling. Johnson's 590-man brigade arrived from Baldwyn, and Forrest posted it on Lyon's right. As soon as the troopers had dismounted, they feigned an attack on Grierson's left.

It was now eleven o'clock, and although Bell's brigade and his artillery had not reached the front, Forrest decided to assault Grierson. He rode along his line and encouraged his men, telling them that he expected everyone to advance when the signal was given. At the sound of the bugle, the dismounted cavalry stormed across the field toward the Federals. Grierson's men held their ground and blazed away. At one point, Rucker's brigade penetrated Grierson's line, but the Federals called up two reserve regiments to close the breach. As the Yanks rushed forward, Rucker shouted to his men, "Kneel on the ground, men, draw your six-shooters, and don't run!" (Each man in Forrest's corps was armed with a rifled musket and two Colt revolvers. Forrest refused to arm his enlisted men with sabers, because he considered them useless in the type of fighting he favored.) After a desperate struggle, the Federals were forced to retire closer to Brice's Cross Roads. By 12:30 Forrest had whipped Grierson's cavalry.[6]

Grierson, on encountering the enemy, had sent a courier galloping to tell Sturgis that he had found the Confederates and needed help. As Forrest brought up fresh units and increased the pressure, Grierson repeated his pleas for reinforcements with greater urgency. It was noon before Sturgis reached the field and after 1:00 P.M. before the advance columns of his infantry arrived, though they had marched as fast as road and weather conditions permitted. They had tramped nine miles since seven o'clock; the "last four miles under a hot sun" and at the "double-quick."[7]

The thirty-two hundred infantry and their three supporting batteries struggled into position covering the cross roads, their left extending well north of Baldwyn Road and their right anchored about two hundred yards west of Guntown Road. West of Tishomingo Creek was Colonel Edward Bouton's reserve brigade of black soldiers and the army's trains. Covered by the infantry, Grierson sought to re-form his exhausted division.

Bell's Tennessee brigade and Captain John W. Morton's artillery now came up. The Confederate cannoneers, having traveled eighteen miles, rode up at a trot, threw their eight guns into battery, and hammered away at Sturgis's masses with telling effect. The Union artillerists replied. Placing himself at the head of Bell's brigade, Forrest rode to the left, dismounted, and formed the newcomers. While Forrest was positioning Bell, the fighting ebbed. The only sounds were the occasional crack of a sharpshooter's rifle-musket, the rustle of underbrush as men moved about, and the hushed commands of officers. The weather was stifling—there was not a cloud in the sky and the air was still. Several men had been felled by sunstroke.

Forrest attacked. Because of the thick undergrowth covering most of the area, the Confederates were able to close to within a few paces of Sturgis's infantry. A crashing volley, however, sent part of Bell's battle line reeling. The Federals called for a charge. Forrest's sixth sense had placed him at the key point. As he dismounted, he shouted for his escort to do likewise. Accompanied by these daring fighters and with revolver in hand, he rushed the Federals. Additional men came up, and the counterattack was repulsed. In the hand-to-hand fighting, the bayonets of the Union infantry were no match for the heavy Colt revolvers. The center of Sturgis's line crumbled, while the Confederate brigades on the right doubled back the Union left upon Ripley Road.

Off to the northwest in the direction of Tishomingo Creek, the 2d Tennessee Regiment, sent out by Bell to attack Sturgis's left and rear, had reached its objective "just as the fighting seemed heaviest in front." To deceive the Federals about their strength, the Confederates made a great commotion and a bugler galloped up and down the line sounding the charge. Not only did this show of force throw Sturgis's reserve brigade and the train guard into confusion, but Grierson sent off most of his cavalry to check the advance of the 2d Tennessee.[8]

Forrest knew that the crisis had come and that now the battle must be won or lost. Riding along behind his line, he told his people that the enemy was starting to give way and that another charge would win the day. He told his young chief of artillery, Captain Morton, to be ready to advance four of his guns, double-shotted with canister, to within pistol range of the Federals at the crossroads. When the bugle sounded, the Confederate battle line pressed forward. At the same time, Morton's cannoneers drove their teams up the narrow country road. At point-blank range, they unlimbered their pieces and fired double-shotted canister into Sturgis's infantry with frightful effect. After a brief but savage fight, the Federals were routed from the crossroads, with the loss of three cannon.

General Sturgis grimly described this phase of the fight:

> I now endeavored to get hold of the colored brigade, which formed the guard to the [wagon] train. While traversing the short distance to where the head of that brigade should be found, the main line began to give way at various points. Order soon gave way to confusion and confusion to panic. . . . Everywhere the army now drifted toward the rear, and was soon altogether beyond control. . . .The road became crowded and jammed with troops, the wagons and artillery, sinking into the deep mud, became inextricable, and added to the general confusion that now prevailed. No power could now check or control the panic-stricken mass as it swept toward the rear.[9]

Several regiments, reinforced by two companies of the 55th U.S. Colored Troops that had crossed to the east side of Tishomingo Creek, attempted to check the onrushing Confederates; but, assailed on the flanks, with Morton's guns sweeping their front with doubled-shotted canister, the Northerners broke. To add to the Federals' embarrassment, a fleeing teamster's wagon overturned on the narrow wooden bridge over rain-swollen Tishomingo Creek and the men were forced to climb over the wreckage. In their frantic efforts to escape, soldiers pushed their comrades aside. Others, seeing it was hopeless to cross the bridge, attempted to wade or swim the creek. Many drowned or were shot as they floundered in the water.

On reaching the bridge, Forrest's men cleared it by pushing the wagons and the dead and wounded teams into the creek. Meanwhile, Forrest's escort forded the stream about four hundred yards south of the bridge and bore down on the flank of the panic-stricken Federals. A number of prisoners were taken, along with some wagons. Although the sun was about to set, Forrest brought up his horse holders and personally took charge of the pursuit. A mile beyond the bridge, some of Sturgis's infantry rallied, but Morton brought up two guns and smashed this pocket of resistance, and as dusk faded into darkness, Forrest and his hell-for-leather troopers overpowered another roadblock hastily manned by black and white Union soldiers. Forrest then halted his command to let his men and their mounts get a few hours rest. During their nighttime crossing of Hatchie Bottom, Sturgis and many of his officers and men panicked, and they abandoned fourteen cannon and most of their wagon train.

Forrest had his people back in the saddle before daybreak and pressed the pursuit relentlessly. The Federals passed through Ripley, twenty-two miles from Brice's Cross Roads. Here Sturgis attempted to re-form his command. But the Confederates came up too soon, and the Federals' retreat was resumed. Not until he reached Salem at dark did Forrest call off the chase. Sturgis's column, which had taken eight days to reach Brice's Cross Roads, retreated to Memphis in sixty-four hours. Union casualties in the fight and retreat were 2,612. Forrest listed his losses as 493 killed and wounded. The Confederates captured 250 wagons and ambulances, 18 cannon, and thousands of stands of arms and rounds of ammunition, as well as the Federals' baggage and supplies.

A noted British soldier, Field Marshal Viscount Garnet J. Wolseley, in commenting on Forrest's victory, called it

> a most remarkable achievement, well worth attention by the military student. He pursued the enemy from the battle for nigh sixty miles, killing numbers all the way. The battle and his long pursuit were all accomplished in the space of thirty hours. When another Federal General was dispatched to try what he could do against this terrible Southerner, the defeated Sturgis was overheard repeating to himself . . . "It can't be done, sir: it c-a-n-'t be done!" Asked what he meant, the reply was, "They c-a-n-'t whip old Forrest!"[10]

Notes

"Bedford Forrest and His 'Critter' Cavalry at Brice's Cross Roads" was originally published in *Leadership During the Civil War: The 1989 Deep Delta Civil War Symposium; Themes in Honor of T. Harry Williams,* ed. Roman J. Heleniak and Lawrence L. Hewitt, 73–84 (Shippensburg, PA, 1992), and is reprinted with permission of White Mane Publishing Company.

1. For more information on Forrest at Brice's Cross Roads, see Edwin C. Bearss, *Forrest at Brice's Cross Roads and in North Mississippi in 1864* (Dayton, OH, 1979).
2. U.S. War Department, *The War of the Rebellion: A Compilation of the Official Records of Union and Confederate Armies,* 128 vols. (Washington, DC, 1880–1901), series 1, vol. 32, pt. 1:697. Hereinafter cited as *OR.* All references are to series 1 unless otherwise indicated.
3. John A. Wyeth, *Life of Lieutenant-General Nathan Bedford Forrest* (1899; reprint, Dayton, OH, 1975), 399.
4. Ibid., 400.
5. *OR,* vol. 24, pt. 2:221.
6. Wyeth, *Forrest,* 414.
7. *OR,* vol. 39, pt. 1:121.

8. Wyeth, *Forrest,* 414–15; R. R. Hancock, *Hancock's Diary: or, A History of the Second Tennessee Cavalry* (Nashville, 1887), 391–93.
9. *OR,* vol. 39, pt. 1:92–93
10. Garnet Wolseley, "General Forrest," *United Service Magazine* 5 (1892): 4.

President Jefferson Davis. Library of Congress.

A Reassessment of Jefferson Davis as War Leader: The Case from Atlanta to Nashville

William J. Cooper Jr.

THE FAILURE OF THE SOUTHERN CONFEDERACY HAS OFTEN BEEN CORRELATED WITH THE failure of Jefferson Davis as war leader. Davis's shortcomings as commander-in-chief of the armed forces, as political head of a new nation, and as chief spokesman to and for the public often lead the list of reasons for southern defeat. For example, Davis pales beside Abraham Lincoln in any comparison of the two presidents as war leaders. Recently Eric L. McKitrick, summing up the views of many, wrote: "It seems apparent that the leadership of Abraham Lincoln was superior to that of Jefferson Davis."[1] Lincoln was flexible; Davis was rigid. Lincoln wanted to win; Davis wanted to be right. Lincoln had a broad strategic vision of Union goals; Davis could never enlarge his narrow view. Lincoln searched for the right general, then let him fight the war; Davis continuously played favorites and interfered unduly with his generals, even with Robert E. Lee. Lincoln led his nation; Davis failed to rally the South.[2] Simply, Lincoln contributed mightily to the Union victory; Davis contributed mightily to the Confederate defeat.

The attack on Jefferson Davis began during the war. Davis, according to the acerbic Richmond editor Edward Alfred Pollard, ran an administration from which neither the Confederate States of America nor its citizens could expect "anything wise, or magnanimous, or unselfish." Editor Pollard scorned "the pitiful personal feelings, of a single man [President Davis]" who wrecked the war effort and the careers of generals "with cold and insolent neglect."[3] Writing in 1863, Pollard viciously indicted Davis for "his intermeddling," which "removed a decisive victory [Antietam] from the grasp of our army, and turned back the war for years."[4]

Final southern defeat failed to temper Pollard's written assaults on the former Confederate president. "Mr. Davis," wrote Pollard in the preface to a biography of Davis he published in 1869, "[was] the prime cause of the failure of the South in the late war."[5] Three years earlier Pollard had attributed to Davis "a peculiar obstinacy, which, stimulated by an intellectual conceit, spurned the counsels of equal minds, and rejected the advice of the intelligent."[6]

Modern historians, while not always adopting the bitterness of Pollard's assessments, have echoed his basic interpretation that Jefferson Davis's failure as a war leader was a crucial factor in the ultimate defeat of the Confederacy. Clifford Dowdey, who found Davis after the fall of Atlanta a man dominated by "a wishful thinking that produced no more than fantasies, daydreams," has published two general studies of the Confederacy which have made major contributions to the preservation of Pollard's estimate of Davis.[7] In a book lauding Lincoln as "a great war president," T. Harry Williams dismissed Davis as a "mediocre one."[8] Perhaps the most damning indictment of Davis's ability by a serious scholar has come from David M. Potter—Davis, in Potter's judgment, was a man "who cared more about proving he was right than about gaining success."[9]

Davis, of course, has had his defenders. The most recent attempt—and the most voluminous—came in Hudson Strode's massive three-volume biography.[10] Strode devotes half of his study to Davis's experiences as president of the Confederate States of America. From Strode's pages Davis emerges as an omniscient hero who failed because of the bitterness, incompetency, and jealousy of "little" subordinates. Although certain of Davis's lieutenants undoubtedly did not give unselfish service to their president, Strode's eulogistic effort to depict Davis as a great man and a great war leader is unsuccessful because his biography is written in a historiographical vacuum and fails to evaluate men and events critically.[11]

Other works viewing Davis more favorably have greater validity. Archer Jones, in a study of early Confederate strategy, concludes that Davis's strategic "method was sound; his application competent."[12] Writing about Davis and his cabinet, Rembert W. Patrick asserted that "Davis's claim to conspicuous ability as a leader is incontestable."[13]

A thoughtful article by Frank E. Vandiver, "Jefferson Davis and Confederate Strategy," stands as probably the most convincing and persuasive attempt to rehabilitate the reputation of Davis. Basic to Davis's strategic planning, according to Vandiver, was the concept of the offensive-defensive, which meant "the Confederates would stand on the defensive because they had fewer men and feebler resources than the Yankees, but they would exploit every chance to counterattack, to take the initiative, to carry the war to the enemy." Vandiver finds "Confederate strategy to have been logical, consistent, and realistic," mainly because it "enable[d] the Confederates to outlast their resources."[14] Unlike Potter, Vandiver found Davis's strategy consistent with reality, and he pictured Davis as striving to institute a comprehensive, overall strategy for the Confederacy.

None of Davis's defenders, however, has caused any significant reassessment of the view initiated by Pollard in the 1860s. To test Davis's performance, this paper examines his actions as war leader during an extremely critical period of Confederate history. Davis functioned energetically as such a leader during the crucial months beginning with the campaigns for Atlanta in the spring and summer of 1864 and the subsequent destruction of the Army of Tennessee in the battles of Franklin and Nashville, which sealed the fate of the Confederate West.

Jefferson Davis lived in Richmond, the Confederate capital. Here, at the northeastern edge of the Confederacy, he found himself in the midst of one of the major theaters of the war. For four years Union and Confederate armies fought over and through the one hundred miles separating Richmond from Washington, the Union capital. The geographical location of the Confederate seat of government could have focused President Davis's mind on Virginia.

Davis, however, did not forget the Confederate West. A Mississippian, he maintained a strong interest in western affairs and in the career of the single most important Confederate force in that vast area, the Army of Tennessee.[15] As its first commander, Davis appointed his friend Albert Sidney Johnston, the man Davis believed to be the best southern general. During his short tenure Johnston showed few signs of military genius. His death at Shiloh in April 1862 required a new commander for the Army of Tennessee, and Davis advanced Pierre G. T. Beauregard, who had been Johnston's deputy at Shiloh. After only two months Davis replaced Beauregard with Braxton Bragg. Initially, it seemed that Davis had acted wisely, for Bragg made a rapid dash into Kentucky—a campaign that marked the northernmost advance made by the Army of Tennessee. But once in Kentucky, Bragg became indecisive and hesitant, and after installing a Confederate governor and fighting an inconclusive battle, he returned to Tennessee.

Although Davis never lost faith in Bragg's abilities as a general, Bragg's subordinates did. Aware of the anti-Bragg feeling in the Army of Tennessee, Davis realized that he had to make some changes. Rather than remove Bragg, he created the Department of the West and placed Joseph E. Johnston in command of Confederate troops from Georgia to the Mississippi. This arrangement failed to work and lapsed after the fall of Vicksburg, but its failure cannot be blamed on lack of support from Davis.

By the autumn of 1863 President Davis faced another crisis in the Army of Tennessee. Bragg won a victory at Chickamauga in September, but he frittered away his opportunity and finally was driven away from Chattanooga two months later. This time Davis knew Bragg would have to go. Certain of Davis's closest advisers, including Secretary of War James A. Seddon, recommended the assignment of Joseph Johnston.[16] Neither Davis nor Seddon had been pleased with Johnston's performance as commanding general of the Department of the West,[17] but as commanding general of the Army of Tennessee he would have only a single field army under his charge. And Johnston, after all, ranked third among Confederate full generals. The president therefore ordered him to Georgia.

In contrast to his situation in Virginia, President Davis found himself constantly searching for the right general in the West. In Virginia Robert E. Lee, who enjoyed Davis's confidence, had military successes unmatched by any other Confederate commander. He and Davis discussed strategy and other matters, and Davis never lost confidence in him as a general or as a loyal subordinate. In the West Davis never found both qualities in one man. Sidney Johnston, probably, and Braxton Bragg, certainly, were loyal, but neither showed consistent ability as a military commander. For Davis, Beauregard and Joseph Johnston proved even less satisfactory. Showing little, if any, more competence than Sidney Johnston or Bragg, neither Beauregard nor Joseph Johnston exhibited much loyalty to Davis as their president and commander-in-chief. When the 1864 campaign opened, Abraham Lincoln had found his general; Jefferson Davis was not at all sure that he had.

Joseph Johnston had had a particularly unsatisfying career as a Confederate general. Piqued initially at being placed fourth on the list of full generals, Johnston seemed to carry this resentment until 1865—and even afterwards. Convinced of his own great ability, Johnston resented directions from his civilian superiors and also their requests for information about his intentions. He commanded at First Manassas, but Beauregard directed the battle. A wound cut short his defense of Richmond in the spring of 1862. Excellent strategical and tactical ideas but no action marked his tenure as commanding general of the Department of the West. The spring of 1864 found Joseph Johnston still an unproven general.[18]

Beset with massive Union offensives in both Virginia and the West, Davis relied on Lee and Johnston to turn back the attacks of Ulysses S. Grant and William Tecumseh Sherman. To the new commander of the Army of Tennessee Davis gave only general instructions. Secretary of War Seddon expressed the hope that Johnston could carry the attack to the enemy, but in the same letter Seddon emphasized that "your own experience and judgment are relied on to form and act on your own plans of military operations."[19]

Johnston's "experience and judgment" counseled against taking the fight to his opponent. Sherman began his move southward from Chattanooga toward Johnston's army and Atlanta on May 4, 1864. Retreating for some two months before Sherman's advance, Johnston abandoned defensive position after defensive position.[20] Unlike Lee in Virginia, who constantly struck out at Grant and who viewed the relentless Federal surge with foreboding, Johnston never hurled his divisions at Sherman.[21] By early July Johnston had his troops in the fortifications protecting Atlanta.

Jefferson Davis had watched Johnston's retreat with increasing concern. As early as June 7 General Bragg, military adviser to Davis, told Johnston that he had informed the president that "The condition of affairs in Georgia is daily becoming more serious."[22] By mid-July an official in the Confederate War Department recorded in his diary: "A very gloomy view of affairs in Georgia prevails in the cabinet."[23] The paucity of communications from Johnston about his plans particularly distressed Davis.[24] No one in Richmond knew his intentions; and after Johnston's continuous retreat of some two months Davis grew anxious to know what they were. Did he plan to fight for Atlanta, the psychological goal of the campaign, the gateway to central Georgia, and an important rail and supply center? Did he intend to withstand a siege in Atlanta? Did he envision continued retreat to Mobile or Savannah (depending upon Sherman's direction)? What chance would Johnston have against Sherman in the plains of central Georgia, if he had not been able to defend against him successfully in the hilly, even submountainous, terrain of north Georgia. Johnston gave Davis reason to expect the third alternative, when on July 11 he "strongly recommended" the immediate removal of prisoners of war from Andersonville, some one hundred miles south of Atlanta.[25]

To ascertain Johnston's intentions, Davis on July 9 ordered Bragg to Atlanta to confer with Johnston about his military plans. Davis instructed Bragg to investigate "dispositions and preparations" of the Army of Tennessee.[26] In Atlanta Bragg found "but little encouraging"; elaborating, he told Davis that "Much disappointment and dissatisfaction" existed in the army. General Johnston, observed Bragg, had "ever been opposed to seeking battle." Bragg implied that only a new commanding general could ensure that the Army of Tennessee would fight Sherman.[27]

Davis never planned to rely solely on Bragg's recommendation in a matter so crucial. Even before he sent Bragg to Georgia, he had Secretary of War Seddon request from Senator Benjamin H. Hill of Georgia an account of the talk he had had with Johnston. After meeting with Johnston on July 1, Hill reported that the general hoped Sherman would attack his trenches, but Hill concluded, as did Bragg, that Johnston had little desire to fight.[28]

Davis next turned to the hard-pressed Lee. "*Gen. Johnston,*" wrote Davis, "has *failed* and there are strong indications that he will *abandon Atlanta.*" Believing "the case [to be] hopeless in present hands," Davis broached the possibility of Johnston's removal and asked Lee, "Who should *succeed* him?"[29] Lee cautioned the president not to act hastily. He thought it "a grievous thing to change [the] commander of an army situated as is that of the Tennessee," but added, "Still if necessary it ought to be done."[30]

Davis's chief cabinet advisers thought a change essential. Secretary of War Seddon, who had been enthusiastic in 1863 about giving Johnston the Army of Tennessee, now advised the president to replace him.[31] From one of his closest confidants Davis heard the same refrain; for some time Secretary of State Judah P. Benjamin had argued that Johnston's strategy could result only in disaster.[32]

Bragg's report, Senator Hill's views, the advice of his most trusted cabinet advisers, Lee's concurrence (although reluctant), and political pressure from Georgia governor Joseph E. Brown taken together built an impressive case against Johnston.[33] Davis turned to Johnston himself. On July 16 he requested from Johnston information about the "present situation, and your plan of operations." Replying that same day, Johnston pointed out that Sherman outnumbered him and that his "plan of operations must, therefore, depend upon that of the enemy."[34] That exchange convinced Davis that his advisers were right. The following day Davis relieved Johnston, not only for failing to stop Sherman but mainly for "express[ing] no confidence that you can defeat or repel him."[35]

To repel Sherman, Davis and the Confederacy turned to John Bell Hood. Hood, thirty-three, had been a corps commander in the Army of Tennessee since April 1864. Best known for his battle leadership as a division commander under Lee, he had had no experience as an army commander. Lee, for one, feared for Hood's success in the larger job and told Davis so.[36] Removal of Johnston left Davis no alternative to Hood. The other full generals and prominent senior lieutenant generals—Lee, Edmund Kirby Smith, Beauregard, Bragg, James Longstreet, William J. Hardee—were for various reasons unavailable. The ability of the new commander to assume control quickly dominated all other considerations because it was believed that the fight for Atlanta could not be delayed. That need in effect limited the choice to someone serving with the Army of Tennessee. Nine months earlier, in a situation not nearly so pressing, Hardee had declined the position; no one remained save Hood.[37] And no one doubted that Hood would fight. Jefferson Davis gave him his chance.

Hood fulfilled Davis's expectations; he fought for Atlanta. And he lost. Hood's critics are legion; they accuse him of destroying the fine army he inherited from Johnston by a series of rashly fought, sanguinary clashes in and around Atlanta that bled the Army of Tennessee almost to death.[38] Voicing a contrary view, the distinguished English military historian Alfred H. Burne pointed out that Hood had lost by an extremely narrow margin. Burne asserted, "It would be difficult to find, in the history of any campaign, a more dazzling series of blows" than Hood delivered in Atlanta. His measured judgment found Hood's fight for Atlanta "sound—even brilliant."[39]

While Hood's fight for Atlanta did not surprise Davis, the fall of the city six weeks after Hood assumed command dealt "a stunning blow" to the president and his advisers.[40] He had gambled with Hood; he had gambled that a strong blow could force Sherman to retreat northward and preserve both Atlanta and central Georgia and Alabama. He accepted the risk of battle because he could see no alternative. When that program received a setback,

Davis hurried to begin anew. Aware of rumbling dissatisfaction with Hood[41] and hoping to renew fervor and patriotism in the citizens of Alabama, Georgia, and South Carolina—the states most immediately threatened by the fall of Atlanta—Davis determined to visit the Army of Tennessee and that section of his country.

Only three weeks after the loss of Atlanta, on September 25, 1864, Davis arrived at Palmetto, Georgia, Hood's headquarters, where he and Hood discussed strategy and command.[42] Hood proposed to move his army into north Georgia, believing that this maneuver would draw Sherman away from Atlanta.[43] If Sherman refused to oblige the Confederates and marched south from Atlanta, then Hood would be close on his heels. A receptive Davis authorized Hood to proceed with these plans. Davis knew he could not approve a retreat by Hood after he had removed Johnston for following such a policy. He also knew that retreat toward the Atlantic Ocean or the Gulf of Mexico could have but one result. Aggressiveness he felt essential, so he supported Hood rather than those who urged a more cautious and static autumn for the Army of Tennessee.[44]

Davis thought talking to the citizens in this invaded and threatened portion of his country was just as important as military discussions with General Hood. On this western trip he made four major speeches—at Macon, Montgomery, Augusta, and Columbia.[45] In each city he exhorted his listeners to rededicate themselves to the success of the Confederate cause. Supreme sacrifices by each individual would result in "final success" for southern arms. Sherman's invasion of Georgia, in Davis's rhetoric, became the counterpart of Napoleon's march into Russia, with the same fate awaiting the new trespasser. Davis was convinced that the innate sense of duty, which he believed dwelled in every southern heart, would ensure southern independence. This faith dominated the four speeches. This personal effort by the president did rekindle flames of confidence; even the hypercritical Edward Pollard admitted that "Confidence was in a measure revived."[46]

In addition to military strategy and public opinion, Davis was thinking about command structure. He knew that elements within and without the Army of Tennessee would have welcomed the dismissal of Hood.[47] To remove Hood meant, of course, to find a replacement, not an easy task in the Confederacy of late 1864. Most generals who had sufficient rank had by that time proved themselves indispensable in their jobs or unavailable, either for military or political reasons. Besides, Hood, even if he had failed to dislodge Sherman, had indeed fought the army; he had done exactly what Davis put him in command to do. Still, Davis had to consider the political climate of the post-Atlanta weeks.

Even before leaving Richmond for Georgia, Davis corresponded with General Lee on the availability and willingness of General Beauregard to accept a western assignment. Because neither Lee nor Beauregard was satisfied with the latter's difficult position in Virginia as Lee's deputy, both generals eagerly welcomed Davis's suggestion.[48] In Augusta, on his return from Hood's army, Davis conferred with Beauregard on the West. After outlining the strategy he and Hood had decided upon and which Beauregard supported, the president appointed him commanding general of the Military Division of the West.[49] This command arrangement, which put Beauregard over everything between Georgia and the Mississippi, harked back to the Department of the West, when Joseph Johnston had held a similar position.

Davis, in giving Beauregard this command, has been accused of "lay[ing] on the shelf a general who was out of favor."[50] Political considerations undoubtedly influenced Davis's decision. Sending the hero of 1861 to the West to support Hood and to lead the citizenry Davis had rallied was a popular move, although an official in the Confederate War Department probably exaggerated public sentiment when he rejoiced at Davis's "response to the universal call of the people."[51] Without doubt, however, Beauregard's appointment met with public approval.

Davis had not placed Beauregard on a shelf. He knew that he could not direct all military activities from Richmond, and he was making another attempt to create a "unified army command on a decentralized basis—in essence, a theater command of the sort so popular in World War II," to give one man the authority, discretion, and mobility to control and coordinate activities over a large area.[52] His first effort with Joseph Johnston had failed, in large part because Johnston would not use the power he had been given.[53] To Beauregard, Davis gave the same general powers "to secure the fullest co-operation of the troops" under his charge.[54] Davis intended Beauregard to exercise command authority in the Military Division of the West; he did not envision Beauregard's becoming tied down in any one place with any particular segment of his command. To enable Beauregard to act as a theater or group commander, his assignment orders authorized him to operate "wherever in your judgment the interests of your command render it expedient."[55]

Davis gave Beauregard sufficient authority to command, but Beauregard, like Johnston before him and like many other Civil War generals, could not conceive of commanding without actually leading troops in the field.[56] General responsibility over a geographical area or theater rather than a specific army made Beauregard uncomfortable. That "he would be without troops directly under him" and "that he was not superseding General Hood" left Beauregard unsure about the new command structure, but Davis's promise of support persuaded him to accept the position.[57]

The device of the Military Division of the West and the selection of Beauregard as its commanding general constituted, in certain ways, a master stroke. It provided advice and restraint for a young field commander, it muted public criticism after Atlanta, and it resurrected an old hero to repel the invader from the southern heartland.

Davis, back in Richmond, watched the strategy he had approved for his western army go into operation. And that strategy functioned as designed. Even before Beauregard's arrival Hood started for northwestern Georgia. Receiving word that the Confederates had advanced north of the Chattahoochee River, Sherman moved his divisions out of Atlanta in pursuit of Hood.[58]

For the better part of a month the two armies played a cat-and-mouse game in northwestern Georgia and northeastern Alabama. Neither commander enjoyed this charade. Although Hood had accomplished his purpose of drawing Sherman out of Atlanta, he wanted a fight, and Sherman made that difficult. Sherman had promised to "make [Hood] suffer" if his "army attempt our rear," but the futility of the chase chafed him.[59] "To pursue Hood is folly," Sherman grumbled, "for he can twist and turn like a fox and wear out an army in pursuit." To Sherman, his army some sixty miles behind Hood, the possibilities in a march to the sea contrasted vividly with the fruitless pursuit of Hood.[60]

While Sherman thought of Savannah, Hood had visions of Confederate divisions in Middle Tennessee. The strategy decided on by Hood and Davis at Palmetto and agreed to by Beauregard at Augusta permitted a northern campaign only so long as Sherman dutifully followed. Should Sherman turn back south to Atlanta and central Georgia, then Hood was to become the hunter. Hood's new idea dropped that contingency. The Army of Tennessee would simply forget about Sherman and invade Tennessee.[61]

Hood made this proposal to Beauregard at Gadsden, Alabama, on October 20. Beauregard, who had returned to Hood's army after visiting Richard Taylor, exhibited little enthusiasm for the plan. After two days of discussion, however, Beauregard, though still dubious, authorized Hood to make preparations for the advance and urged on his subordinate the absolute necessity for speed. Hood had to get into Middle Tennessee before the Federals were ready for him.[62]

His decision made, Beauregard informed the War Department on October 22 and 24 of the change in operations.[63] A week later he sent President Davis a letter outlining the new strategy.[64] Nothing came from the president or the War Department countermanding Hood's invasion.

On October 31 Hood arrived at Tuscumbia, Alabama, the place he and Beauregard had decided to use as a jumping-off point for the invasion. Once in Tuscumbia, Hood failed to heed Beauregard's admonishment to make haste and get into Tennessee as quickly as possible. For three weeks, in spite of repeated pleas from Beauregard to "Push on active offensive immediately," Hood waited in Tuscumbia.[65] The dashing, energetic, even impetuous Hood sat down. Not until November 21, for reasons not yet convincingly explained, did he advance into Tennessee.[66]

This three-week delay greatly decreased the chances for Confederate success in Tennessee. While Hood marched between Gadsden and Tuscumbia, Sherman turned back toward Atlanta and began the march to the sea.

He placed Major General George H. Thomas in charge of defending Tennessee against attack by Hood, but Thomas did not begin to get his troops positioned until November 4. Disarray pervaded Federal concentration efforts to such an extent that Hood, when he finally did begin his advance, came close to cutting off and isolating from Thomas a sizable portion of his force. Once in Tennessee ill fortune and poor execution plagued Hood every bit as much as Federal opposition.[67]

Hood's Tennessee campaign was crucial for Davis and the Confederacy. On its accomplishments depended the western war. When the battles of Franklin and Nashville crushed Hood and the Army of Tennessee, the Confederacy had no military force of consequence left in the West. Sherman, opposed only by militia units and paltry remnants of the Army of Tennessee, gutted Georgia and the Carolinas.

Jefferson Davis backed Hood all the way in the autumn of 1864. He did not countermand Beauregard's authorization of the invasion. Before Hood left Tuscumbia, Davis wrote him a long, vague letter expressing the hope that Hood might beat the Federals in detail and "advance to the Ohio River."[68] To Beauregard he spoke of Hood's "reach[ing] the country proper of the enemy" from whence he could "change the plans for Sherman's or Grant's campaigns."[69]

After the war Davis tried to exempt himself from complicity in Hood's project. Claiming in his memoirs that the invasion was "ill-advised," he asserted that he had no knowledge until after the fact that Hood had decided not to follow Sherman.[70] In the bitter battles the Confederate chieftains fought long after Appomattox, Davis denied that Hood conceived the idea of the Tennessee invasion and insisted that Beauregard forced it on Hood.[71] Neither of these claims has any basis in the contemporary documents. Davis knew from the beginning what Hood was about. He hoped, as did Beauregard and Hood, that Hood would succeed in Tennessee. He hoped that Sherman would be forced out of Georgia. He hoped that the invasion of Tennessee would rejuvenate the Confederate military effort in the West.

Supporting Hood's invasion meant taking a great risk for Davis; not to support it meant certain disaster. The evidence from the Atlanta campaign and from the battles for Atlanta itself gave no indication that the Army of Tennessee could do more than delay, for brief moments, the inexorable southward advance of Sherman. But resounding victory in Tennessee might make a substantial difference. Davis gambled—with higher stakes than those before Atlanta because the times were more desperate—and according to Beauregard and Federal commanders it might have paid off had Hood acted quickly.[72] He did not.

This essay suggests a somewhat different Jefferson Davis from the one portrayed over a century ago by Pollard and brought down to the present. In the late period of the western war Davis did not exhibit a rigidity that refused advice. He showed more than a little political sense and made a strong effort to gain the support of a threatened public. He did not play favorites nor needlessly interfere with his generals in the field. Davis's every act in those months was calculated for victory.

Notes

"A Reassessment of Jefferson Davis as War Leader: The Case from Atlanta to Nashville" was initially published in the *Journal of Southern History* 36, no. 2 (May 1970): 189–204, and is reprinted with permission of the Southern Historical Association.

1. Eric L. McKitrick, "Party Politics and the Union and Confederate War Efforts," in *The American Party Systems: Stages of Development,* ed. William N. Chambers and W. Dean Burnham (New York, 1967), 119.
2. Many of the works cited below give part of the Lincoln-Davis story.
3. *Richmond Examiner,* Sept. 5, 1864; Edward A. Pollard, *The Second Year of the War* (Richmond, VA, 1863), 302.
4. Pollard, *Second Year of the War,* 302.
5. Edward A. Pollard, *Life of Jefferson Davis, with a Secret History of the Southern Confederacy* (Philadelphia, 1869), vi.
6. Edward A. Pollard, *The Lost Cause; A New Southern History of the War of the Confederates* (New York, 1866), 91.
7. Clifford Dowdey, *Experiment in Rebellion* (Garden City, NY, 1946) and Clifford Dowdey, *The Land They Fought For: The Story of the South as the Confederacy, 1832–1865* (Garden City, NY, 1955). The quotation comes from *Experiment in Rebellion,* 372.
8. T. Harry Williams, *Lincoln and His Generals* (New York, 1952), 7.
9. David M. Potter, "Jefferson Davis and the Political Factors in Confederate Defeat," in *Why the North Won the Civil War,* ed. David Donald (Baton Rouge, LA, 1960), 106.
10. Hudson Strode, *Jefferson Davis: American Patriot, 1808–1861* (New York, 1955); Hudson Strode, *Jefferson Davis: Confederate President* (New York, 1959); Hudson Strode, *Jefferson Davis, Tragic Hero: The Last Twenty-five Years, 1864–1889* (New York, 1964).
11. See the superb review by David Donald in the *New York Times Book Review,* Sept. 27, 1964, p. 6.
12. Archer Jones, *Confederate Strategy from Shiloh to Vicksburg* (Baton Rouge, LA, 1961), 239.

13. Rembert W. Patrick, *Jefferson Davis and His Cabinet* ([Baton Rouge LA], 1944), 44.

14. Frank E. Vandiver, "Jefferson Davis and Confederate Strategy," in *The American Tragedy: The Civil War in Retrospect,* ed. Avery O. Craven and Frank E. Vandiver (Hampden-Sydney, VA, 1959), 19–32. The quotations come from pages 20 and 32.

15. The only general history of the Army of Tennessee is Stanley F. Horn, *The Army of Tennessee: A Military History* (1941; reprint, Norman, OK, 1953). Two excellent volumes treat the earlier years of the war—Thomas L. Connelly, *Army of the Heartland: The Army of Tennessee, 1861–1862* (Baton Rouge, LA, 1967) and Jones, *Confederate Strategy*—and form the basis for the generalizations on that period. *Editors' note:* Thomas L. Connelly's *Autumn of Glory: The Army of Tennessee, 1862–1865* was not published until 1971, a year after this article was first published.

16. Horn, *Army of Tennessee,* 307–8.

17. Jones, *Confederate Strategy,* passim.

18. The most recent biography of Johnston is Gilbert E. Govan and James W. Livingood, *A Different Valor: The Story of General Joseph E. Johnston, C.S.A.* (Indianapolis and New York, 1956). Jones, in *Confederate Strategy,* and Alfred H. Burne, in *Lee, Grant and Sherman: A Study in Leadership in the 1864–65 Campaign* (New York, 1939), discuss various aspects of his generalship. *Editors' note:* Craig L. Symonds's *Joseph E. Johnston: A Civil War Biography* was published in 1992, twenty-two years after this article was first published.

19. U.S. War Department, *The War of the Rebellion: A Compilation of the Official Records of the Union and Confederate Armies,* 128 vols. (Washington, DC, 1880–1901), series 1, vol. 31, pt. 3:842–43. Hereinafter cited as *OR.* All references are to series 1 unless otherwise indicated.

20. For the Atlanta campaign see Burne, *Lee, Grant and Sherman,* 70–100; Horn, *Army of Tennessee,* 305–40; Lloyd Lewis, *Sherman, Fighting Prophet* (New York, 1932), 375–87.

21. For an absorbing account of Lee's fight against Grant, see Clifford Dowdey, *Lee's Last Campaign: The Story of Lee and His Men Against Grant—1864* (Boston, 1960).

22. *OR,* vol. 38, pt. 4:762. Thomas R. Hay, "The Davis-Hood-Johnston Controversy of 1864," *Mississippi Valley Historical Review* 11 (June 1924): 54–84, has a minutely detailed discussion of the relief of Johnston.

23. Robert G. H. Kean, *Inside the Confederate Government: The Diary of Robert Garlick Hill Kean,* ed. Edward Younger (New York, 1957), 165.

24. Telegrams did arrive at the War Department from the Army of Tennessee between May and July 1864, but they reported minor actions, personnel problems, and similar matters. Never did Johnston talk about the course of his campaign, and he admitted as much. *OR,* vol. 38, pt. 4:795–96.

25. Ibid., pt. 5:876.

26. Jefferson Davis, *Jefferson Davis, Constitutionalist: His Letters, Papers and Speeches,* 10 vols., ed. Dunbar Rowland (Jackson, MS, 1923), 6:286.

27. *OR,* vol. 38, pt. 5:878; ibid., vol. 39, pt. 2:712–14 (the two final quotations are on page 713).

28. *OR,* vol. 52, pt. 2:693–95, 704–707. Hill had a free and open conversation with Johnston. Govan and Livingood, *A Different Valor,* 299–300.

29. Davis, *Jefferson Davis, Constitutionalist,* 6:291–92.

30. Robert E. Lee, *The Wartime Papers of R. E. Lee,* ed. Clifford Dowdey and Louis H. Manarin (Boston, 1961), 821–22.

31. Jefferson Davis, *The Rise and Fall of the Confederate Government,* 2 vols. (New York, 1881), 2:560–61; Roy W. Curry, "James A. Seddon, a Southern Prototype," *Virginia Magazine of History and Biography* 63 (Apr. 1955): 140.

32. Benjamin to Davis, Feb. 15, 1879, Walter L. Fleming Papers, 1685–1932, New York Public Library (hereinafter cited as Fleming Papers).

33. Louise B. Hill, *Joseph E. Brown and the Confederacy* (Chapel Hill, NC, 1939), 182–83.

34. Both letters are in *OR,* vol. 38, pt. 5:882–83.

35. Ibid., 885.

36. Lee, *Wartime Papers,* 821–22.

37. Hay, "The Davis-Hood-Johnston Controversy of 1864," 55, 76.

38. Among Hood's critics are W. Birkbeck Wood and James E. Edmonds, *Military History of the Civil War with Special Reference to the Campaigns of 1864 and 1865* (New York, 1960), 160–73, and Thomas R. Hay, "The Atlanta Campaign," *Georgia Historical Quarterly* 7 (Mar. and June 1923): 32–43, 100–18.

39. Burne, *Lee, Grant and Sherman,* 110; see also Burne, "General J. B. Hood," *Army Quarterly and Defence Journal* 79 (Jan. 1960): 220–23.

40. John B. Jones, *A Rebel War Clerk's Diary at the Confederate States Capital,* 2 vols., ed. Howard Swiggett (New York, 1935), 2:277; in addition, consult Mary B. Chesnut, *A Diary from Dixie,* ed. Ben Ames Williams (Boston, 1949), 443.

41. Jones, *Rebel War Clerk's Diary* 2:288; Horn, *Army of Tennessee,* 371; Richmond *Examiner,* Sept. 5, 1864; Josiah Gorgas, *The Civil War Diary of General Josiah Gorges,* ed. Frank E. Vandiver (University, AL, 1947), 140.

42. Horn, *Army of Tennessee,* 372–73; John Bell Hood, *Advance and Retreat: Personal Experiences in the United States and Confederate States Armies* (New Orleans, 1880), 253–55.

43. *OR,* vol. 39, pt. 2:862.

44. Both Bragg and Lieutenant General Richard Taylor, the commander of the Department of Alabama, Mississippi, and East Louisiana whom Davis visited when

he left Hood's army, objected to an offensive campaign. They argued that Hood's lack of strength and Sherman's overwhelming force made remote the chance of Confederate success. Of course, adopting the defensive in no way guaranteed success either. Richard Taylor, *Destruction and Reconstruction: Personal Experiences of the Late War* (New York, 1879), 205–6; Bragg to Davis, Sept. 29, 1864, quoted in Don C. Seitz, *Braxton Bragg: General of the Confederacy* (Columbia, SC, 1924), 462.

45. Davis, *Jefferson Davis, Constitutionalist,* 6:341–44 (Macon), 345–47 (Montgomery), 356–61 (Augusta), 349–56 (Columbia).
46. Edward A. Pollard, *Southern History of the War: The Last Year of the War* (New York, 1866), 92.
47. See note 41 above.
48. Alfred Roman, *The Military Operations of General Beauregard in the War Between the States, 1861 to 1865,* 2 vols. (New York, 1883), 2:273–75; Lee to Davis, Sept. 19, 1864, Letterbook, June 7, 1863—Oct. 12, 1864, Robert Edward Lee Papers, Virginia Historical Society, Richmond; T. Harry Williams, *P. G. T. Beauregard: Napoleon in Gray* (Baton Rouge, LA, 1955), 238–40.
49. Undated manuscript on military operations in P. G. T. Beauregard Papers, Manuscript Division, Library of Congress (hereinafter cited as Beauregard Papers); Roman, *Beauregard* 2:275–79. The second major segment of Beauregard's new command was Richard Taylor's Department of Alabama, Mississippi, and East Louisiana.
50. Williams, *Beauregard,* 242; Thomas R. Hay, *Hood's Tennessee Campaign* (New York, 1929), 28–29.
51. Jones, *Rebel War Clerk's Diary* 2:300.
52. Frank E. Vandiver, *Rebel Brass: The Confederate Command System* (Baton Rouge, LA,1956), 35; Vandiver, "Jefferson Davis and Confederate Strategy," 31.
53. Vandiver, *Rebel Brass,* 34–35, 57–59; Jones, *Confederate Strategy,* 89–197 and passim.
54. Davis, *Jefferson Davis, Constitutionalist,* 6:344–45.
55. *OR,* vol. 39, pt. 3:782; Davis, *Jefferson Davis, Constitutionalist,* 6:368.
56. Johnston's performance as commanding general of the Department of the West illustrates this attitude. Also consult Williams, *Lincoln and His Generals,* 245.
57. Roman, *Beauregard* 2:279.
58. Burne, *Lee, Grant and Sherman,* 125–38; Jacob D. Cox, *Atlanta* (New York, 1882), 225–26.
59. William T. Sherman, *Home Letters of General Sherman,* ed. M. A. De Wolfe Howe (New York, 1909), 310.
60. Quoted in Freeman Cleaves, *Rock of Chickamauga: The Life of General George H. Thomas* (Norman, OK,1948), 245; also see Burne, "General J. B. Hood," 225.

61. Hood, *Advance and Retreat,* 264–69; John P. Dyer, *The Gallant Hood* (Indianapolis and New York, 1950), 281–82.

62. Roman, *Beauregard* 2:287–88; *OR,* vol. 45, pt. 1:647–48; Beauregard to Charles Jones, Apr. 1, 1875, Beauregard Papers.

63. *OR,* vol. 39, pt. 3:841; ibid., pt. 1:796–98.

64. Actually, he sent Davis a copy of a new dispatch to the War Department. *OR,* vol. 39, pt. 3:870.

65. For the quotation, see *OR,* vol. 45, pt. 1:1226. See also *OR,* vol. 45, pt.1:1215; *OR,* vol. 39, pt. 1:800.

66. *OR,* vol. 39, pt. 3:913, and Hood, *Advance and Retreat,* 272–74, contain Hood's reasons for his long delay. The closest student of Hood's activities found most of Hood's explanations unconvincing; he suggested that Hood procrastinated but really put forth no new and persuasive reasons for the long stay in Tuscumbia. Hay, *Hood's Tennessee Campaign,* 61–66.

67. Cleaves, *Rock of Chickamauga,* 244–50; Wood and Edmonds, *Military History,* 230n; Jacob D. Cox, *The March to the Sea: Franklin and Nashville* (New York, 1884), 7, 14; Hay, *Hood's Tennessee Campaign,* 50–51, 83–102.

68. Davis, *Jefferson Davis, Constitutionalist,* 6:398–99.

69. Ibid., 413.

70. Davis, *Rise and Fall* 2:569–70.

71. Davis to Lucius Northrop, Apr. 9, 1879, Fleming Papers.

72. See notes 62 and 67 above.

General John Bell Hood. Courtesy of the Museum of the Confederacy, Richmond, Virginia.

General Hood as Logistician

Frank E. Vandiver

General John B. Hood was a fighter. This had been a major consideration influencing his appointment to succeed Joseph E. Johnston in command of the Confederate Army of Tennessee on July 18, 1864. But aside from this attribute, what qualifications did the general have for army command? He was undeniably quick in battle, had shown a grasp of objective in action and certainly could move troops where he wanted them to fight. He had, at the beginning of his independent command good and sufficient confidence in himself and advocated the offensive. But, as time went on certain gaps appeared in his proficiency.

President Davis knew that the choice of Hood was not the most ideal he could have made, but felt that what he lacked in professional finesse he might make up in action. Davis knew, too, that General Lee was unsure of Hood's overall qualifications.[1]

It is not surprising that neither the President nor the commander of the Army of Northern Virginia worried about whether Hood had a keen sense of logistics. Surely this was an obvious requisite of any capable field commander, and need hardly be questioned. Hood was, after all, a graduate of West Point. But no complete analysis of his military capacity can overlook his attention to what Frederick the Great has called "the primary duty of a general"—supply.[2] An examination of efforts to provide ordnance for Hood's Tennessee campaign will serve perhaps to measure him from this standpoint.

The Army of Tennessee was adequately supplied with ordnance when Hood took command. Confederate ordnance officers had pushed efforts to furnish ample ordnance stores from the outset of Johnston's campaign to retard Sherman's advance from Dalton to Atlanta, Georgia, in May, 1864. By April the Ordnance Bureau had been able to issue about 120 rounds of small arms ammunition per man in the ranks, along with adequate artillery ammunition. Even so, the Bureau seemed compelled to apologize for not being able to do better.[3] An apology was out of order—the Bureau had done well.

Supplying such a volume of ammunition had strained resources considerably. Cooperation among all the ordnance establishments alone made the achievement possible. Under a logistical plan set up in March, 1863, the Army of Tennessee was to be supplied by the arsenals, armories and depots nearest to it.[4] This threw the main distribution responsibility upon Atlanta Arsenal and its supporting installations. Columbus Arsenal provided a large portion of the small arms ammunition to Atlanta Arsenal. From here it was, in turn, sent to the army. The production capacity of the cartridge laboratory at Columbus was increased during May and June, so that by June 28, its commanding officer reported a weekly fabrication of from 100,000 to 120,000 bullets. All of these were destined for the Army of Tennessee.[5]

In the emergency created by the summer campaign Atlanta Arsenal received aid from almost all of the ordnance installations in the deep South. From Savannah, for instance, as well as from Macon, came assistance in the form of cartridges.[6] No city contributed more toward supplying the army's needs than Macon. Here were located the Confederate States Central Laboratories for Ordnance, an arsenal, a cannon foundry and a National Armory.[7]

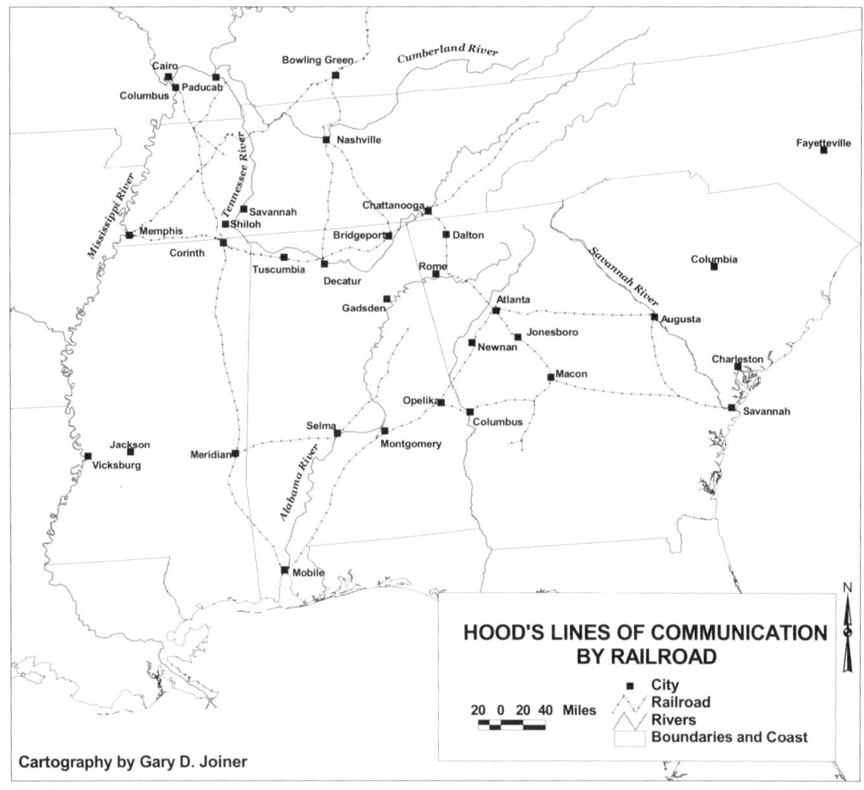

The Ordnance Bureau naturally expected Macon to carry a major portion of the supply load for the Army of Tennessee.

As the army retreated closer toward Atlanta, fear increased concerning the safety of the ordnance plants in Georgia. The loss of Atlanta Arsenal would be a severe blow to the Ordnance Bureau. But, in addition, the presence of Sherman beyond Atlanta meant danger to all deep South installations. Although generally not mentioned probably for security reasons, the retention of Atlanta was almost essential to the Bureau. This should have been obvious to even the most unmilitary onlooker.

With this tactical necessity as a spur, all possible measures were taken to sustain the army backing into Atlanta's entrenchments. In mid-July, after the army had occupied the city's defense lines, the percussion cap factory there was hastily moved to Macon.[8] Colonel Josiah Gorgas, the Confederate Chief of Ordnance, had decided to concentrate his Bureau's resources in Macon to sustain the army now under Hood's charge. Atlanta Arsenal was too exposed to rely on for other than distribution functions. Macon, Columbus, and Augusta arsenals were picked to provide Hood's wants, though the latter arsenal was all but isolated from army by rail. The ammunition laboratory in Atlanta was united with that in Macon and put under the command of Colonel John W. Mallet, the Confederacy's Superintendent of Laboratories. Gorgas directed Mallet to organize the workers of the two laboratories into a single force and rush production—"time is chief consideration."[9]

Macon Armory was sorely taxed to repair arms for Hood, but by July 26 it was able to return 200 arms a day to the army.[10] The Ordnance Bureau encountered many hindrances in keeping up Hood's ordnance supply. Negro labor was an essential part of the working force at all Macon ordnance plants. As the war moved into Georgia it became increasingly difficult to persuade slaveowners to rent their slaves to the various installations close to the theater of war. In late June Mallet and James H. Burton, at Macon Armory, had been forced to seek authority to impress slave labor in order to keep going.[11]

The little relief this expedient offered was short lived. On July 1, Burton complained to Gorgas that an armed guard had appeared at the armory and laboratories with instructions to impress one-third of the Negro labor for work on fortifications.[12] Unfortunately this was to be the first of several similar interruptions.

Realizing that confusion might result from the changed logistical plan forced by Atlanta's situation, Gorgas sought to keep matters in hand. In an attempt to prevent decentralization from degenerating into chaos, he sent Colonel Moses H. Wright, trusted commander of Atlanta Arsenal, to Macon on August 4 to take charge of supplying Hood's needs. Wright was soon moved to the command of Columbus Arsenal when it became apparent that no confusion would develop.[13]

Hood received all kinds of cooperation from ordnance officers. Arms came to him from Richmond, even though General Lee needed almost all on hand in that city;[14] gunstocks came with difficulty to Macon for him from North Carolina.[15] Charleston Arsenal, basking in a period of rare quiet, was called upon to supply everything possible.[16] Percussion caps, a basic need, were in urgent demand by late August. Mallet, doing all he could with the combined Atlanta and Macon machinery, asked Gorgas if an additional supply could be brought through the blockade. This despite the fact that a million pistol caps and a million and a quarter musket caps had come through during June and July. Two million were sent to Macon from Richmond.[17]

Apparently unaware of all that was being done to keep his troops supplied with ammunition and ordnance, Hood settled down to the siege of Atlanta. He made his first sally into the realm of ordnance logistics on August 1. On that day he telegraphed Colonel Richard M. Cuyler and Mallet, in Macon: "General Bragg directs that you send me at once all the negroes employed on public buildings at your post." Two days later James H. Burton got substantially the same message.[18]

This maneuver crippled operations at Macon, but Hood had not yet learned his lesson. On August 31–September 1, during General William J. Hardee's desperate action at Jonesboro on the Central railroad, all of the skilled and unskilled workers in the numerous ordnance works at Augusta were sent to reinforce him. Badly as they were needed at Jonesboro, the move was unwise. George W. Rains, commanding the Augusta works, complained that as a consequence "all the works here were stopped for some days . . . " He wrote Colonel James M. Kennard, chief ordnance officer of Hood's Army, that "I think it would be well for the General to give directions that the employees of the *Small Arm Cartridge Laboratory* . . . and Powder Works at this place should be exempted from the local duties, or in other words that they should remain under my control at all times: I think this very important for the public interests."[19]

As Hood's position in Atlanta became obviously insecure, uncertainty gripped ordnance officials. No one had any sound knowledge of what place would be safe. Burton did not wait for instructions; he assumed the Macon machinery would have to be moved and had several flat boats constructed—60 feet long and 14 feet wide—to transport it down the Ocmulgee River to an undetermined point in Georgia. Burton was excited and his action premature. Gorgas calmed him down and later suggested Columbia, South Carolina, as the place to send his machinery, if, indeed, it must be moved at all.[20]

Ordnance Bureau officers at Augusta, Columbus and Macon—indeed everywhere in the south, anticipated bad news from Atlanta. But they could hardly have conceived how bad that news would be. Mallet was the first to realize what had happened, and on September 5 expressed it to Gorgas in stark words: "Gen. Hood has blown up his reserve Ordnance train. Can any cartridges of calibre fifty-four and fifty-eight or rifle shell—three inch, ten pounder Parrott, and two and half inch Blakely—be had from North of Augusta?"[21]

There it was. Atlanta fell on September 2, and with it the arsenal, shops, railroad connection and the reserve ammunition of the Army of Tennessee. The ordnance was destroyed at about two o'clock that morning.

Reports were confused. All seemed to agree that some eighty-one cars and from three to five engines had been blown up, and it was generally assumed that all of the cars contained ordnance stores.[22] This was not quite true, but the truth was sickening enough. Twenty-eight of the eighty-one cars did contain ordnance supplies. This constituted all of the available reserve, particularly of artillery ammunition.[23]

The repercussions of Hood's explosive withdrawal from Atlanta were immediately felt in the Ordnance Bureau. Hood could hardly have picked a more inopportune time to retreat, from an ordnance standpoint. Initial provision of his reserve ammunition had required every exertion, and now that Atlanta Arsenal was gone, replacement was doubly difficult.

George Rains at Augusta, unruffled by events, asked ladies in Augusta and nearby communities to volunteer for work in his cartridge factory. With these patriotic assistants, he was able to reach a daily production of 75,000 cartridges during the critical days following the loss of Hood's ordnance.[24] But Rains' enterprise accounted for only one deficiency.

Suddenly, as if at a given signal, all kinds of shortages appeared. Colonel Wright, at Columbus, summed up their general nature in a telegram to his friend Mallet on September 6: "I need beeswax, twine, thread, gum arabic, sulphur, mealed powder, woolen yarn, lead & percussion caps as well as powder."[25] Mallet, whose wary eye surveyed the whole ordnance scene, told Gorgas of an even more serious need a few days later. Rains' supply of lead was running out, he said, and since all the other southern arsenals depended upon him for it a general shortage was in sight.[26]

Once out of Atlanta Hood came face to face with his own supply problems. Concentrating near Lovejoy's Station, on the Central railroad, he thought in terms of grand strategy. On September 6 he voiced an idea to President Davis. He felt that as soon as his army had rested, and after the Federal prisoners at Andersonville, Georgia, had been moved, he should strike Sherman's attenuated line of communications. This might force Sherman to follow him toward Chattanooga and might offer a favorable chance for battle.

Considering the disparity in size of the opposing armies, Hood's strategy was probably sound. To put this plan into effect, Hood realized, he had to shift his position from the Macon railroad and he informed Davis of his intention to draw supplies from the West Point and Montgomery railroad after he had changed his location. Hood described his move:

> Causing the iron to be removed from the several railroads out of Atlanta for distances of forty miles, and directing railroad stock to be restored to the West Point railroad, the movement to the left toward that road began on the 18th of September. Arriving at that road the army took position with the left touching the Chattahoochie River and covering that road, where it remained several days to allow the accumulation of supplies at Blue Mountain and a sufficiency with which to continue the movement.[27]

Hood's chief of staff, Brigadier General Francis A. Shoup, recorded the change in position in his daily journal of the army's movements. He noted that army headquarters were at Palmetto, on the West Point railroad on September 19. On the 20th he observed that Hood's orders concerning the removal of track were being executed:

> The telegraph wire and railroad iron between Lovejoy's and Griffin, on Macon railroad, and the iron above Newnan on West Point railroad, also on the Georgia railroad between Oconee River and Stone Mountain, have been ordered to be taken up at once and saved for future use.[28]

Ensconced at Palmetto, Hood proceeded to demand supplies. On September 20 he suggested to General Bragg that powder mills for his use be established at Cahaba, Alabama, or somewhere else in that state. This suggestion was the first indication that he was aware of any difficulties involved in providing ordnance to his army. Even in this instance he was uninformed, perhaps excusably so. A new powder mill was almost ready to begin operation at Selma, and would produce enough for Hood's needs—if niter production in Alabama was not interrupted by conscription.[29]

Having paid lip service to basic logistics, Hood's attention focused on more obvious matters. Arms were badly needed to refurbish the army and supply returning troops. Where could they be found? Colonel Kennard, possessed of commendable directness, suggested that since the Georgia militia had recently become inactive, their arms be taken for the Army of Tennessee. The Chief of Ordnance, in giving approval to this scheme, pointed out that the militia could soon be resupplied from the stock of arms being repaired at Macon. And he told the harassed Kennard: "Every exertion will be made to assist you, but the drain is simultaneous and difficult to meet."[30] Negotiations with Georgia began.

Meantime, Kennard tried other sources which he hoped might provide faster assistance. He queried Colonel Hypolite Oladowski, now commanding Columbus Arsenal, about how many arms could be drawn from there. The answer was discouraging. Oladowski had not a single rifle or musket to send. He thought, though, that Captain W. D. Humphries, at Hood's intermediate base, West Point, could send some 2,000 arms. With authorization for Gorgas another 600 could be had from Columbia Arsenal.[31]

Kennard's anxious search for harness and saddles for artillery horses brought Gorgas to his rescue. The Chief of Ordnance ordered saddles sent him from the arsenal at Mount Vernon, Alabama, along with 1,000 sets of harness from Richmond.[32]

Hood himself telegraphed Gorgas for ammunition. Envisioning the need for a change of base, he asked that future supplies be sent in quantity to Selma. Gorgas telegraphed Kennard immediately, asking what stores he needed and where he wished them sent.[33] Colonel Rains, instructed to aid Kennard, started shipping stores to Selma on September 26—4,000 rounds of fixed ammunition (mostly for 12-pounder Napoleon guns) and 500,000 rounds of .54 and .57 caliber small arms ammunition. He told Kennard that he could probably ship him 350,000 cartridges and 1,200 rounds of artillery ammunition a week. These figures could be raised considerably, given the proper conditions. Rains thought Hood should be told of the interruptions occurring in the operations at Augusta. The forcible removal of bullet moulders and wood agents had to stop, since

> such interruptions are likely to be disastrous. I write you these facts in order that General Hood may see to what interruptions I am liable to continually in preparing supplies, and hence cannot say as to the amount of stores I can send him but presume it will be as above stated.[34]

Finally Hood's negotiations for the Georgia militia rifles produced an answer. General Gustavus W. Smith, commanding the militia, would not surrender the arms without authority from Governor Joseph E. Brown. Colonel Cuyler, at Macon, through whom Hood was negotiating, found himself caught between Hood and Smith. He appealed to Brown on September 27, for a release of the arms. Brown said no, as certainly the arsenal commander must have expected. Hood asked that Cuyler seize the arms, stored at Macon Arsenal, and send them to him immediately. Cuyler's position was awkward, to say the least. Faced with an order from a general in the field he was almost duty bound to comply. But apparently he was more forcefully aware than was Hood of the wrath such an action could produce in Milledgeville.

He had to telegraph Kennard on the 28th: "Governor Brown will not give up the arms. I cannot undertake to take them. . . . " Later that same day, anguished in his dilemma, Cuyler again telegraphed Kennard: "Governor Brown has refused to give up the arms.—I must have high authority before

I conflict with him. If Col. Gorgas my immediate superior, or the President order me, I will take them."

Brown again was told that Hood needed 1,000 of the militia rifles for the defense of Georgia, and the colonel hoped he would reconsider. Again, no. Cuyler gave up, and the matter was headed for court decision on October 1, when General Howell Cobb managed to persuade Brown to release the arms if they were later replaced. But this came a little late, since Hood had already started operations without the rifles.[35]

Gorgas did not wait to hear of the outcome of the Brown-Hood altercation. On September 28 he directed Major John T. Trezevant, at Columbia Arsenal, to send Kennard 1,000 Enfield rifles and 1,500 accoutrements. Captain Humphries, at West Point, told Kennard on the same day that he would ship arms and accoutrements on the 29th, but he had received no ammunition to send forward. The delay was in rail transportation, frequently monopolized by Commissary officials. Ammunition shipments could be made only when cars were released.[36]

Hood had at least one bit of good luck. On September 28, the day he moved his army out of camp at Palmetto, General Robert C. Tyler telegraphed him from West Point: "Have just found four hundred thousand percussion caps shall hold them subject to your order."[37]

Supplying Hood's needs at Palmetto had not been easy. He still used Jonesboro as his main depot, which involved somewhat complicated rail connection with his army.[38] Now that he was moving, he decided, after talking with General Beauregard, commanding the geographical department in which Hood operated, to change his base, as he had anticipated doing. The new base would be Jacksonville, Alabama, the railhead of the Meridian to Blue Mountain road. Ordnance was to be accumulated at Selma Arsenal for shipment to the army. As his plans developed, Jacksonville became progressively less attractive as a base.

Hood's itinerary took him north to Dalton and then southwest to Gadsden, which was reached on October 20. Here a day was consumed in issuing the supplies received from Selma and Jacksonville. These supplies had had to be transported from the railroad at Jacksonville to Gadsden by wagon, a distance of some eighteen to twenty miles.[39]

Having distributed his supplies, Hood appeared ready to cross the Tennessee River at Guntersville, Alabama. At any rate, this is what Beauregard had been led to believe in a conference with Hood at Gadsden on October 21. As a result, Beauregard began logistical preparations to sustain an offensive. Since Sherman's pursuing forces had come within fifteen miles of Gadsden and were thus all too close to Jacksonville, Beauregard made hurried plans to change Hood's base.

Hood himself had attempted to anticipate such a possible shift. As early as October 8 he had requested that the Memphis and Charleston Railroad be put in working condition from Corinth to Decatur. He and Beauregard together decided that Tuscumbia would be the best base, and it was so designated. Much trouble was encountered in getting supplies there. From Selma, the main assembly point, supplies went to Meridian, Mississippi, then up the Mobile and Ohio as far as Corinth; were there transferred to the Memphis and Charleston, and sent to Tuscumbia. Despite the fact that these lines ought to have been in working shape, they were not. The Memphis and Charleston was not put in running order until November 21, and Hood, who moved his army to Tuscumbia October 30, had to wait. His movement from Gadsden to Tuscumbia consumed most of what he had obtained at Gadsden—there was no alternative but to wait for stores.[40]

He was able to accumulate enough supplies by this route before the rail line was repaired, and despite muddy roads, to move across the Tennessee on November 21.

Rations and shoes were Hood's concern at Tuscumbia.[41] His main supplies of ordnance had already been sent to him at Newnan and Jacksonville. Reserves would come from Selma. After he had obtained supplies at Gadsden on October 20–21, the Ordnance Bureau could feel reasonably sure that it had done its best by him.

Hood's campaign was a failure—costly and bloody. His only chance of success lay in a speedy movement into Tennessee. And his fateful delay at Tuscumbia allowed the enemy ample time to prepare for his coming. Although want of cavalry and anxiety over Sherman's movements have been adduced as reasons for the delay,[42] Hood's logistical problems certainly loom large as a cause.

What of Hood's part in these logistical problems? Was he a victim of circumstances?

He blundered in August, 1864, by requisitioning the Negroes employed in the Macon ordnance plants. This mistake might well be excused on the grounds of ignorance, but the assignment of skilled ordnance technicians as reinforcements for Hardee at Jonesboro merits no such charity. He had been told of the havoc created in Macon and other ordnance cities by such interruptions.

Confusion might be offered as a reason for the loss of the Army of Tennessee's reserve ordnance at Atlanta. Nevertheless, this is no excuse. Hood, writing long after the event, placed the responsibility on his chief quartermaster.[43] This scapegoat had been provided by a Court of Inquiry appointed to assess blame for the loss of stores at Atlanta. After sifting the evidence, this body, Colonel M. B. McMicken, Hood's chief quartermaster, and Kennard, exonerated Kennard completely. The principal guilt was dumped in McMicken's lap—he failed "to comply with the specific and repeated instructions from the chief of staff . . . had at his disposal sufficient cars and engines to move all trains as ordered, and they were not so moved because proper instructions were not given by him to the railroad agents." The Court slightly censured Shoup for failing to see that his instructions to move the stores, issued on August 30, were carried out. Hood exonerated him in an indorsement to the Court's findings.[44] Hood was not mentioned in the proceedings. He is not blameless. Perhaps he was too busy with pressing matters to pay specific attention to the removal of the stores, but this hardly seems a legitimate excuse. Granted he trusted his chief of staff to carry out this operation, still it would not seem unreasonable to expect the army commander to inquire specifically about such an important detail. A recent biographer of Hood observes that the loss of these stores was "another example of the poor staff work in evidence throughout the Atlanta campaign."[45] Agreed, but it must be added that Hood's apparent indifference contributed to this situation. Closer supervision from him would have insured at least a bit more energy in his staff.

Once having committed himself to the Tennessee venture, Hood does not appear to have grasped the overwhelming difficulties involved in supplying his needs for this campaign. The shift of the army from Jonesboro to Palmetto was perhaps tactically correct. Logistically it was awkward. Hood did not change his base immediately; consequently his line of communication ran from Jonesboro to Macon, thence to Columbus, Georgia, and Opelika, Alabama. The West Point and Montgomery railroad ran through the latter town, which connected, in turn, with the Atlanta and West Point at West Point, Georgia. Here was established Hood's intermediate base after the move to Palmetto. Ordinarily this would have been ideal, since the Atlanta and West Point ran through Palmetto. But Hood ordered the track torn up for a distance of forty miles from Atlanta. As a result his railhead was Newnan, fourteen miles southwest of Palmetto. The gap was bridged by wagon trains.[46]

When Hood did change his base he chose Jacksonville, Alabama. This seemed the proper place if he contemplated operating on Sherman's communications as far as Chattanooga, or if he planned a swift thrust across the Tennessee River at Guntersville. Jacksonville, as the head of the Meridian to Blue Mountain line, was the nearest railhead, but it was far from ideal. A decision to rely on this railroad involved the transfer of the depot supplies from Jonesboro, as well as transportation of future supplies from the Georgia arsenals. This would not have been too difficult had the rail connections from Georgia been continuous to Selma. Such was not the case. Selma, which must serve as the new assembly base for the army, was connected to Montgomery by a steamboat on the Alabama River. This gap, of course, necessitated breaking bulk and reshipping—and delay. Montgomery was connected with Georgia by rail well enough, but available rolling stock really was inadequate for rapid transportation, even had there been no break in the route.[47]

Hood came no nearer Jacksonville than Gadsden. The eighteen to twenty miles between towns were bridged again by the army's overworked wagons. Jacksonville became unprofitable as a base after Hood decided to move west toward Decatur and Tuscumbia. A new base was located at the latter town. This could only be stocked by devious rail connections. The plan was to collect supplies at several places along the Mobile and Ohio railroad in Mississippi and Alabama. They were to be transported to Corinth in north Mississippi, put on the Memphis and Charleston road and sent east to Tuscumbia. Even under the most ideal conditions a poor arrangement, particularly since the arsenals at Selma and Demopolis lay to the east of Meridian, on the Meridian to Blue Mountain, and not on the Mobile and Ohio. Conditions, moreover, were not ideal. Hood had, on October 8, asked that the railroad from Corinth be repaired to Decatur, anticipating his possible needs, and as an alternative route should it become necessary. And though Forrest was able to protect the line as far east as Cherokee Station, its serviceability ended there. A fifteen mile expanse of wrecked track lay between Cherokee Station and Tuscumbia, which was not fully replaced until November 21. "It was thus necessary to transfer all shipments to wagons, which then had to be hauled over a country road, which in clear weather was none too good and which became a quagmire as soon as the rains began."[48]

In fairness to Hood it should be stressed that he had given indication of a need to use the Memphis and Charleston road and he cannot be blamed for the repairs not being completed in time.

But Hood is perhaps guilty of an even greater misjudgment. Beauregard conferred with him at Decatur in late October and hoped he would move immediately into Tennessee, not far from there.[49] But he went on to Tuscumbia, pleading that he had not supplies enough to go into Middle Tennessee. These he must accumulate at his base.[50]

This seems indeed a valid reason for delay, but Hood was not thus making his problem easier. Although it is generally considered advantageous for an army to operate close to its base, Hood's position was peculiar in that this was untrue in his situation. The longer he remained at Tuscumbia, the more taxing it was on the Confederate supply bureaus to keep him equipped. His attenuated line of communications to his sources of supplies was unequal to a prolonged effort, and certainly to a rapid and efficient effort.[51]

An authority on Hood's Tennessee campaign has observed that his delay at Tuscumbia is militarily puzzling; it was "not due so much to lack of supplies, to the absence of Forrest, and to the necessity for repairing the railroad, as it was to anxiety concerning what Sherman would do." Without this anxiety to hold him back, he "would have advanced sooner than he did, at least by the 7th of November."[52] It is equally true that had he been thoroughly conscious of his logistical position there can be scarcely any doubt that he would have advanced sooner. Sherman had returned to Atlanta, and the Jacksonville line of communication would have been relatively safe. Hood might well have risked advancing at some point east of Tuscumbia to relieve the strain on the railroads.

Perhaps politically and tactically Hood's march into Tennessee was the best possible maneuver. But from a supply standpoint it left the heart of the Confederate Ordnance Bureau brutally exposed while imposing unnatural strain on the arteries carrying equipment to the army. The laboratories at Macon suspended operations in the face of Sherman's troops and were evacuated by November 17;[53] by December Columbus Arsenal was closed down,[54] and Savannah was lost, with all its ordnance and stores. Columbia Arsenal was destroyed in February, 1865, and Fayetteville, North Carolina, Arsenal and Armory followed in March. A relatively unopposed Sherman had wrecked the Ordnance Bureau.

From a logistical standpoint the Tennessee campaign was a catastrophe. It would have been infinitely better to keep the Confederate Army between Sherman and the arsenals and close to its supplies. Yet, militarily this appeared impossible.

Hood cannot be censured alone for the decision to invade Tennessee. Others in higher places shared that decision. In carrying out the plan, however, it must be concluded that although determined and reckless in battle, he was, sadly enough, an irresponsible logistician.

Notes

"General Hood as Logistician" first appeared in *Military Affairs* 16 (Spring 1952): 1–11, and is reprinted with permission of the *Journal of Military History* (formerly *Military Affairs*).

1. For the background to Hood's appointment see John P. Dyer, *The Gallant Hood* (Indianapolis and New York, 1950), 243–44.
2. Frederick II, *Instructions for his Generals,* translated by Brig. Gen. Thomas R. Phillips (Harrisburg, PA, 1944).
3. See Col. Julius A. de Lagnel to Col. Hypolite Oladowski, Richmond, Apr. 4, 1864, Manuscripts Collection, series 50, vol. 111, Mississippi Department of Archives and History, Jackson. Hereinafter cited as Mississippi Manuscripts Collection. In this letter de Lagnel, assistant to the Confederate Chief of Ordnance, observed to the chief ordnance officer of the Army of Tennessee that ammunition distribution among the army's different corps was unequal. See also de Lagnel to Col. M. H. Wright, Richmond, Apr. 4, 1864, in Personal Service File of J. A. de Lagnel, Compiled Service Records of Confederate General and Staff Officers, and Non-Regimental Enlisted Men, National Archives Microcopy 331, Roll 74 (hereinafter cited as CSR M331). This letter includes a chart of ammunition distribution in the Army of Tennessee.
4. Ordnance Circular, Richmond, Mar. 31, 1863, in Mississippi Manuscripts Collection, series 50, vol. 111.

5. De Lagnel to Col. M. H. Wright, Columbus, GA, June 22, 28, 1864, Personal Service File of De Lagnel, CSR M331, Roll 74; De Lagnel to Wright, June 24, advising of shipment of 190,000 cartridges, ibid.
6. Col. John W. Mallet to Col. Josiah Gorgas, telegram, Macon, July 18, 1864, Telegrams Sent, Superintendent of Laboratories, Oct. 1863–Apr. 1865, chap. 4, vol. 52:15, Records of Ordnance Establishments, Record Group 109, War Department Collection of Confederate Records, National Archives and Records Service, Washington (hereinafter cited as RG109); James H. Burton to Gorgas, Macon, July 26, 1864, Letters Sent by the Superintendent of the Armory, June 1862–Apr. 1865, chap. 4, vol. 31:511, ibid.
7. Josiah Gorgas, *The Civil War Diary of General Josiah Gorgas,* ed. Frank E. Vandiver (University, AL, 1947), 90, 91.
8. Mallet to Gorges, Macon, July 19, 1864, Telegrams Sent, chap. 4, vol. 52:16, RG109.
9. Gorgas to Mallet (telegram), Richmond, July 22, 1864, in Personal Service File of John W. Mallett, CSR M331, Roll 162. Mallet was instructed to suspend all other work in favor of ammunition production.
10. Burton to Gorgas, Macon, July 26, 1864, Letters Sent, chap. 4, vol. 31:511, RG109.
11. Burton to Gorgas, Macon, June 24, 1864, ibid., 460.
12. Burton to Gorgas, Macon, July 1, 1864, ibid., 471. The Negroes were returned on July 2 by order of Brig. Gen. Marcus J. Wright, commanding the Post of Atlanta. See ibid., 476.
13. Wright to de Lagnel (telegram), Macon, Aug. 4, 1864, Telegrams Sent, Arsenal, chap. 4, vol. 101:309, RG109. For Wright's leaving Macon see "Circular Order," Macon, Aug. 19, 1864, Letters and Telegrams Received, Superintendent of Laboratories, chap. 4, vol. 37:60, ibid.
14. U.S. War Department, *The War of the Rebellion: A Compilation of the Official Records of the Union and Confederate Armies,* 128 vols. (Washington, DC, 1880–1901), series 1, vol. 52, pt. 2:590. Hereinafter cited as *OR.* All references are to Series 1 unless otherwise indicated.
15. Burton to Gorgas, Macon, Aug. 12, 1864. Letters Send, chap. 4, vol. 31:528, RG109.
16. Gorgas to Maj. Nathaniel R. Chambliss, Richmond, Aug. 31, 1864, Mississippi Manuscripts Collection, series 50, vol. 111.
17. Mallet to Gorgas (telegram), Macon, Aug. 26, 1864, Letters Sent, chap. 4, vol. 52:20, RG109; Col. W. L. Broun to Mallet (telegram), Richmond, Aug. 26, 1864, Letters and Telegrams Received, chap. IV, vol. 37:97, ibid.; Maj. J. T. Trezevant to Mallet, Richmond, July 25, 1864, vol. 37:53, ibid. The last reference is a list of "leading" ordnance stores received at Columbia, SC, Arsenal from Wilmington, NC, in June and July, 1864.
18. *OR,* vol. 38, pt. 5:939; Burton to Gorgas, Macon, Aug. 3, 1864, Letters Sent by the Superintendent, chap. 4, vol. 31:520, RG109.

19. In Mississippi Manuscripts Collection, series 50, vol. 111.

20. Burton to Gorgas, Macon, Aug. 29, Sept. 5, 10, 1864, Letters Sent by the Superintendent, chap. 4, vol. 29:27, 36–37, 41, RG109.

21. Mallet to Gorgas (telegram), Macon, Sept. 5, 1864, in Personal Service File of John W. Mallet, CSR M331, Roll 162. Another, slightly different version, is in Letters Sent, chap. 4, vol. 52:21, RG109.

22. This idea seemed to grow with time. See Dyer, *Hood,* 270. Contemporary reports of the disaster reflected uncertainty by equivocation. The Chief of Ordnance apparently was aware that all the cars did not contain ordnance. See Gorgas, *Diary,* 140. The southern press made the distinction also. See Mobile *Advertiser and Register,* Sept. 21, 1864. General Hood probably contributed to this distortion in his memoirs. He wrote that on the night of September 2 his chief quartermaster "grossly neglected to send off a train of ordnance stores and five engines. . . . This . . . entailed the unnecessary loss of these stores, engines and about eighty cars." See Hood's article, excerpted from his memoirs, in Robert U. Johnson and Clarence C. Buel, eds., *Battles and Leaders of the Civil War,* 4 vols. (New York, 1884–1888), 4:344.

23. *OR,* vol. 38, pt. 3:992. For a partial list of the stores lost at Atlanta, see ibid., 685–86. Hood's regular ordnance train was under Hardee's care at Jonesboro. See ibid., 701.

24. See Frank E. Vandiver, *Ploughshares into Swords: Josiah Gorgas and Confederate Ordnance* (Austin, 1952), 216.

25. Letters and Telegrams Sent by W. H. McMain, Military Storekeeper of Ordnance, Apr. 1862–Apr. 1865, chap. 4, vol. 27:137, RG109.

26. Mallet to Gorgas (telegram), Macon, Sept. 12, 1864, Telegrams Sent, vol. 52:23, ibid. Rains had only enough lead to satisfy Augusta's wants for five weeks. See also Rains to Kennard, Augusta, Sept. 27, 1864, Mississippi Manuscripts Collection, series 50, vol. 111.

27. *OR,* vol. 39, pt. 1:801. See also Dyer, *Hood,* 271–72.

28. *OR,* vol. 39, pt. 1:805. For accounts of Brig. Gen. Edward M. McCook's raid against the Atlanta and West Point railroad, see ibid., vol. 38, pt. 3:688, 689, 955, 962–63, 972–73.

29. *OR,* vol. 39, pt. 2:847–48. Also endorsements by Bragg, James A. Seddon, Gorgas, and Col. Isaac M. St. John.

30. De Lagnel to Kennard, Richmond, Sept. 21, 1864, Mississippi Manuscripts Collection, series 50, vol. 111; Gorgas to Kennard (telegram), Richmond, Sept. 22, 1864, ibid., series E, vol. 64.

31. Oladowski to Kennard, Columbus, Sept. 22, 1864, Mississippi Manuscripts Collection, series E, vol. 64.

32. Gorgas to Kennard (telegram), Richmond, Sept. 23, 1864, ibid.

33. Gorgas to Kennard, Richmond, Sept. 26, 1864, ibid.

34. Rains to Kennard, Aug., Sept. 27, 1864, Mississippi Manuscripts Collection, series 50, vol. 111.

35. Cuyler to Brown (telegrams), Macon, Sept. 27, 28, 1864, Telegrams Sent, Arsenal, chap. 4, vol. 101:363, 368, RG109; Cuyler to Kennard (telegrams), Macon, Sept. 27, 28, 1864, ibid., 364, 365, 366; Cuyler to Gorgas (telegrams), Macon, Sept. 28, Oct. 3, 1864, ibid., 367, 369; Vandiver, *Ploughshares into Swords,* 219–20,

36. Trezevant to Kennard (telegram), Columbia, Sept. 28, 1864, Mississippi Manuscripts Collection, series E, vol. 64; Humphries to Kennard, West Point, Sept. 28, 1864, ibid.; Cuyler to Humphries (telegram), Macon, Sept. 28, 1864, ibid.

37. Tyler to Hood, West Point, Sept. 28, 1864, Mississippi Manuscripts Collection, series E, vol. 64.

38. See *OR,* vol. 39, pt. 1:796, 801, 805; ibid., vol. 45, pt. 1:659.

39. *OR,* vol. 39, pt. 1:802, 807; ibid., vol. 45, pt. 1:659.

40. *OR,* vol. 39, pt. 1:796–97; ibid., vol. 45, pt. 1:651; Thomas R. Hay, *Hood's Tennessee Campaign* (New York, 1929), 59, 60, 61, 62.

41. Hay, *Hood's Tennessee Campaign,* 61, 62, 64.

42. Ibid., 65.

43. See *supra,* note 22.

44. *OR,* vol. 38, pt. 3:992.

45. Dyer, *Hood,* 270.

46. *OR,* vol. 39, pt. 1:801, 805.

47. Ibid., ser. 4, vol. 3:733–34.

48. Ibid., ser. 1, vol. 39, pt. 1:797, 802; Hay, *Hood's Tennessee Campaign,* 61, 62.

49. The Decatur crossing of the Tennessee was too well defended and Hood hoped to cross at Lamb's Ferry or Bainbridge—these failing, then at Tuscumbia. *OR,* vol. 45, pt. 1:648; Hay, *Hood's Tennessee Campaign,* 61.

50. Hay, *Hood's Tennessee Campaign,* 61.

51. Hood's ordnance replacements were being drawn from Georgia as late as mid-November. See Wright to Mallet, Columbus, Nov. 15, 1864, Letters and Telegrams Received, chap. 4, vol. 38:328, RG109.

52. Hay, *Hood's Tennessee Campaign,* 65.

53. Mallet to Gorgas (telegrams), Macon, Nov. 17, 21, 1864, Letters Sent, chap. 4, vol. 52:30, 31, RG109. Some of the Macon Armory machinery was sent away at the same time. See Burton to Gorgas, Dec. 7, 1864, Inventory of Tools and Machines at the Arsenal, chap. 4, vol. 39:165, ibid.

54. *OR,* vol. 45, pt. 2:704.

Lieutenant General William Joseph Hardee. Library of Congress.

Hardee's Defense of Savannah

Nathaniel C. Hughes Jr.

AMONG THE EPISODES OF THE CIVIL WAR WHICH HAVE BEEN CONSIGNED TO LIMBO stands the capture of Savannah by Sherman in the winter of 1864. Chronologically fixed midway between the fall of Atlanta and the surrender at Appomattox, the capture of Savannah lacks the desperation of the former and the heart-twisting pathos of the latter. Yet this episode merits continuing attention because it gives insight into the strategic and operational mind of one of the greatest Union commanders. Moreover, for the student of Confederate history, the siege of Savannah provides an opportunity to observe an able Confederate commander, Lieutenant General William Joseph Hardee. In this action Hardee acted not in his customary subordinate capacity, but in direct operational control of a sizable number of troops. His assignment was to hold the city of Savannah against Sherman advancing from Atlanta with 60,000 experienced troops.

In a postwar letter to the Georgia historian, Charles C. Jones, Hardee stated that "there is no part of my military life to which I look back with so much satisfaction."[1] How could this veteran of Shiloh, Murfreesboro, and Atlanta make such a statement? At Savannah he evacuated one of the most important Confederate ports (and Georgia's largest city) with a considerable loss of war material. This study will endeavor to answer this question and to offer an appraisal of Hardee's performance.

In September 1864, following his release from the Army of Tennessee, Hardee assumed command of the Department of South Carolina, Georgia, and Florida. He spent that October and November strengthening the defenses of Charleston and organizing his 12,000 troops, scattered along the south Atlantic seaboard. While General Hardee adjusted himself to the problems and personnel of his geographic command, Hood and Sherman maneuvered across North Georgia. When Hood finally broke off contact with the enemy and moved his army into Alabama, Sherman reacted by dividing his army, assigning the able George H. Thomas to watch Hood with one-third of the command, while he, himself, grouped the four remaining corps around Atlanta.

On November 16, 1864, Sherman's army departed Atlanta and disappeared into central Georgia. Sherman had decided to "strike out for Savannah," destroying as he went the military resources and the morale of the Confederate "heartland." With a base established at Savannah, he felt confident that a new vista of military possibilities would be opened. Sherman's objective was unknown to the Confederate commanders and they spent the following month ineffectually opposing his advance. In desperation the Confederate War Department combined the inactive forces in Augusta, Georgia under Bragg with Hardee's troops in front of Sherman. Bragg was given command of both forces and ordered to halt Sherman. Bragg accepted the task but declared "that no practicable combination of my available men can avert disaster."[2] With Bragg in command of the theater, Hardee's activities and responsibilities were confined to the Savannah area. At Savannah Hardee would act independently until he came under P. G. T. Beauregard's control in December.

Panic gripped the Savannah area during the closing days of November 1864. Places that had not seen a blue uniform since 1860 and that would not see another until after the surrender were reported occupied. Rumors had Sherman's columns crossing every stream and menacing every town in southeastern Georgia. Hardee found it trying to sift information under such circumstances. Even Joseph Wheeler, his reliable source of information in the first three and a half years of the war, seemed unable to penetrate the heavy curtain of foragers that surrounded Sherman's army. Behind these "bummers" ranged Union cavalry patrols that insulated the main columns and concealed the actual strength behind the screen. The general direction of Sherman's advance, however, could not be hidden for long.

Anticipating an attack on Savannah, Hardee set to work preparing the city's defenses. As early as November 20, he had transferred guns from the coastal fortifications to the western side of the city.[3] He now ordered all able-bodied men in the city to report for duty. Soldiers on furlough and assignment together with convalescents were organized into battalions under Major General Lafayette McLaws. A few units from the north, including companies of the veteran 1st North Carolina Artillery, arrived and were put to work on the trenches. Hardee arranged with his navy commander, W. W. Hunter, to use the gunboat *Macon* to patrol the Savannah River. He assigned Wheeler the task of destroying everything the enemy could use between the Ocmulgee River and the city.

As soon as all the roads leading to Savannah had been obstructed, Hardee released Wheeler from his position in front of Sherman and ordered him to operate on the enemy's flanks and rear. Thus Hardee sacrificed the buffer between Sherman's army and his own infantry, but it meant that Wheeler would not be confined in the narrow, marshy terrain around Savannah.[4]

Amid the calamity reports pouring into headquarters came one dispatch that gave Hardee serious concern. A lonely Confederate sentinel on the bank of the Broad River sat eating sweet potatoes that he had just fried for his breakfast. The river that he guarded lay just above Savannah and led directly inland to the Savannah & Charleston Railroad that constituted Hardee's line of communications with the rest of the Confederacy. As the sentinel peered into the heavy fog that had hung over the river several large enemy steamers suddenly loomed out of the haze. They headed up river. The astonished picket hurried off leaving his breakfast untouched.[5] These steamers brought the division of Brigadier General John P. Hatch, part of Major General John G. Foster's Union forces at Port Royal. Hatch disembarked about 5,000 men at Boyd's Landing and advanced inland toward the railroad. To oppose them was only one cavalry regiment.[6]

Hardee responded to this threat to his communications by ordering Major General Sam Jones at Charleston to Grahamville and Pocotaligo with two Georgia infantry regiments.[7] After dispatching Jones to the threatened point, Hardee wrote to Secretary of War Seddon, "As railroad and telegraphic communications may soon be cut with Charleston I desire you to know that I have, including the local troops, less than 1,000 men of all arms. General [Gustavus W.] Smith is expected with 3,200 men, but has not yet arrived. If railroad communication is cut with Charleston, which is threatened by ten gunboats and barges, of course no reinforcements can be sent from Augusta."[8] The fate of Savannah depended upon blocking Hatch's advance. Hardee wired Smith to take the first two train loads of Georgia militia through Savannah to Grahamville, and to "drive the enemy back to their gun-boats."[9]

General Smith and his sleeping soldiers pulled into Savannah from Macon at 2:00 A.M., November 30. Smith received Hardee's dispatch at the railroad depot. He left his men asleep and went directly to Hardee's headquarters. Smith awakened his commander and said, "If you can satisfy me that it is absolutely necessary that my command shall go into South Carolina [the enemy landing had occurred across the river] I will endeavor to carry out your orders. If you do not satisfy me . . . I will be under the disagreeable necessity of withdrawing the State forces from your control." With the help of a map and a handful of dispatches, Hardee quickly satisfied Smith that it was necessary.[10] While Smith's troops hurried into South Carolina, Hardee impatiently wired Jones asking why his troops had not left Charleston. Where were the troops from Augusta?[11]

At midmorning on November 30, Smith reported to Hardee that he had engaged the enemy at Honey Hill and that he needed reinforcements. The Charleston troops had not arrived. Later in the day Hardee anxiously opened another dispatch from Smith. It proved to be one of the most welcome reports that he received during the war. Smith sent the news that the enemy had been repulsed.[12]

As soon as he learned that reinforcements were nearing Grahamville, Hardee left his headquarters and went to join Smith. He found the enemy "badly whipped" and allowed Smith to take his troops back to Savannah. The enemy advanced again on December 2, but retired without a battle. Hardee left the defense of the railroad to Jones and returned to Savannah.

To keep the situation under control Hardee needed about 3,000 troops. Bragg sent word on December 3 that he was dispatching 10,000 men to Hardee's assistance. If he had sent 10,000, Hardee could have forgotten about his exposed communications. The few thousand that Bragg did send were needed in the lines about Savannah while the Pocotaligo and Grahamville situation remained in Jones's hands. Jones expressed dismay when he inspected his motley command of two Georgia regiments, some reserves, and several battalions from other units.[13]

Jones might well have been dismayed, but the situation could not have been remedied without more assistance from Augusta or Richmond. Hardee worried more about Sherman's 60,000 troops already around Millen and the Augusta Railroad.

To slow Sherman's advance Hardee sent McLaws with a brigade of Georgia militia and Brigadier General Laurence S. Baker's North Carolina troops to Station Number 4 1/2 on the Georgia Central Railroad. McLaws found the position untenable and fell back, with Hardee's approval, to Station Number 1 1/2, about four miles from the city.[14]

Sherman drew closer to Savannah on December 5–7. The enemy who had crossed the Ogeechee River forced the abandonment of the outposts on the Augusta Railroad.[15] Hardee now brought the navy into play, ordering the *Macon* and a floating battery to protect the railroad bridge across the Savannah River.[16] In Hardee's rear at Grahamville trouble also developed as the enemy secured a lodgment near the railroad. Jones attempted to drive the enemy from their new positions but failed.[17] Hardee urged Jones to encourage his men and to try to drive the enemy back. He needed a fighting commander at Grahamville as badly as he needed more troops.

Beauregard in Charleston displayed alarm at the deterioration of the situation as revealed in Hardee's telegrams. He asked both Hardee and Jones to come up for a conference. Hardee replied that he and Jones could not come "without injury to the service" and suggested that Beauregard come to Savannah. Later on December 8, Hardee wired Beauregard stating, "I hope you will not fail to come here tonight. It is all important that I should confer with you."[18]

Before leaving Charleston for Savannah, Beauregard sent Hardee a wire that set forth the guiding principle for the defense of Savannah: "Having no army of relief to look to, and your forces being essential to the defense of Georgia and South Carolina, whenever you shall have to select between their safety and that of Savannah, sacrifice the latter, and form a junction with General Jones, holding the left bank of the Savannah River and the railroad to this place as long as possible."[19]

While waiting for Beauregard's arrival Hardee received from Wheeler an intercepted enemy message disclosing Sherman's planned dispositions for the direct investment of Savannah. This relieved Hardee's worries about Sherman swinging north to cut his communications with Charleston. Now certain of the enemy's objective and dispositions, Hardee immediately ordered all able-bodied men in Savannah to the trenches and began moving his troops to the main defense line located about two and a half miles from the city. He urged the people of the city to send him their spades, axes, and other tools. The mayor of Savannah strongly backed the request.[20]

Beauregard arrived early on December 9 and spent most of the day conferring with Hardee who informed him that the enemy was about six miles from his intermediary line which protected the Charleston & Savannah Railroad and its bridge across the Savannah River. Sherman's army was advancing in separate columns down the Middle Ground Road, the River Road, and the Augusta Railroad, with 35,000 to 40,000 men, the bulk of his army of 60,000. The main body of Hardee's 10,000 troops occupied the main works while Brigadier General Hugh W. Mercer and Adjutant General Henry Wayne delayed Sherman. Hardee expected the enemy to strike the main line on December 9 or 10. As for the defenses on the north side of the Savannah River, Hardee did not place troops there, believing that the navy would deter the enemy. If the Federals did succeed in crossing the river the boggy rice fields would prevent a rapid strike in force. Beauregard asked Hardee about his plans for evacuation. Hardee told him that none had been made since he relied on the gunboats in Savannah harbor to ferry his troops across to South Carolina. Beauregard thereupon directed Hardee to begin at once the construction of pontoon bridges across the Savannah. After reiterating his order giving the safety of the army priority over that of the city and cautioning Hardee to look to his communications, Beauregard left the city to visit Jones at Pocotaligo.[21]

The intermediate line of which Hardee spoke collapsed before Beauregard left the city. Old and infirm General Hugh Mercer, who had commanded one of Hardee's brigades at Atlanta, had no sooner occupied this line of detached works behind the Monteith Swamp than the enemy outflanked him and rushed toward the main defense line.[22] The two armies had at last made contact.

The armies of Sherman and Hardee confronted each other on a peninsula about thirteen miles wide, bordered on the north by the Savannah River and on the south by the Little Ogeechee River. Below the Little Ogeechee existed "a natural barrier, consisting of small rivers, creeks, and impassable swamps."[23] The peninsula was cut and cross-cut by innumerable creeks and marshes. Hardee availed himself of the advantages of the terrain. His line followed a series of creeks from the Savannah south to the Little Ogeechee.

All approaches to Savannah from the west, and there were only five, must cross this line. Hardee first obstructed these approaches, then he had all canals and rice dikes cut. When they had been cut, he opened the sluices and flooded the fields. "Thus the entire front of the Confederate line . . . was submerged to a depth varying from three to six feet."[24]

The advantage of terrain could only assist, not replace, the human barrier. Opposed to Sherman's 60,000 troops were only 10,000 Confederates divided into divisions under Smith, McLaws, and Major General A. R. "Rans" Wright. Wright had replaced Mercer on December 10, when the siege proper opened. Wright's presence promised experienced, capable combat leadership. As President of the Georgia Senate, he had taken command of the Georgia forces east of the Oconee River after the fall of Milledgeville. Wright held the left of Hardee's line with a scrambled force of veterans, workers, clerks, militia, and locals. His line, mounting thirty-two guns, extended for seven miles from the Little Ogeechee River to Shaw's Dam.

McLaws held the center. Major General McLaws also had ample combat experience in Virginia, but "his record as a divisional commander had not been one of uniform promptness and of average success."[25] This professional soldier and fellow Georgian had come into the department after he had been relieved of command at Knoxville. At Savannah he commanded the best troops and held the key position. Among the 4,000 tried troops holding his four miles of trenches from Shaw's Dam to the Georgia Central Railroad crossing was the famous Orphan Brigade from Kentucky. These men knew Hardee well, and he knew that they could be relied on. As their spokesman Johnny Green put it, "We have seen a great deal of service with [Hardee] and the men all admire him."[26]

Gustavus W. Smith held the right. Smith and Hardee had been associated during the battles around Atlanta when Smith commanded a division of Georgia militia. Smith was known in the South as a one-time engineer, manufacturer, politician, wing commander, and acting secretary of war. His military competence was suspect. His part of the line at Savannah consisted of two and a half miles of trenches stretching from the railroad crossing to the Savannah River. To hold this line he had about 2,000 militia and twenty guns.[27]

Generally Hardee's defense line centered about a tandem of earth redoubts that commanded likely avenues of approach. One or more forts had been advanced to give enfilade fire down the line. As much as possible Hardee had these forts connected with rifle trenches, two or more in depth. At points where the five main causeways entered his defenses Hardee placed twenty-four- and thirty-two-pound cannon that easily outmatched the light Union artillery. He had drawn this heavy ordnance from the wealth of guns protecting Savannah's seaward side.[28] Savannah, like Charleston, had successfully withstood attack throughout the war. Its coastal defenses were the work of the Confederate Department of Engineers and contained an enormous amount of firepower from three mutually supporting defense lines. The entrance to the Savannah River was impassable because of the obstructions and network of guns.[29]

When both wings of Sherman's army closed in on his works, Hardee withdrew the outposts defending the Savannah & Charleston Railroad bridge and had the bridge destroyed. By this action he cut off Savannah from her normal and most important means of communication. Hardee reported to Beauregard that the enemy was all along his front and that skirmishing had begun in earnest. "I have not a reserve."[30] Hardee tried to manufacture some corps reserves by having McLaws withdraw a Georgia regiment from his line hoping to replace it with more artillery.[31] Seeking a more realistic reserve Hardee requested that a Georgia regiment serving with Sam Jones be sent to him.[32] Beauregard, who had taken over supervision of Jones's command, tried to procure the troops for Hardee in South Carolina, but could not take away Jones's troops at the moment for he was planning an attack on Hatch, an attack which never materialized. The following day, December 11, Hardee again called on Beauregard for troops: "I have been obliged to extend my lines. It is impossible to hold it without immediate reinforcement."[33] Hardee did receive some help at this time by the acquisition of E. C. Anderson's detached cavalry regiment and by ordering Brigadier General S. W. Ferguson's cavalry into the trenches.[34]

On December 10 and 11, skirmishing continued and minor attacks were repelled at Shaw's Dam, Fort Hardeman, and at Williamson's. To strengthen his right flank at Fort Hardeman and to provide cover in the event of evacuation over the river, Hardee ordered back the *Macon* and the *Sampson* from their patrol duty. On their return down the winding Savannah, however, Federal batteries surprised them, disabled and captured the tender *Resolute* and forced the two larger ships back up the river.[35] This action disquieted Hardee. It deprived him of the *Sampson*, which he intended to station at the right flank of his line to deliver supporting fire, but more important it meant that Sherman had challenged his control of the river and might send troops to the north bank of the Savannah. In anticipation of this move Hardee ordered Wheeler to shift his cavalry from Sherman's rear to the South Carolina side of the Savannah and establish his headquarters at Hardeeville, South Carolina. Thus situated Wheeler somehow was to prevent the enemy from crossing in force.[36]

By this time little doubt existed that Savannah would ultimately have to be evacuated but no threat as yet rendered immediate evacuation imperative. The nature of Sherman's previous campaigns indicated that he would adopt the surest methods first. At Savannah this meant that he would open a line of communications with his naval forces lying off the coast and with Foster's infantry at Hilton Head. An attack against Hardee's lines of breastworks, arranged in depth, seemed imprudent without an established base.

Sherman felt little apprehension that his quarry could escape quickly. The city had been isolated from the rest of the Confederacy except for a tenuous route by water across the Savannah, then by way of rice dikes to Hardeeville. Major General Henry Slocum had broken the Savannah and Charleston rail communications when he moved up against the right of Hardee's line. His position on the Savannah River also halted traffic on the river. Major General Oliver O. Howard had penetrated south to Flemingo, breaking the Gulf Railroad at that point.

The immediate obstacle preventing Sherman from opening his communications with the sea was Fort McAllister, an isolated bastion dominating the Great Ogeechee River. The army already felt the need for supplies. In the Union camps even hard crackers were selling for a dollar apiece when they could be found.[37]

Anxious to open his communications with the fleet, Sherman decided to move against Fort McAllister immediately. He directed Major General William B. Hazen with his division to attack the fort from the land side. About 250 men garrisoned Fort McAllister. They expected attack from the sea and their guns on barbette mounts pointed out over the river. When Hazen attacked on December 13 the defenders were quickly overcome. That evening the army opened communications with Admiral John Dahlgren's naval forces. Sherman was jubilant. One of his staff officers who disdained his commander's "cockiness" wrote, "[Sherman] says the city is his sure game and stretches out his am and claws his bony fingers in the air to illustrate how he has his grip on it."[38]

With Fort McAllister gone and with heavy Union guns coming ashore, Hardee began to feel those "bony fingers." He wrote President Jefferson Davis emphasizing the seriousness of the situation. Foster's force under Hatch menaced his flimsy line of communication and Sherman was now ready to strike. "Unless assured that force sufficient to keep open my communications can be sent me, I shall be compelled to evacuate Savannah."[39] Davis replied that because of the critical state of affairs in Virginia no troops could be sent.[40] Hardee also communicated his anxiety to Beauregard: "Our occupation of Savannah depends on your ability to hold the railroad. Whenever you are unable to hold the road I must evacuate. . . . Inform me instantly if Foster is reinforced by Sherman or otherwise. I feel uneasy about my communications."[41]

A new threat began to materialize north of the Savannah. Federals along the south bank of the Savannah began to jump in small numbers from island to island in the middle of the river searching for rice and better flank protection. A regiment was soon on Argyle Island and Wheeler, after repelling several sorties on the north bank, admitted reluctantly that a lodgment had been made. If this lodgment grew to brigade or division strength Hardee's last lifeline could be cut. Hardee also had serious trouble at his main defense line. A mutiny had occurred in the Foreign Battalion that he himself had recruited with care. Fortunately the mutiny exposed itself prematurely and was easily put down. Hardee dealt with the situation summarily, executing seven ringleaders on the spot and shipping the battalion immediately to Florence, South Carolina. In exasperation Hardee urged that similar efforts to enlist foreign troops be prohibited.[42]

On December 16 Hardee called his generals together for a council of war. Most of the generals agreed that Savannah should be evacuated as soon as the pontoon bridge had been completed. Brigadier General P. M. B. Young, fresh from the Army of Northern Virginia, was ordered by Hardee to collect rice flats along the Savannah. They would be used in the construction of the bridge. Young reminded Hardee that he had intended to attack the enemy lodged on the north side of the river the next morning. Hardee replied that the assault "was of no importance" compared to procuring the rice flats.[43] Young with difficulty gathered the flats and also managed to deliver an effective cavalry attack.[44]

While the Confederates attacked enemy fragments that had roamed across the river, Federal commanders prepared for the assault of Hardee's main line. They constructed fascines, stringers, and ladders, all the while skirmishing constantly to develop weak points in Hardee's line. Larger guns brought by boat from Hilton Head and the navy assured additional artillery support. Fully confident of success, Sherman demanded Hardee's surrender on December 17. He told Hardee that he had a supply base established and guns large enough to reduce the city. "I have for some days held and controlled every avenue by which the people and garrison of Savannah can be supplied." He would be willing to grant reasonable terms, but if forced to resort to assaults or a prolonged siege, he would "make little effort to restrain [his] army burning to avenge."[45]

Hardee waited a day before replying to Sherman. Beauregard had arrived and the two discussed plans for the evacuation and the subsequent disposition of the troops.[46] On December 19, Hardee in that frozen language that he could command so well, wrote to Sherman refusing to surrender. He stated that his two lines of defense still remained intact, that Sherman did not have any troops nearer than four miles to the city, and that in spite of Sherman's efforts, he still kept in touch daily with Charleston. In reply to Sherman's threat Hardee answered, "I have hitherto conducted the military operations entrusted to my direction in strict accordance with the rules of civilized warfare, and I should deeply regret the adoption of any course by you that may force me to deviate from them in the future."[47]

December 19, 1864, should have been evacuation day for the Confederate forces. Everything, however, depended on the completion of the pontoon bridge which had been delayed by fogs, ships running aground, and many other unpredictables. During the day the enemy attacked on A. R. Wright's front, but was repulsed.[48] On the north side of the river the enemy became aggressive as they grew in numbers. The Union regiments there attacked and drove back a Confederate cavalry brigade, establishing themselves at Izards—much closer to the Confederate line of retreat. Hardee crossed the river to observe the fighting. The situation looked critical and he wired General William Taliaferro in Charleston to come to Hardeeville with what men he had.[49] Hardee also sent Wheeler about 700 men and 6 guns. He cautioned Wheeler that "the road to Hardeeville must be kept open at all hazards."[50]

The successful evacuation of Savannah depended upon the pontoon bridge. Until the moment the formal siege began, or perhaps even later, Hardee seems to have entertained the idea of evacuating the garrison by boat.[51] The limited number of craft, accentuated by Major General Henry W. Slocum's cutting off the *Macon* and the *Sampson,* relegated this plan to an alternate status and Hardee gave priority to the completion of the bridge. The engineers fastened about thirty rice flats, seventy to eighty feet long, end to end and covered them with timber ripped from the Savannah wharves.[52] To supplement the inadequate engineer companies, Georgia militia and Confederate sailors provided the necessary working parties.[53] The bridge spanned the Savannah River in three sections. The first segment ran from the city to Hutchinson's Island, then across that island by causeway to the second bridge which spanned the middle of the river to Pennyworth Island. A road across Pennyworth Island connected with the third span which reached across the Black River to Screven's Ferry.[54] On December 13, when Fort McAllister fell, Hardee had completed the bridge only as far as Hutchinson's Island. When the enemy appeared in increasing numbers on the north side of the river, Hardee decided to build a floating dock on the north side of Hutchinson's Island and complete the crossing by ferry. He discarded this plan when Sherman did not press his riverhead advantage on the north side of the Savannah. Throughout the period of construction the Confederate navy protected and concealed the pontoon bridge from the enemy. It is plausible to assume, however, that Sherman knew that it was being built or at least expected it to be built. The engineers completed the bridge about nine o'clock on the night of December 19. Hardee would have evacuated Savannah that night if the bridge had been completed a few hours earlier.[55]

December 20, 1864, was a day of suspense for the Confederates. The men knew the city would be evacuated and anxiously awaited Hardee's order. The Confederates on the main defense line kept the Union soldiers pinned down in their trenches all day, with surplus ammunition they knew would have to be destroyed if it were not expended.[56] Early in the day Hardee visited the South Carolina shore and found that Colonel Ezra A. Carman's Union brigade had crossed the river and engaged Wheeler's cavalry. With his communication line in jeopardy, Hardee ordered Taliaferro to furnish Wheeler all the men he called for.[57] To further agitate the Confederate commander news arrived that advanced enemy units had penetrated to within three-fourths of a mile of the

Savannah & Charleston Railroad at Pocotaligo.[58] Fortunately for Hardee, the advance of Foster's troops lacked the finality that Sherman wished. On this very day Sherman had decided to reinforce Foster's troops with one of his divisions. The united force would then move on the Savannah & Charleston Railroad and place themselves athwart Hardee's escape route.[59]

Hardee knew only too well that he must act quickly to avert destruction of his command. His orders, based on the experience of the Atlanta Campaign, called for the customary night withdrawal. Prior to moving his troops and artillery, Hardee directed that all the army wagons and caissons be sent across the bridges.[60] A confidential circular outlined the troop movement. The light artillery would be pulled by hand to the bridges and sent across before the men left their trenches. Troops at outlying posts, such as Forts Jackson and Bartow, would assemble at 9 P.M. and move by steamer to Screven's Ferry. At the nearer forts, such as Rosedew and Beaulieu, the men would withdraw at dusk, march into the city and over the bridge. Hardee would begin the uncovering of the main line from the left with the troops having the longest distance to cover. This force, Wright's division, would pull out of the trenches at 8 P.M., McLaws at 10 P.M., and Smith at 11 P.M. The division skirmish lines would be strengthened and left in position for about two hours after their parent units had left. After the skirmishers had crossed, Colonel John G. Clarke of Hardee's staff would destroy the pontoon bridge. All powder to be abandoned would be destroyed by dousing it with water; ammunition would be dumped into the river, and the numerous heavy guns that must be left would be spiked at the exact time their division withdrew from the line.[61]

Hardee had worked out the plans for the disposition of the naval force with Beauregard on December 18. The *Isandiga* and *Firefly* would proceed up the river to join Commodore Hunter near Augusta. The *Georgia* would be scuttled; the *Savannah* would cover the evacuation, wait two days to protect the rear of the army and the stores at Screven's Ferry, and then proceed to sea.[62]

During the morning of December 20, Union troops on the South Carolina side of the Savannah witnessed a scene that they never forgot. Over the pontoons, as it seemed, came the entire civilian population of Savannah, "wagons, family carriages, men and women on foot. . . . The stream of fugitives and number of carriages and wagons increased as the day wore on."[63] To discourage Carman's troops from molesting this procession, Hardee strongly reinforced Wheeler's line just to the west of the line of retreat. Next he subjected the Union soldiers to a bombardment by the *Savannah* which steamed up to their flank.[64]

While the cavalry and navy held Carman in check, the infantry in the main line began their withdrawal on schedule. Preceded by forty-nine pieces of artillery, the three divisions retired without incident. The last, Smith's division, began crossing at 1 A.M. When Smith had passed over the bridge, rockets were fired to inform the forts down by the sea that the city had been evacuated.

Hardee, himself, with most of his staff crossed the river on the steamer *Swan* about 9 P.M. He left one staff officer behind with a detail "to preserve order" in Savannah until the last possible moment and to be certain that no skirmishers had been left behind. Hardee also left behind many of his sick and wounded because he could not provide transportation for them.[65]

Despite the precautions taken to insure a minimum of noise, the wind carried the grinding and the rumbling of the wagons up the river to the Union positions. Colonel Carman informed his superiors "but was instructed not to risk interference with the movement, as [the Confederates] would be cut off from above [by Foster]."[66] Many of the Confederates also expected to be cut off. There seems to have been some confusion, some units became tangled, and everyone appeared to be ill humored. It was a bitter moment. Many of the Savannah garrison were leaving not only their long-held military post but their homes as well. The fires of the burning ships gave the whole column a garish appearance. One Confederate noted that "the constant tread of the troops and the rumblings of the artillery as they poured over those long floating bridges was a sad sound, and by the glare of the large fires at the east of the bridge it seemed like an immense funeral procession stealing out of the city in the dead of night."[67]

The solemnity of the Confederate evacuation contrasted sharply with the exuberance of the triumphant Federals. Sherman demonstrated the attitude of his troops when he presented Savannah to President Abraham Lincoln by telegram as a Christmas present.[68] And the Confederates left some presents for Sherman. Although most of the movable war materials had been carried away, the heavy guns and equipment alone represented a rich prize. Over 200 guns fell into Union hands.[69] To their amazement Sherman's soldiers uncovered an even richer prize: at least 35,000 bales of cotton, negotiable anywhere at the best prices. This cotton caused the pulses of merchants and speculators throughout the North to quicken and brought droves of "seekers" to Savannah.[70]

Why did Hardee fail to destroy the cotton? This the Confederate House of Representatives demanded of Davis, and this in turn Davis demanded of Hardee.[71] The cotton was of great economic benefit to the enemy and the general who neglected to destroy it had a brother, Noble Andrew Hardee, who was a prominent Savannah cotton merchant. Hardee replied:

> The cotton was distributed throughout the city in cellars, garrets and warehouses, where it could not have been burnt without destroying the city. It had not been sent off by railroad previous to the cutting of the road, because railroad transportation was monopolized for removal of ordnance, commissary, and other important Government stores. From the cutting of the road to the

> evacuation of the city—twelve days—every man was required to work in the lines, and every wagon, dray and cart that could be impressed was needed to keep the troops in a line of twelve miles long supplied with ordnance and commissary stores. Not a man nor a woman could have been spared to collect the cotton in a place where it could have been burnt.[72]

When one reflects on the harried use of trains by the Confederates during the last few weeks when the road was in operation and the necessity of preventing the destruction of the town without alerting the enemy, Hardee's answer seems self-evident. That the financial interests of his brother influenced him is inconsistent with the scrupulous integrity with which Hardee had conducted his life heretofore. Furthermore, the commissary and ordnance stores salvaged from Savannah did more to prolong the life of Confederate military resistance than any amount of cotton could have done.

Hardee's role in the capture of Fort McAllister merits closer scrutiny as well. Brigadier General Josiah Gorgas and others criticized him for the sacrifice of this garrison. Why did Hardee leave this fort exposed and fail to recall the garrison in time? Or if Hardee knew, as he should have known, that the Great Ogeechee was Sherman's most feasible line of communication, why did he fail to strengthen the garrison and make a determined effort to hold the position? Charles C. Jones, the historian of the Savannah siege, believed that Hardee hoped that a "bold retention" of the fort might lead Sherman to hesitate, perhaps to give up the idea of establishing a base near Savannah or even of attacking Savannah. Jones admits, however, that this thin reasoning hardly justifies the loss of the garrison.[73] Two other factors should be considered. First, Hardee, who had an intimate knowledge of the terrain probably believed, as did others, that Sherman could not maneuver a large assaulting body through the supposedly impassable area around the Great Ogeechee. Here, as later, Hardee appears to have underestimated the ability of Sherman's men in overcoming natural obstacles. Second, and most important, Hardee probably had little choice. If he abandoned the fort, Sherman got his line of communication gratis; if he reinforced the fort he would weaken a line already bare of reserves and hardly strong enough to withstand a determined attack by a Union corps.

Could Sherman have captured Hardee's army in Savannah? It is doubtful that Sherman could have cut off this body of 10,000 men unless he had planted at least a strong army corps across Hardee's line of retreat after the siege had opened. To have attempted the movement earlier would have led to a rapid and fairly easy evasion by Hardee up the coast line by rail while Bragg applied pressure against Sherman's left flank. Sherman hesitated to place a body of men across Hardee's communications once the siege had begun for he feared that Hardee might crush an isolated force with his advantage of interior lines. Sherman also worried about the Confederate ships that commanded the vital stretch of the Savannah River and the soft, marshy ground just across the river on the South Carolina side.[74] Actually Sherman's primary objective appears to have been not Hardee's army but a base on the Atlantic. After he secured this base he steadily prepared for the destruction of his opponent. The real criticism of Sherman comes in the question of timing. Surely he and Foster could have arranged to attack the railroad at Pocotaligo in force sooner than they did. Hardee, however, with a large segment of his pontoon bridge completed and a sizable number of steamers on hand probably could have evacuated most of the garrison as early as December 17. The methodical Sherman risked the escape of his foe so that he might establish his base. Once it had been established he would be able to close in on Hardee with the number of variables reduced to a minimum.

Although credit for the successful evacuation must be given to Hardee, Beauregard's contributions should not be overlooked. Beauregard remained in general control of the operation throughout. He made the basic decision placing the army's welfare above the city's. He sketched the general plan for the line and method of retreat, checking constantly with Hardee to insure that the details had been worked out properly. Most important, perhaps, he succeeded in giving Hardee additional troops when they were so desperately needed and dispatched the men necessary to keep the line of retreat open.

In light of Beauregard's contributions and the conclusion that perhaps Sherman could not have cut off all of Hardee's garrison if he had made a supreme effort, how can Hardee be justified in stating, "Tho' compelled to evacuate the city, there is no part of my military life to which I look back with so much satisfaction."

Beauregard's contributions and Sherman's slim hope of success both were predicted on the assumption that Hardee had immediate operational control of the Savannah forces and would conduct the withdrawal in his proven excellence in this type of endeavor. As the commander of this garrison he went beyond his traditional successful retreat in the face of the enemy. He had to. Never had he operated so close to the enemy with such limited means in regard to the number and quality of his troops. He displayed a signal aptitude in improvisation and in the utilization of terrain and resources. Sherman's men marveled at the Confederate battery mounted on a railroad car that skipped from one front to another with celerity, firing accurately into exposed groups of men and generally harassing the Union army. To Hardee should be given credit for the novel technique of employing ships not only to hinder Union troop movements, but also to guard land features and to serve as mobile forts on the Confederate right flank. This close integration of naval and land forces represents one of the finest displays of joint operations by a Confederate commander during the war. The nature and location of

Hardee's main defense position demonstrates his skill in the advantageous use of terrain. He used the flooded rice fields and canals effectively to offset the disparity in numbers. Hardee's main line evoked admiring comments from friend and foe alike and gave great assurance to many untried defenders who badly needed it.

In assembling the Savannah garrison and giving it a combat efficiency quite disproportionate to its surface capabilities, Hardee displayed once again his talent for getting the maximum yield from his human resources. Through two weeks of constant skirmishing and periodic attacks both officers and men handled themselves effectively. Smith and Wright exhibited steadiness and consistency. McLaws and Wheeler behaved in a manner that rivaled better performances in the salad days of their careers. In building this effective garrison, a great deal must be attributed to Hardee's personal leadership. His confident and decisive manner pervaded all ranks, giving his militia, laborers, clerks, dismounted cavalry, and the others faith in themselves and the troops next to them.

To remove the large civilian group, the war material, and 10,000 men from the presence of 60,000 veteran troopers led by one of the ablest Civil War commanders, represents one of Hardee's finest achievements. After 10:30 P.M. December 20, one enemy assault in only division strength might have disrupted the evacuation and resulted in the capture of most of the garrison. To minimize this risk Hardee insured secrecy by the silent destruction of ordnance, muffled cannon wheels, and strong skirmish lines. The orders upon which the withdrawal was based were detailed and precise, providing for the most expeditious movement of troops. This scrupulous regard for proper planning prevented panic, disaster, and blundering that in one form or another occurred in the evacuations of Corinth, Nashville, New Orleans, and other cities. Thus in getting the maximum performance out of his nondescript force, in utilizing terrain well, and in conducting a skillful retirement in the face of an overwhelming opponent, Hardee's accomplishment could justifiably be a source of satisfaction to him.

After pulling his forces out of the city, Hardee marched them across a slender line of South Carolina rice dikes in sight of the enemy, but protected by a heavy force entrenched along the threatened side of the road. While the infantry column labored on its way to Hardeeville the navy remained behind with the army details to forward the stores deposited at Screven's Ferry. The gunboat *Savannah* discouraged any Union attempt to molest the removal of these stores. When the last supplies had been started inland Hardee ordered the wharf, the steamer *Firefly,* the cotton-clad *Isandiga,* and the *Savannah* burned. For the second night in succession the water around Savannah reflected the lurid glare of burning vessels.[75]

The Confederate army followed by the navy detachments on foot arrived in Hardeeville, disgruntled and half-frozen. The first to arrive mounted the water tanks, smashed the thick layers of ice, and formed relays to pass the water over to the locomotives' boilers. Hardee assigned priorities to the infantry units and dispatched them toward their destinations as quickly as he could. At this time Hardee released the volunteer battalions of workers from the Macon and Augusta war factories and all the men that he had forced into the army at Savannah.[76]

Hardee departed on December 22, leaving McLaws in command and instructing him to expedite the shipment of the supplies and troops to Charleston.[77] With the garrison and supplies safe in South Carolina, the troops distributed, and a new defense line being drawn behind the Combahee River, Hardee made his way to Charleston to set that house in order before the uninvited arrived.

Notes

"Hardee's Defense of Savannah" was previously published in *Georgia Historical Quarterly* 47 (Mar. 1963): 43–67, and is reproduced by permission of the Georgia Historical Society.

1. Hardee to Charles C. Jones, May 14, 1866, Georgia Portfolio, Manuscript Division, Duke Univ. Library, Durham, N.C.; Mrs. Howard Bowen, interview with author, Aug. 5, 1957, Birmingham, AL.
2. U.S. War Department, *The War of the Rebellion: A Compilation of the Official Records of the Union and Confederate Armies,* 128 vols. (Washington, DC, 1880–1901), series 1, vol. 44:901. Hereinafter cited as *OR.* All references are to series 1 unless otherwise indicated.
3. Ibid., 877; Hardee to Beauregard, Nov. 21, 1864, Confederate States of America Archives, 1861–1865, Rare Book, Manuscript, and Special Collections Library, Manuscript Division, Duke Univ. Library, Durham, N.C. (hereinafter cited as CSA Archives); Charles Colcock Jones Jr., *The Siege of Savannah in December, 1864 . . .* (Albany, NY, 1874), 96.
4. *Savannah Daily Morning News,* Nov. 28, 29, 1864; Hardee to W. W. Hunter, Nov. 27, 1864, Savannah Squadron Papers, 1862–1865, Manuscript Division, Emory Univ. Library, Atlanta (hereinafter cited as Savannah Squadron Papers); *OR,* vol. 44:906.
5. Madeleine Vinton Dahlgren, *Memoir of John A. Dahlgren* (Boston, 1882), 479–80.
6. *OR,* vol. 44:420; Charles Colcock Jones Jr., *The Battle of Honey Hill . . .* (Augusta, GA, 1885), 11; Richard Taylor, *Destruction and Reconstruction: Personal Experiences of the Late War,* ed. Richard B. Harwell (New York, 1955), 262.
7. Hardee to S. Jones, Nov. 29, 1864, Sam Jones Papers, 1861–1864, War Department Collection of Confederate Records, Record Group 109, National Archives (hereinafter cited as Sam Jones Papers).

8. *OR,* vol. 44:905.
9. T. B. Roy to G. W. Smith, Nov. 29, 1864, Department of South Carolina, Georgia, and Florida, Letters Sent, 1863–1864, War Department Collection of Confederate Records, Record Group 109, National Archives.
10. Robert U. Johnson and Clarence C. Buel, eds., *Battles and Leaders of the Civil War,* 4 vols. (New York, 1884–88), 4:666–67; *OR,* vol. 44:415.
11. Hardee to S. Jones, Nov. 30, 1864, Sam Jones Papers.
12. *OR,* vol. 44:911, 914, 920. Contributing to Smith's creditable and fortunate victory were the facts that Hatch had been misguided upon landing and thus prevented him from breaking the railroad, and that his troops were not battle seasoned.
13. Jones, *Siege of Savannah,* 92; *OR,* vol. 44:929; Hardee to S. Jones, Dec. 5, 1864, Sam Jones Papers; S. Jones to T. B. Roy, Jan. 11, 1865, Confederate Miscellany, Manuscript Division, Emory Univ. Library, Atlanta (hereinafter cited as Confederate Miscellany).
14. *OR,* vol. 53:35; Jones, *Siege of Savannah,* 55.
15. Jones, *Siege of Savannah,* 51–52; *OR,* vol. 44:938.
16. U.S. Naval War Records Office, *Official Records of the Union and Confederate Navies in the War of the Rebellion,* 31 vols. (Washington, DC, 1894–1927), series 1, vol. 16:472 (hereinafter cited as *ORN;* all references are to series 1 unless otherwise indicated.); T. D. Roy to W. W. Hunter, Dec. 7, 1864, Savannah Squadron Papers.
17. To Jones's credit, he never allowed the enemy to approach any closer to the railroad. Although the enemy could and did shell the line, only one locomotive and one car were damaged during the next two and a half weeks as the trains continued to run. S. Jones to T. B. Roy, Jan. 11, 1865, Confederate Miscellany; Jones, *Siege of Savannah,* 93–94; *OR,* vol. 44:420, 443–44.
18. Hardee to S. Jones, Dec. 8, 1864, T. B. Roy to S. Jones, Dec. 8, 1864, Sam Jones Papers; Hardee to Beauregard, Dec. 8, 1864, CSA Archives.
19. *OR,* vol. 44:940.
20. *OR,* vol. 44:410; Hardee to Beauregard, Dec. 9, 1864, CSA Archives; *Savannah Daily Morning News,* Dec. 8–9, 1864.
21. Johnson and Buel, *Battles and Leaders* 4:680; Hardee to Beauregard, Dec. 9, 1864, CSA Archives; *OR,* vol. 53:381–82.
22. George Anderson Mercer, Diary, 1851–1865, Dec. 8, 1864 entry, Southern Historical Collection, Wilson, Library, Univ. of North Carolina at Chapel Hill (hereinafter cited as Mercer Diary).
23. Charles Seton Henry Hardee, "Reminiscences and Recollections of Old Savannah," n.d., Southern Historical Collection, Wilson Library, Univ. of North Carolina at Chapel Hill.

24. *OR,* vol. 44:277; Berry G. Benson, Manuscript, Berry Benson Papers, 1845; 1865–1922, Southern Historical Collection, Wilson Library, Univ. of North Carolina at Chapel Hill (hereinafter cited as Benson Manuscript); Jones, *Siege of Savannah,* 78.

25. Douglas Southall Freeman, *Lee's Lieutenants: A Study in Command,* 3 vols. (New York, 1942–44), 1:435.

26. John William Green, *Johnny Green of the Orphan Brigade . . . ,* ed. A. D. Kirwan (Lexington, KY, 1956), 179.

27. Jones, *Siege of Savannah,* 86, 112–13; Ezra A. Carman, *General Hardee's Escape from Savannah,* Military Order of the Loyal Legion of the United States, Commandery of the District of Columbia, War Papers, No. 13, Washington, DC, 1893, 4.

28. *OR,* vol. 44:57; Jones, *Siege of Savannah,* 80–85; Hardee,"Reminiscences."

29. Jones, *Siege of Savannah,* 98; *OR,* vol. 44:13; Dahlgren, *Memoirs,* 488–89.

30. Hardee to Beauregard, Dec. 10, 1864, CSA Archives.

31. *OR,* vol. 44:951–952.

32. Hardee to Beauregard, Dec. 19, 1864, CSA Archives.

33. Hardee to Beauregard, Dec. 11, 1864, ibid.

34. E. C. Anderson to T. B. Roy, Dec. 26, 1864, George Wayne Anderson Papers, 1758–1896, Southern Historical Collection, Wilson Library, Univ. of North Carolina at Chapel Hill; S. W. Ferguson, Reminiscences, Heyward and Ferguson Family Papers, 1806–1923, Southern Historical Collection, Wilson Library, Univ. of North Carolina at Chapel Hill.

35. Frobel, Notes of the Siege of Savannah, Charles Colcock Jones Papers, 1763–1893, Manuscript Division, Duke Univ. Library, Durham, NC (hereinafter cited as C. C. Jones Papers); Jones, *Siege of Savannah,* 117–18; Hardee to Beauregard, Dec. 12, 1864, CSA Archives; Carman, *Hardee's Escape,* 12–13; Hardee to W. W. Hunter, Dec. 12, 1864, Savannah Squadron Papers.

36. Hardee to Beauregard, Dec. 12, 1864, CSA Archives.

37. William Tecumseh Sherman, *Memoirs of General William T. Sherman,* 2 vols.(New York, 1875), 2:193; John C. Van Duzer, Diary, 1864, entry of Dec. 13, 1864, Rare Book, Manuscript, and Special Collections Library, Duke Univ. Library, Durham, NC.

38. John Chipman Gray and John Codman Ropes, *War Letters 1862–1865 of John Chipman Gray and John Codman Ropes* (New York, 1927), 42.

39. *OR,* vol. 44:10, 61.

40. Jefferson Davis, *Jefferson Davis, Constitutionalist: His Letters, Papers and Speeches,* 10 vols., ed. Dunbar Rowland (Jackson, MS, 1923), 6:421.

41. *OR,* vol. 44:960.

42. Carman, *Hardee's Escape,*13ff; Mercer Diary, Dec. 15, 1864, entry; Jones, *Siege of Savannah,* 138; *OR,* ser. 2, vol 7:1268; ibid., ser. 1, vol. 44:966.

43. P. M. B. Young to C. C. Jones, n. d., C. C. Jones Papers; *OR,* vol. 44:962–63.

44. P. M. B. Young to C. C. Jones, n. d., C. C. Jones Papers; Carman, *Hardee's Escape,* 20.

45. Jones, *Siege of Savannah,* 139–40.

46. *ORN,* vol. 16:484; *OR,* vol. 44:964–65.

47. Jones, *Siege of Savannah,* 141–42.

48. A. R. Wright to T. B. Roy, Jan. 20, 1865, Confederate Miscellany.

49. Jones, *Siege of Savannah,* 145; Hardee to Beauregard, Dec. 19, 1864, CSA Archives; Carman, *Hardee's Escape,* 23–25; *OR,* vol. 44:968–69.

50. *OR,* vol. 44:967–69.

51. Frobel, Notes on the Siege of Savannah; Hardee to J. G. Clarke, P. G. T. Beauregard Papers, 1844–1893, Rare Book, Manuscript, and Special Collections Library, Duke Univ. Library, Durham, NC.

52. This unusual and unstable method was adopted because the Confederates could obtain only half the number of boats necessary to complete the bridge with pontoons perpendicular to the bridge as customary.

53. Frobel, Notes on the Siege of Savannah; *ORN,* ser. 2, vol. 16:484.

54. Frobel, Notes on the Siege of Savannah; J. G. Clarke to P. G. T. Beauregard, Apr. 6, 1875, quoted in P. G. T. Beauregard to C. C. Jones, Apr. 13, 1875, C. C. Jones Papers; Jones, *Siege of Savannah,* 133–34.

55. Jones, *Siege of Savannah,* 134.

56. Benson Manuscript.

57. P. M. B. Young to C. C. Jones, n.d., C. C. Jones Papers; *OR,* vol. 44:973.

58. *OR,* vol. 44:450–51.

59. Gray and Ropes, *War Letters,* 42; *OR,* vol. 44:6–7; D. H. Poole to McLaws, Dec. 19, 1864, Lafayette McLaws Papers, 1861–1865, War Department Collection of Confederate Records, Record Group 109, National Archives (hereinafter cited as McLaws Papers, NA).

60. T. B. Roy to McLaws, Dec. 20, 1864, McLaws Papers, NA; *OR,* vol. 44:973; Carman, *Hardee's Escape,* 26–27.

61. *OR,* vol. 44:967.

62. Ibid., 965; *ORN,* ser. 2, vol. 2:xvi, 482. The *Savannah* found she could not escape to sea through the obstacle choked waterways and had to be burned. The *Isandiga* ran aground and had to be burned. The *Firefly* met the same fate after doing good service in ferrying material. The several ships under construction in the city had to be burned also.

63. Carman, *Hardee's Escape,* 27.

64. Ibid.

65. Frobel, Notes on the Siege of Savannah; Hardee,"Reminiscences"; Jones, *Siege of Savannah,* 162.

66. Carman, *Hardee's Escape,* 29.

67. J. B. Elliott to his mother, Jan. 10, 1865, Habersham Elliott Papers, 1863–1885, Southern Historical Collection, Wilson Library, Univ. of North Carolina at Chapel Hill.

68. Sherman, *Memoirs* 2:231.

69. *OR,* vol. 44:63–65.

70. George Winston Smith,"Cotton from Savannah in 1865,"*Journal of Southern History* 21 (Nov. 1955): 495–96.

71. *OR,* vol. 53:413.

72. Hardee to S. Cooper, Feb. 6, 1865, Adjutant and Inspector General's Office, Record of Telegrams Received, 1862–1865, War Department Collection of Confederate Records, Record Group 109, National Archives.

73. Jones, *Siege of Savannah,* 107–8.

74. Sherman, *Memoirs* 2:216; Gray and Ropes, *War Letters,* 42; *OR,* vol. 44:6–7.

75. Hardee to Beauregard, Dec. 21, 1864, CSA Archives; *ORN,* ser. 1, vol. 16:484; William Douglas Pickett, *Sketch of the Military Career of William J. Hardee, Lieutenant-General C.S.A.* (Lexington, KY, 1910), 42.

76. Hardee,"Reminiscences"; Hardee to Beauregard, Dec. 21, 1864, CSA Archives; *OR,* vol. 53:38.

77. T. B. Roy to McLaws, Dec. 22, 1864, McLaws to wife, Dec. 27, 1864, Lafayette McLaws Papers, 1836–1897, Southern Historical Collection, Wilson Library, Univ. of North Carolina at Chapel Hill.

Major General James Patton Anderson. Courtesy of the U.S. Army Military History Institute, Carlisle, Pennsylvania.

Patton Anderson: Major General, C.S.A.

Richard M. McMurry

JAMES PATTON ANDERSON WAS ONE OF THOSE QUIET MEN WHO PLAYED A SECONDARY role in the Civil War. Yet, he was a man whose life story is interesting for it illuminates some little-known facets of the conflict while providing concrete examples of several larger themes that run through both antebellum and Confederate history. A successful lawyer, militia officer, politician, and planter before the war, the tall, dark-haired, dark-eyed Anderson proved himself to be a capable leader whose abilities carried him to high rank in the Confederate army.

Anderson was born on February 16, 1822, in Winchester, Tennessee. His father, William Preston Anderson, had been a lieutenant in the United States Army in the 1790s and a federal district attorney in Tennessee. He returned to military service in the War of 1812 as colonel of the 24th United States infantry Regiment. Patton Anderson's mother, Margaret L. Adair, Colonel Anderson's second wife, was the daughter of a prominent Kentucky family. Patton's early childhood was spent at "Craggy Hope," his father's farm near Winchester.

Soon after Colonel Anderson's death in April 1831, the widow and her six children moved to her father's place in Mercer County, Kentucky. For several years Patton lived with uncles, aunts, grandparents, and other relatives. His education came from "country schools" and private tutors,

In October 1836, young Anderson was sent off to Jefferson College in Cannonsburg, Pennsylvania (now Washington and Jefferson College in Washington, Pennsylvania), where he studied for a year until what he later termed "pecuniary misfortunes"[1] forced the family to withdraw him from school. He spent the winter of 1838–1839 near Hernando, De Soto County, Mississippi, assisting his stepfather, Dr. J. N. Bybee, "in building cabins, clearing land, &c., for the comfort of the family"[2] which planned to relocate in that newly opened area.

In the spring of 1839 Anderson returned to Jefferson College, from which he graduated in October 1840. He then settled in Hernando, hoping to practice law. There were, however, many able attorneys in that prosperous town in the northwestern corner of the state, only a few miles from Memphis, and Anderson found the overcrowded legal profession less than lucrative. He accepted a position as deputy sheriff—his brother-in-law was the sheriff—and practiced law when the opportunity offered. In the summers of 1844 and 1845 he returned to Kentucky to study law in a school run by one of his uncles. Admitted to the bar in 1843, he did not begin full-time legal practice until 1846.

Meanwhile, Anderson was elected colonel of De Soto County's militia regiment. In October 1847, at the "earnest appeal"[3] of Governor Albert G. Brown, Anderson raised a company from his regiment for Mexican War service and was elected its captain. The company became part of a battalion sent to garrison the Mexican town of Tampico. In early 1848 Anderson was elected lieutenant colonel to command the battalion. The unit was mustered out of service in the following summer, and its commander resumed his law practice. Although he saw no combat in Mexico, Anderson had acquired valuable experience commanding and administering a military unit on active duty, and he had had a bout with malaria that seriously impaired his health then and weakened it for the remainder of his life.

In 1849 Anderson, a Democrat, was elected to the Mississippi Legislature where he allied himself with Governor John A. Quitman and Senator Jefferson Davis in opposing the Compromise of 1850 as a "great injustice and wrong"[4] to the South. In the bitter state elections of 1851 Anderson, like Davis and most other Southern rights extremists, went down to defeat. In Anderson's case the loss was by about 100 votes out of 1,800, and poor health prevented him from making a vigorous campaign.

In 1853 Jefferson Davis's appointment as Secretary of War in the administration of Franklin Pierce gave Anderson an opportunity to escape the "miasmatic climate"[5] of Mississippi. After an effort to procure for Anderson a commission in a new infantry regiment failed because the bill creating the regiment did not become law, Davis secured for him the appointment as United States marshal for the newly created Washington Territory.

Anderson went to Memphis to consult his first cousin and fiancée Henrietta Buford Adair. She agreed to accompany him to the Northwest, and they were married on April 30, 1853. With funds given by Etta's father and money borrowed from friends, the couple left immediately after the wedding. They traveled by way of Nicaragua and arrived in Astoria near the end of June.

Anderson's years in the territory were busy and happy. His chief duty was to make a census, and for months he traveled through the area by foot, horseback, and canoe, carrying his blankets, food, and papers on his back. The active outdoor life soon improved his health. Anticipating that Olympia would become the territorial capital, the Andersons moved there. Whenever he could do so without conflict with his duties as marshal, Anderson practiced law. As he prospered, he purchased real estate in Olympia, Steilacoom, and Seattle.

In 1855 Washington's Democrats nominated Anderson to be the territorial delegate to Congress, and he was elected in June. That fall he and his wife returned to the East by way of Panama and were in Washington, D.C., for the opening of the Thirty-fourth Congress in December. Anderson played no important role in Congress's deliberations, but he was surely influenced by the growing sectional hostility that was frequently on display in the legislative halls.

During the Christmas recess of 1855, the Andersons visited their aunt Ellen Adair White Beatty on her plantation "Casa Bianca," near Monticello, in Jefferson County, Florida, just east of Tallahassee. Mrs. Beatty, a distinguished, very wealthy, twice-widowed woman, was seeking someone to manage her large plantation, with its 130 slaves, its looms, and its cobbler shop. She and Anderson soon reached a mutually profitable agreement under which Anderson would manage the enterprise. Mrs. Beatty would be relieved of the burden of directing a large plantation, and her nephew and niece would have a comfortable, healthy place to live. Declining appointment from President-elect James Buchanan as Governor and Superintendent of Indian Affairs in Washington Territory, Anderson moved to Florida when his congressional term ended in 1857. Later he would assert that he and his wife had concluded that the United States was breaking up and they wanted to be in the South when it did.

For four years Anderson managed his aunt's property. He was active enough in public affairs to be elected a delegate from Jefferson County to Florida's 1861 secession convention where, he later wrote, the state's ordinance of secession "received my hearty approval."[6] He was, in fact, a member of the committee that prepared the ordinance, and there is some evidence indicating that he may have been its principal author. While the convention was completing its work, Governor Madison Perry decided to attempt to seize the United States forts and navy yard at Pensacola. A volunteer company, the "Jefferson Rifles," had been organized in the county, and Anderson had been elected its captain. At the request of his men, Anderson left the convention to lead the company to Pensacola.

Anderson took charge of both the "Jefferson Rifles" and another volunteer company and had gotten them as far as Chattahoochee when the attack was called off and the force ordered to disband. Meanwhile, Anderson had been chosen as one of Florida's three delegates to the Provisional Confederate Congress in Montgomery. He was named to the Committee on Military Affairs and the Committee on Public Lands. Perhaps his most notable legislative proposal was the suggestion to use slaves as army cooks, teamsters, nurses, and pioneers.

When Congress adjourned, Anderson returned to Casa Bianca and was there in March when he learned that Perry planned to send a regiment to Pensacola for Confederate service. The "Jefferson Rifles" reorganized and reported to Chattahoochee where the regiment was to be formed and its field officers chosen. On April 5, Anderson, without opposition, was elected colonel of the 1st Florida Infantry (the "Magnolia Regiment"), and that night he left with his men for Pensacola to report to Brigadier General Braxton Bragg.

The 1st Florida was part of the reinforcements being sent to Bragg in the hope that he would be able to capture Union-held Fort Pickens on Santa Rosa Island, 1 1/2 miles offshore. Although Bragg wanted to take the fort, it proved impossible to do so. For eight months Anderson was with Bragg's army which was camped in a semicircle around Pensacola Bay, drilling, building fortifications, and observing the enemy on the island. For much of the time, Anderson commanded a brigade. The only notable combat was an October 8–9 raid against the island launched by Bragg in retaliation for Federal sorties against Confederate positions on the mainland.

To undertake the raid, Bragg organized an *ad hoc* force of 1,060 men divided into three battalions, one of which was led by Patton Anderson. Anderson's battalion, made up of 400 men detailed from several regiments, formed the left of the Rebel force that crossed the bay in small boats and stormed, captured, and burned the camp of the 6th New York Infantry—"a band of thieves and cut-throats,"[7] Anderson called the New Yorkers. The battalion commander, furious because some of his men who had been posted to protect a Federal hospital were made prisoners, was nonetheless pleased with the expedition which, he announced, "fully and completely"[8] accomplished its purpose.

By the spring of 1862 it was clear that the Florida Gulf Coast would not become the scene of major operations, and Bragg, with most of the Pensacola army, went to join the force that Generals Albert Sidney Johnston and P. G. T. Beauregard were assembling in northern Mississippi. Anderson, who had been promoted to brigadier general on February 10, was assigned to command a brigade in Brigadier General Daniel Ruggles's division of Bragg's Second Corps in what was to become the ill-starred Army of Tennessee.

For the next two years Anderson served with the Army of Tennessee, commanding a succession of brigades and divisions as he was tumbled about in the kaleidoscope that was that army's command structure. In the first day's fighting at Shiloh, April 6, 1862, his brigade formed the reserve for Ruggles's division on the left of the second line of battle. As the Confederates advanced on the Federal camps, Anderson's brigade "worked . . . [its] way into [an interval in] the front line." The men crawled across "a boggy ravine, the narrow swamp of which was thickly overgrown with various species of shrubs, saplings, and vines, so densely interwoven as to sometimes require the use of the knife to enable the footman to pass." They helped capture a Northern artillery battery and to overrun the camps of Major General William T. Sherman's division. Following Bragg's vague order to "go wherever the fight is thickest," Anderson's brigade then drifted into the poorly coordinated but ultimately successful assaults on the "Hornet's Nest," near the center of the battle, where the remnants of a Federal division were struggling to hold back the attacking Rebels. Anderson led his men throughout the battle, often signaling them by waving his hat since his voice could not be heard. The brigade suffered heavily before the beleaguered Northerners in the "Hornet's Nest" surrendered.[9]

Anderson helped cover the Confederate position on the second day at Shiloh and on the subsequent retreat. In the two-day battle his brigade lost 434 of its 1,633 men. It had captured several Federal guns and generally fought well. Anderson had performed about as capably as any of the brigadiers on the field, and he was praised by Bragg who wrote that "[Anderson] was . . . among the foremost where the fighting was hardest, and never failed to overcome whatever resistance was opposed to him. With a brigade composed almost entirely of raw troops his personal gallantry and soldierly bearing supplied the place of instruction and discipline." A short time later Bragg went out of his way to praise his friend Anderson as a "true and reliable" officer.[10]

In the two months after Shiloh Anderson's brigade underwent three changes in organization as regiments were shuffled in and out. When the turmoil subsided in late May, the brigade consisted of the 1st Florida Battalion (the shrunken 1st Florida Regiment); the 4th, 17th, and 25th Louisiana regiments; and a company of the Washington Artillery of New Orleans. Meanwhile, Bragg had become commander of the army and Ruggles had been succeeded as division commander by Major General Thomas C. Hindman. When Hindman was transferred in late May, Anderson took command of the division. His first stint as division commander, however, was brief. Illness soon forced him away from the army for a short time, and when he returned he found that Major General Samuel Jones had been assigned to the division. Anderson resumed command of his brigade.

Jones's division took part in Bragg's Kentucky Campaign of 1862. As the army passed through Chattanooga on its way to the Bluegrass, Jones was detached to take charge of the base of operations established in that city and to command Rebel forces in East Tennessee. Anderson was back in command of the four-brigade division which was assigned to Major General William J. Hardee's wing of the army. He led the division at Perryville, where two of his brigades covered the army's extreme left, and the other two were detached and sent off to fill a gap in the line to the right. They participated in the bloody attack on the Federal left. Owing to the division of his command, there was little opportunity for Anderson himself to accomplish much in the battle.

After Perryville the army returned to Tennessee where on December 12 it underwent yet another reorganization. The division that Anderson had been leading was broken up, and he was assigned to command an Alabama–South Carolina brigade in Major General Jones M. Withers's division of Lieutenant General Leonidas Polk's First Corps. On December 27, however, as the army maneuvered in the campaign that would soon result in the Battle of Murfreesboro, Anderson was given a new assignment. Another of Withers's brigade commanders, Brigadier General Edward Cary Walthall, was too ill to lead his brigade, and his troops successfully petitioned that Anderson command them in the approaching battle. Most of the regiments of the brigade—45th Alabama; 24th, 27th, 29th, and 30th Mississippi; and 39th North Carolina—had previously served with Anderson.

Murfreesboro, or Stones River, fought over the period from December 31, 1862, to January 1, 1863, was Anderson's finest battle. "No brigade occupied a more critical position, nor were the movements of any invested with more important consequences," reported Withers. Advancing against the center of the Northern line at about 9:00 A.M. on December 31, Anderson's brigade made a savage assault on Major General James S. Negley's division of the Fourteenth Corps. Negley's position was an excellent one, especially for artillery. Repulsed again and again, Anderson's men, supported by Alexander P. Stewart's brigade, returned to the attack. Federal artillery raked the cleared field across which the Confederates came, and Anderson's regiments suffered terribly before they poured over the Yankee line, forced the enemy back, and captured nine pieces of artillery and many prisoners. The 30th Mississippi lost 201 men in an acre of ground in front of the Yankee guns; brigade casualties were 783—131 killed, 638 wounded, and 14 missing. "About 216 more than [in] any other Brigade in the fight," Anderson wrote a week later.[11]

After a quiet New Year's, Anderson's brigade was detached and rushed to the extreme Confederate right on January 2 to cover the wild retreat of Major General John C. Breckinridge's men who had been repulsed in an assault on the Federals. Anderson arrived after most of the fighting was over and went into position to help protect the Rebel right.

Bragg praised Anderson's performance in the battle, bestowing too much credit for his role on January 2 and not enough for his leadership on December 31. Polk and Withers also lauded Anderson and his brigade, especially for their work in capturing the Union artillery. Anderson was proud of his men. "They behaved most gallantly as Mississippians have always done in this war," he wrote his wife. "They took *nine* pieces of artillery but lost many of their best officers and men."[12]

Withers was ill after Murfreesboro, and Anderson commanded the division. When Withers returned, Anderson was assigned to lead a brigade consisting of the 7th, 9th, 10th, and 41st Mississippi regiments, Blythe's Mississippi Regiment, the 9th Battalion of Mississippi Sharpshooters, and Garrity's Alabama Battery.

Withers was replaced by Hindman whose illness elevated Anderson to division command on September 17, as the army maneuvered for the Battle of Chickamauga. On September 19, however, Hindman returned to the command, and it was as a brigade commander that Anderson participated in the great battle of the Western Theater. Hindman's division was part of Lieutenant General James Longstreet's Left Wing of the army. On September 20, Anderson's "fearless Mississippians" moved up on the extreme Confederate left to replace a unit that had taken heavy losses. The brigade was caught up in the great Confederate breakthrough that routed the right of the Federal army, and it was credited with the capture of eight pieces of artillery, several flags, hundreds of prisoners, and nine ordnance wagons. The brigade was then sent to the right to assist in the effort to complete the victory by defeating the rest of the Union army. Its "gallant and impetuous charge" there won high praise from other Rebels, but the Northerners held until dark and then withdrew. Anderson reported 568 casualties among his 1,865 men. At 11 o'clock that night, Hindman, "suffering much pain from an injury received about midday," turned the division over to Anderson.[13]

Bragg's bumbling campaign against Chattanooga in late 1863 was Anderson's most humiliating period of military service. It was his misfortune to be in command on November 25, when the division was overrun at Missionary Ridge.

Scanty evidence makes it impossible to assess fairly the blame for the Rebel disaster that day. Anderson's division was placed on Missionary Ridge in the center of Bragg's army. The position was a weak one. Lateral communication along the ridge was difficult, and there were many sheltered ravines by which attackers could approach the summit. Anderson's men did not move into their position until a short time before the battle. Troop deployment was faulty, and tactical dispositions were unsound. Engineer officers had placed the main Southern line on the geological crest rather than the military crest of the ridge, and that poor deployment made it impossible for fire from the main line to reach large areas in front of the position. Gaps were left in the line, and one of these was clearly attributable to Anderson, who expected it to be filled as troops posted lower on the ridge fell back.

Plans were not clearly explained, and many men in advanced positions were unsure whether their mission was to make a real effort to stop the enemy or merely to slow his advance before falling back. The Confederates were physically weakened from exposure, unhealthy camps, and inadequate rations. Anderson's men were aware of their thinned ranks—the men were seven to eight feet apart—and many knew that earlier Federal successes in some preliminary movements threatened the left flank and rear of the Missionary Ridge position even as vast hordes of Northerners were deploying to assail the right and center.

Morale had also been hurt when Bragg had announced a few days earlier that the exchange of prisoners had been halted. Therefore any Rebel who was captured could look forward to a long incarceration in a Northern prison where his chances of death from disease were higher than the chance that he would be killed in battle. Bragg's intent was doubtless to inspire his men to fight rather than surrender, but the message conveyed to many Confederates was that they might escape imprisonment by running away if their situation appeared at all doubtful.

The outcome was predictable. When the Northerners advanced on November 25, some Rebels at the base of the ridge attempted to hold their position; others fired a few shots and scrambled up toward the main line. The Northerners found that they were more protected as they went on and followed the retreating Confederates. Southern troops atop the ridge could not fire without hitting their own men. Quickly the Blue infantrymen swarmed up the ridge and poured into the gap Anderson had left in the line. The Rebel position disintegrated, and Bragg's left and center fled the field and streamed south to safety in Dalton, Georgia. Anderson's men were among the first to break. The division lost 1,676 men—76 killed, 476 wounded, 1,124 missing.

Whatever the reasons for the disaster, Bragg blamed Major General John C. Breckinridge who, he asserted, had been drunk. Breckinridge, who had commanded the corps to which Anderson's division was attached, was soon reassigned. Hindman returned to duty and took command of the corps. Bragg soon relinquished command of the army. No official blame for the rout was ever attached to anyone. Anderson, on December 18, announced himself "mortified at the conduct of some of our troops," but he believed himself without blame because he had protested against the deployment, calling it "the worst I have ever seen."[14]

The army spent the winter of 1863–1364 in Northern Georgia where it had come to rest after its flight from Missionary Ridge. Anderson was at army headquarters in Dalton on January 2, 1864, when Major General Patrick R. Cleburne read to the assembled corps and division commanders a proposal to augment the Confederate armies by using black slaves as soldiers and granting them freedom for their services. This proposal, which was made as a sincere effort to replenish the thinned ranks of the Southern armies, quickly became embroiled in the labyrinthine politics of the Army of Tennessee. The army had long been divided into bitterly squabbling pro-Bragg and anti-Bragg factions. Cleburne and the supporters of his proposal were members of the anti-Bragg faction. Although Bragg had left the army, his malevolent influence in its affairs lingered. From his new post as adviser to President Jefferson Davis, Bragg continued to intervene in army politics, and he and his supporters among the generals sought to use Cleburne's proposal as a weapon against their rivals who could now be branded as abolitionists and traitors.

Like many officers of the old Pensacola army, Anderson was a supporter of Bragg—although his relatively low position in the army's hierarchy kept him on the periphery of the quarrel—and he was quick to denounce Cleburne's proposal as a "monstrous proposition . . . so revolting to Southern sentiment, Southern pride, and Southern honor." Unlike some of Cleburne's critics who stood to gain promotion if their rivals could be disgraced, however, Anderson seems to have been motivated by sincere convictions that such a policy would negate what the Confederacy was fighting for and lead to disaster. Had the Confederate government not suppressed discussion of the proposal and had not Anderson himself soon received another assignment, he might well have taken an active public stand in opposition to Cleburne and those who supported the idea of using blacks in the Rebel armies.[15]

In February 1864 Anderson was promoted to major general. Only ninety-two men ever held the grade of Confederate major general (twenty of them were subsequently promoted to higher grade). Of the sixty-six promoted to major general before 1864, fifty-three (80%) were professional or professionally trained officers and most of the others were politically motivated appointees. Of the twenty-six elevated to major general in 1864 and 1865, only eight (35%) were professionals. Casualties had exhausted the supply of professional officers, and a new source of leaders had to be found. Anderson was one of the new type of Rebel major general—men, usually with some college training, who had been lawyers, businessmen, and/or planters before 1861 but who had often had some military experience in the Mexican War or militia service and who had demonstrated an ability to lead in battle as regimental and brigade commanders.

Anderson's promotion carried with it a new assignment. Shortly after being made a major general he was transferred to command Confederate forces in Florida. The assignment and the promotion seem to have been dictated by the need for "an officer of superior rank and reputation" in Florida and by Anderson's status as a citizen of the state whose prewar political contacts would prove useful. They may also have owed something to pressure from Anderson's friends Senator Augustus E. Maxwell and General Braxton Bragg.

At the time Anderson assumed his new duties in early March the military situation on the peninsula had stabilized. The Federals had long held several coastal towns in East Florida. In February 1864 they had made an unsuccessful advance inland, and, after their defeat at Olustee on February 20, they had fallen back toward the coast. The Saint Johns River then became the *de facto* line between the two forces. The Confederates sought to keep the enemy east of the river and to protect the people and agricultural resources in the rest of the state.

Anderson took charge of the 12,000 Rebel troops in Florida and set about his mission. There was an occasional skirmish with the Yankees and raids by both sides, and Anderson's men managed to destroy several enemy gunboats in the Saint Johns. As the year wore on, however, both sides withdrew troops from the state and sent them to more important areas. Anderson soon found himself devoting most of his time to chasing deserters; protecting civilians; attempting to control the activities of disloyal persons, outlaws, "villains," "skulkers," "plunderers," runaway slaves, and bandits; and organizing and administering the scanty "Reserve Forces" of the state, of which he was made commander on April 30. Perhaps more congenial to his instincts as a lawyer, although frustrating to him as a military commander, was the long legal ruckus that arose that summer over taking up track from one railroad to use on a more important route. Local interests objected to the removal of the track and sought a court injunction to block the Confederate government from acting. Anderson made an effort to prevent a clash between state and Confederate authorities, but the matter had not been resolved in late July when he was ordered elsewhere.

Command in Florida was not a happy experience for Anderson. His health was not good, and, as an ardent Southerner, he was frustrated by the relative unimportance of his assignment. In May he broke his longstanding rule "*not* to *apply* for *anything*" and suggested to the War Department that, since almost all of his troops had been sent to other areas, he be given an active command. Not until late July was his wish granted. By that time the Army of Tennessee was near Atlanta, Georgia, attempting to defend that crucial city. Anderson's friend Braxton Bragg, then *de facto* chief of staff to President Davis, ordered Anderson back to command his old division in place of Hindman whose recent wound and perennial sickness had rendered him permanently unfit for field duty.[16]

Anderson reached Atlanta and assumed command on July 30. His division, then part of Lieutenant General Stephen D. Lee's corps, was posted on the extreme Confederate left southwest of the city. There were 304 officers and 3,459 enlisted men present for duty in the division.

For the next several weeks Anderson's men were engaged in indecisive skirmishing along the fortified line. On August 30 the division, along with other troops, was ordered to Jonesboro, twenty miles below Atlanta, to drive away Northerners who threatened to seize the only railroad that supplied the Rebel army. After a "toilsome" night march, Anderson's jaded men reached Jonesboro at 11:00 A.M., August 31. Orders were to attack the Federals who were building breastworks on a ridge some 1,200 yards to the west.

At 3:00 P.M. the Confederates were ready to assault. Anderson's division was on the right of the Rebel force, three brigades in its front line and one in reserve. When the signal to advance came, Anderson started forward, his men driving away the enemy skirmishers and moving up a "moderately steep"[17] incline toward the Yankee works. Slowly but resolutely the Gray ranks moved to within pistol range of the enemy. By that time, the fire had become so heavy that the advance lost momentum and then stopped, and some men began to waver. Anderson, believing that with support he could push on, called up the second line. Casualties, however, were so heavy that the division could not continue. Still, Anderson held his men to their work and sent to Lee for reinforcements. It was then about 4:30. Federal Major General John A. Logan described what happened:

> The assault raged severely in front of . . . [my line], the enemy approaching their lines at the average distance of 50 and 100 paces. . . . The most terrible and destructive fire I ever witnessed was directed at the enemy, and in less than one hour he was compelled to retire. . . . The rebel general, Patton Anderson, and his staff, rode fearlessly along his lines . . . and did all that a commander could do to make the assault a success. But few of those who rode with him in that perilous performance of duty returned from the field. Himself, with many of his staff, were seen to fall. . . . I could not help but admire his gallantry, though an enemy.[18]

Anderson was badly wounded. A bullet had smashed through the lower portion of his head, breaking the jawbone on both sides and cutting through the tongue. Bleeding was heavy, and there were fears that the wound would prove mortal. He was taken to a hospital, and doctors who worked through the night managed to staunch the flow of blood and save his life. He was soon sent to Monticello, where he spent about seven months recuperating. As late as March 1865 he was still forced to live on a liquid diet.

Despite his suffering, Anderson planned to rejoin the army in December, but a bone fragment in his jaw abscessed and forced him to remain in Florida. In January 1865 he wrote Stephen D. Lee, "I am tired—tired to death of looking out of my door daily, yea, almost every hour of the day" and seeing men who used their wounds as an excuse to stay home. He was annoyed at the croakers who had lost faith in the Southern cause and especially hurt by the activities of prominent political leaders who would not give full cooperation to the government in its hour of need. He yearned to rejoin the army and to continue the fight. "This is a dark day in the history of the present war," he wrote on February 28, "but I believe a brighter day will soon dawn upon us. If dissension and faction does not distract us, we will certainly achieve our independence."[19]

Although not fully recovered, Anderson, over the objections of his doctors, rejoined the army in the Carolinas in March 1865. In the reorganization of April 9, he was assigned to command a two-brigade division of Georgia and South Carolina troops. At the surrender on April 26, at Greensboro, North Carolina, his command numbered 185 officers and 2,441 enlisted men.

Neither Anderson's life nor his role in the Civil War was either theatrical or decisive, but both are of some interest as illustrations of several aspects of mid-nineteenth-century American history. For one thing, nineteenth-century Americans were a remarkably mobile people, willing frequently to change both their homes and their occupations in an attempt to better their circumstances. Anderson—who lived in five states, one territory, and the District of Columbia for significant portions of his life and practiced several professions—was a personification of the Americans' tendency to move about.

From their beginnings, Americans have often debated whether a full-time career military force would meet the nation's needs better than would a system under which a small professional army served as a cadre for an essentially civilian military organized as a militia, reserve, or national guard. In the decades prior to the Civil War the latter approach was favored by most Americans, and Anderson's career is a strong argument in its favor. Like Jacob D. Cox, John A. Logan, Wade Hampton, John B. Gordon, and dozens of others, he demonstrated that native intelligence, sharpened by discipline, broadened by education, seasoned by experience and training, and applied to military problems is at least as useful as the more narrow experiences that shaped such professional officers as John Bell Hood, Braxton Bragg, and John Pope.

On a more specialized level, Anderson's career also illustrates the constantly shifting command structure of the Army of Tennessee. In 1862 and 1863 he commanded six different brigades in the army—when he was not busy commanding two different divisions. One problem that hampered the Army of Tennessee throughout its history was the turmoil that resulted from its constant reorganizations. The command structure simply never had a chance to jell. Anderson's service with that army amply illustrates that fact.

The end of the war found Anderson in desperate straits. He never fully recovered from his Jonesboro wound, and he suffered from occasional relapses into his old "malarial symptoms." He, his wife, and their five children, were poor—Casa Bianca had been sold early in the war for payment in Confederate money, and Anderson had invested much of his own capital in slave property. After the war, he refused to seek a pardon and therefore he could neither practice law nor hold public office. He eventually settled in Memphis and, until his death on September 20, 1872—the ninth anniversary of the Army of Tennessee's greatest victory—he earned a living selling insurance and editing an agricultural magazine.

For Anderson, the Confederacy's struggle for independence had been "the most momentous earthly crisis which men were ever called upon to meet." Anderson met the crisis with everything he had—he had even objected to the surrender in 1865. He was truly an unreconstructed Rebel.

Notes

"Patton Anderson: Major General, C.S.A." first appeared in *Blue & Gray Magazine,* vol. I, issue 2 (Oct.–Nov. 1983): 10–17, and is reprinted with permission from Blue & Gray Enterprises, Inc., Columbus, OH.

1. "Autobiography of Gen. Patton Anderson, transcribed by Mrs. Anderson," *Southern Historical Society Papers* 24 (1896): 58.
2. "Autobiography of Gen. Patton Anderson," 58.
3. Ibid., 59
4. Ibid., 60.
5. Ibid.
6. Ibid., 65.
7. U.S. War Department, *The War of the Rebellion: A Compilation of the Official Records of the Union and Confederate Armies,* 128 vols. (Washington, DC, 1880–1901), series 1, vol. 6:299. Hereinafter cited as *OR.* All references are to series 1 unless otherwise indicated.
8. Ibid., 299.
9. Ibid., vol. 10, pt. 1:496–97.
10. Ibid., vol. 10, pt 1:468, pt. 2:538.
11. Ibid., vol. 20, pt. 1:755; Anderson to wife, Jan. 8, 1862, James Patton Anderson Collection, P. K. Yonge Library of Florida History, Univ. of Florida, Gainesville (hereinafter cited as Anderson Collection).
12. Anderson to wife, Jan. 8, 1862, Anderson Collection.
13. *OR,* vol. 30, pt. 2:303, 305, 462.
14. "Autobiography of Gen. Patton Anderson," 69.
15. Quoted in Edward Turner Sykes, *Walthall's Brigade: A Cursory Sketch . . .* (Jackson, MS, 1916), 552–54.
16. Anderson to wife, May 3, 1864, Anderson Collection.
17. *OR,* vol. 38, pt. 3:772.
18. *OR,* vol. 38, pt. 3:109, 773.
19. "Autobiography of Gen. Patton Anderson," 70.

Brigadier General Daniel Chevilette Govan. Courtesy of the Alabama Department of Archives and History, Montgomery.

No Better Officer in the Confederacy: The Wartime Career of Daniel C. Govan

Daniel E. Sutherland

Sixteen Arkansans served the Confederacy as general officers. Most of the sixteen remain obscure figures in the annals of history, largely because they compiled lackluster records or because they served in the low-profile Trans-Mississippi theater. Yet neither of these limitations explains why historians have ignored the military career of Daniel Chevilette Govan. Govan never held independent command, but he rose rapidly from company commander to brigade commander. Promotion came partly from his Arkansas connections with Thomas C. Hindman and Patrick R. Cleburne. Yet Govan also possessed qualities—personal courage, loyalty, compassion, and common sense—prized in military leaders by both superiors and subordinates. He left few personal papers to reveal his thoughts or explain his actions, yet his wartime career, like those of many regimental and brigade commanders, may be explored and evaluated through the story of his men. The brigade he led for nearly half the war formed part of the "best fighting division in the Army of Tennessee" and the "hardest-hitting division in the Confederate army." John M. Harrell, himself a Confederate colonel and one of the earliest historians of Arkansas's role in the war, maintained, "No officer of the army of Tennessee enjoyed to a greater degree than General Govan did, the esteem of his men and of his superior officers."[1]

Govan did not settle in Arkansas until 1854, at the age of twenty-seven. He had by that time already led an adventurous life. Born July 4, 1827, in Northampton County, North Carolina, he moved west as a child with his father, Andrew R. Govan, a planter and politician, and his mother, Mary Pugh Jones. The Govans emigrated first to Tennessee, but by 1830, they had made their home in Mississippi. Govan studied with a private tutor in his youth before going to South Carolina, his father's native state, to attend Columbia College (now, the University of South Carolina), his father's alma mater. The younger Govan drilled as part of a military company while at college, but he returned for unexplained reasons to his father's Mississippi plantation, Snowdown, in 1848 before being graduated.[2]

The following year, Govan joined the California gold rush with his cousin Ben McCulloch, another future Confederate general. A dozen men—eight whites and four blacks (probably slaves)—departed Austin, Texas, on the southern overland route to California through northern Mexico. "A small party of gentlemen," McCulloch said of the white men, "bent on getting wealth . . . for the glorious privilege of being independent." They attracted additional followers en route, so that a rugged band of some twenty men finally reached Mazatlan on Mexico's Pacific coast. From there, they traveled by sea to San Francisco, where the group concluded a journey of nearly three months in December.[3]

Govan met with only fair success panning for gold along the San Joaquin and other rivers. So when McCulloch won election as the sheriff of Sacramento in 1850 (following the shooting death of the previous sheriff), Govan accepted his offer to serve as deputy. But neither McCulloch nor Govan was keen on California. The weather did not suit them, the mines were not as rich as reputed, and the population, with its antislavery leanings, was not entirely congenial. When McCulloch went home, so did Govan, returning to Mississippi in 1852 to engage in planting. The following year, he married eighteen-year-old Mary F. Otey, the daughter of the first Episcopal bishop of Tennessee.[4]

The Govans had been operating a plantation in Phillips County, Arkansas, for six years when the war came. They had four children by that time, two boys and two girls, the eldest being five and the youngest under a year. The eldest was named for Ben McCulloch, and a proud grandmother back in Mississippi liked to boast about how little Ben "is outgrowing his clothes and is very pretty, smart and sweet." The Govan homestead was worth $12,000 in real property and $22,000 in personal property in 1860. The personal property included twenty-five slaves, eleven male and fourteen female, who ranged in age from a few months to fifty years. Govan owned a respectable amount of livestock, including 2 horses, 6 mules, 8 cows, 5 oxen, 15 cattle, and 50 hogs, the whole valued at $2,000. But he made his money from cotton. He grew nearly 28,000 pounds of cotton (69 bales) on 200 acres in 1860. This amount could not compare with the state's largest cotton producers, but it was more than that grown by about thirty percent of Arkansas planters.[5]

No record remains of Govan's position on the heated political debates of the 1850s, but a planter with Govan's background—South Carolina birth and Mississippi boyhood—undoubtedly sympathized with the secession movement. Phillips County's slave population expanded tremendously in the decade before the war, and the white population strongly supported secession by 1861. When Arkansas seceded in May, Govan helped raise a volunteer company in Phillips County. The enthusiastic band, evenly divided between "plain country folk" and planters, elected Govan, a man of wealth and influence with at least marginal military experience, as their captain. The men trained initially at Hopefield, Arkansas, where, on June 5, they mustered into Confederate service as Company F, Second Arkansas Infantry. Commanded by the recent United States congressman from Phillips County, Colonel Thomas C. Hindman, the Second Arkansas traveled to Memphis. The regiment spent nearly two months drilling and equipping itself, and soon numbered seventeen companies. On July 3, Brigadier General William J. Hardee assumed command of this expanded "legion." Hardee, a no-nonsense Georgian who graduated from West Point in 1838, had been sent to defend Northeast Arkansas and Southeast Missouri against a possible Union advance from St. Louis. Arkansas was in military and political turmoil early in the war, far beyond the ability of a man with Hindman's limited military experience in the Mexican War to put right. The situation required a professional soldier.[6]

Govan and his men chaffed for action by early July. At first, many in the regiment had refused to serve outside of Arkansas. They had even threatened to resign from the army should they be ordered eastward. In Tennessee, rumors circulated about a Union invasion from Missouri. The "whole militia" of Arkansas would be called out to defend the Arkansas-Missouri border, Govan informed his wife, "and he who lags behind must be either traitor or coward." On July 10, the men marched into Northeast Arkansas to establish Camp Hardee at Pittman's Ferry, a "stone's throw of the Missouri line." Their leader, however, despaired of the Arkansas troops assigned to him. One officer reported Hardee as saying, "These men have no manners and I am afraid we can't make much out of them."[7]

Captain Govan and his company now found themselves part of a brigade that included five infantry regiments, a battalion of cavalry, and four batteries of artillery. Colonel John R. Gratiot commanded the Second Arkansas, Hindman having been sent to Northwest Arkansas to persuade mutinous troops there—unhappy with the lack of clothing and pay—to stay in Confederate service. One of Govan's fellow Phillips County residents, Patrick R. Cleburne, commanded the First Arkansas Infantry. Govan also witnessed the grim side of army life at Camp Hardee. By late July, the men still lacked "clothing of every description," and they had been struck by a catastrophic outbreak of measles. Govan estimated some two hundred cases of measles in the brigade, with most companies suffering two to three deaths in the epidemic.[8]

However, spirits rose significantly on July 27. Govan had been visiting friends in Cleburne's regiment when he heard thunderous cheering in his own neighboring camp. He hurried back to learn of a great Confederate victory at Manassas, Virginia. War whoops and rebel yells sounded a "general jubilee in which men and officers mingled together . . . without restraints."

Govan liked such occasions. He never expressed disdain for enlisted men; he was never seduced by the authority of command. And on this July evening, he, like many southerners, expressed the naive hope that the quick, decisive Confederate triumph in Virginia meant that he could soon return home to his wife and children.[9]

On the contrary, the war soon began in earnest for Govan when Hardee led his men north to Greenville, Missouri, in early August. Lack of coordination among Confederate units aborted a possible military action against the Federals, but the expedition gave Hardee an opportunity to drill his undisciplined force and mold a unit that would form, in time, the nucleus of the famous Hardee-Cleburne division. Hardee roused his men daily at 4:00 A.M., Govan informed Mary. Missouri-born West Pointer Colonel John S. Marmaduke instructed the men in regimental drill for two hours before breakfast. Then came two hours of company drill to finish out the morning, followed at 4:00 P.M. by two more hours of company drill. The brigade returned to Pittman's Ferry on August 28, and by mid-September, as more regiments continued to assemble there, Hardee commanded a force of nearly six thousand men.[10]

Hardee moved his brigade to Columbus, Kentucky, in mid-September. The move was a direct response to Brigadier General Ulysses S. Grant's seizure of Paducah, Kentucky, on September 6, which encouraged Union and Confederate armies to jockey for position in Missouri and Kentucky. The brigade passed through Bowling Green to join forces with Brigadier General Simon B. Buckner, with both Hardee and Buckner under the command of the new Western Department commander Albert Sidney Johnston. The Arkansans bivouacked near Cave City, so called, Govan informed his wife, "from its proximity to a considerable cave within a hundred yards of our Camp." He also noted that the army was within eight miles of the famous Mammoth Cave, and that a large hotel had been built at Cave City for tourists.[11]

By mid-October, Hardee had been promoted to major general and given command of a division. His promotion meant elevation for three of his regimental commanders—Cleburne, Hindman, and R. G. Shaver—who now commanded three brigades containing eight Arkansas, two Tennessee, and one Mississippi regiments. Hindman was even promoted to brigadier general. From his company camp, Captain Govan had no notion of being a general or of what the generals were planning. He only knew that from 15,000 to 30,000 Confederates were assembling in the region and that they must soon be tested in battle.[12]

Govan's company along with Hindman's entire brigade stayed in nearly constant motion during September and October as it screened Hardee's division, conducted reconnaissance, and formed raiding parties. Govan's men had by this time earned a reputation as "the best company in the Reg." How they earned that reputation is unclear, although, given Govan's college experience, they likely were a well-drilled and well-disciplined group when compared to most green troops. They nearly engaged the enemy in October when assigned to "special duty" with a battalion of cavalry. In an odd mode of advance, Govan and each of his men was mounted behind a cavalryman. The expedition had hoped to launch a surprise attack on a nearby Union encampment. Alas, when heavy rains made the rivers impassable, the Confederates retired.[13]

Such missed opportunities frustrated Govan. The nation had been at war for half a year, yet he and his men seemed to spend all their time marching aimlessly through the countryside. Like many a junior officer, he questioned the wisdom and ability of his superiors. "A battle is so uncertain and distant," he lamented to Mary, "that I cannot form any idea as to when it will occur. Military operations are so slow that our military leaders seem lacking in enterprise and energy." Equally disheartening to Govan, sickness continued to plague the army. Most of the problems, he was convinced, came from exposure, sleeping on the ground, and the failure of many men to take "proper care of themselves." Besides battling measles and scurvy, men from Texas, Mississippi, Tennessee, and Central Arkansas had difficulty adjusting to the damp climate of Kentucky's cave region. When his orderly sergeant died from sickness, the solemnity of war and finality of death weighed heavily on the sensitive Govan. He traveled personally to Bowling Green to arrange shipment of the sergeant's remains to Arkansas.[14]

The longer the war continued without resolution, and the longer he remained away from home, the more Govan worried about his wife and children. Nothing is known of the arrangements Govan made for his family or his plantation while in service. Supposedly, like many planters, he left an overseer in charge of the latter, and other members of the Govan and Otey clans who had settled nearby likely watched out for Mary. The separation made Govan "far sadder" than he had been in "many, many years," and "more depressed" then ever in his life. He would try to visit in December, he promised Mary, but for the moment he took satisfaction in serving his country and resisting "Northern fanaticism" and the Yankee "war of extermination upon the South." Should men leave the army now, they would expose their families to "the mercy of a brutal soldiery, ready to do the bidding of their abolition master." "Can I or ought I to resign in the very face of the enemy?" the introspective captain asked his wife. "These are questions which it is difficult for me to answer."[15]

Govan's worries and mixed emotions stemmed, no doubt, from scenes he witnessed daily in the Bluegrass state. "The condition of things in Ky is lamentable and heartrending," he assured Mary. "Many families houseless and homeless, driven by the abolition incendiary or the fanatic soldier to put themselves at the mercy of every pitiless storm." People were being murdered, their property destroyed, and everyone lived "in constant dread." "They know not when they lie down at night but that the torch or the musket may rouse them before morning. This is civil war in earnest—with all its terrible realities."[16]

Govan finally saw action at Rowlett's Station, on December 17. When Brigadier General Don Carlos Buell advanced with the intention of seizing Bowling Green, an important railroad center in the struggle for Eastern Kentucky and Eastern Tennessee, Hindman, directed by Hardee to be aggressive in holding the region, attacked. It was not much of a fight, only a minor action, really. Hindman moved forward cautiously with his force of 1,100 infantry, 250 cavalry, and four pieces of artillery in response to reports that the Thirty-second Indiana Infantry was attempting to repair a railroad bridge across the Green River, near Woodville. After his cavalry drove back a handful of Union infantry skirmishers, the general decided to decoy the main Federal force up a small hill and into the teeth of his combined infantry and artillery. The plan went awry when a Federal force of 300 men suddenly appeared on Hindman's right flank. Colonel Benjamin F. Terry of the Eighth Texas Cavalry threw 75 of his Texas Rangers against the Hoosiers. The Texans attacked with "lightening speed" and "infernal yelling," but Terry fell mortally wounded. More Federals moved onto the flanks. The First Arkansas Battalion under Marmaduke drove back the enemy's right. The Second Arkansas and an artillery battery moved up in support, but by that time, Hindman realized he was outnumbered two to one and withdrew. Total Confederate losses numbered 4 killed and 10 wounded; the Federals lost 11 killed and 22 wounded.[17]

Govan considered the contest "severe." He was most impressed by the action of the Texans, and much saddened by the death of Terry, a "gallant and heroic" officer whom he had met in California through Ben McCulloch. However, it does not appear that his company or even the Second Arkansas came under direct fire. The regiment's job, Govan reported to his wife, had been to support the artillery. The Rangers and two infantry companies from the First Arkansas did all the fighting. Happily, though, his men had at least been prepared for combat. They had recently exchanged their flintlock muskets for Enfield rifles, and all the training, drill, and forced marches of October and November had made them fit for battle.[18]

On January 28, 1862, while the army sat impatiently in winter camp at Bowling Green, Hindman promoted Govan to colonel and gave him command of the Second Arkansas. No record tells why Hindman promoted Govan in preference to his other company commanders. The fact that they both came from Phillips County may have worked in Govan's favor, as would the high reputation of his company. In any event, Govan suddenly found himself in an entirely different military circle. He likely joined the "Shynnedygge Militaire" organized that night by Hindman's staff, an evening of dancing and joking for the regiment's officers and local civilians. Otherwise, the Confederates in Kentucky and Tennessee did not fare well that winter. They suffered severe setbacks at Logan's Cross Roads (or Mill Springs) on January 19 and Fort Henry on the Tennessee River fell to Grant on February 6. As a result, the Army of Central Kentucky evacuated Bowling Green on February 11. Its original destination was Nashville, where Johnston hoped to concentrate his disparate forces, but when Fort Donelson fell to Grant on February 16, western and central Tennessee seemed on the verge of collapse. Hardee's army pushed on to Mississippi, by way of Decatur, Alabama, to arrive at Corinth in late March. By then, word of Confederate defeat at Pea Ridge and the death of Ben McCulloch had reached the army.[19]

Pittsburg Landing would be the next test for Govan and the Arkansans in Johnston's newly created Army of the Mississippi, "our first lesson in the stern realities of the battlefield," as Govan put it. Alas, Colonel Govan would be too ill to lead his new regiment—assigned to Hindman's First Brigade of Hardee's Third Corps—for much of the fight. After having boasted of his good health all through the sickly time in Kentucky, Govan succumbed to a severe case of "flux," or diarrhea, around the first of April. Even though in a still weakened condition when Johnston caught the Federals napping near Shiloh church, Govan led his men all through the horrendous first day of the two-day battle (April 6–7). Hindman's 2,000-man brigade moved hesitantly during the initial assault of April 6, as though reluctant to engage the enemy. Finally, however, the men methodically pushed forward across Seay Field, and when other, less disciplined, brigades lost their alignment and broke ranks, Hindman's men held their ground and saved the morning's earlier success with a bayonet charge. They remained in the thick of the action and continued to push back the Federal right flank all morning, "exposed all the while to the fire of the enemy" and sustaining heavy casualties. Not until 11:00 A.M., physically spent and nearly out of ammunition, did the brigade fall back to rest and rearm.[20]

All this was but prelude to the Hornet's Nest. Hindman had withdrawn from the fight due to painful injuries suffered when his horse was shot from under him earlier in the day. Colonel Shaver replaced him, and after allowing the brigade to rest for about an hour, he led it back to the field. As the men rejoined the struggle at about 3:00 P.M., they anchored the center of a Confederate line that sought repeatedly to dislodge the last enclave of Federal defenders. As the brigade charged "with great vigor" directly against three Iowa regiments, the Second Arkansas faced not only superior numbers but a withering crossfire by cannon and rifle. Govan's men remained pinned down for an hour while "taking advantage of every possible cover." They fought "desperately," enemy bullets whizzing all around them, until darkness came, the Federals withdrew, and they collapsed in exhaustion. Govan was so ill by then that he could not file the regiment's battle report.[21]

After Shiloh, the army retreated to Corinth and reorganized once more. Govan took five days leave of absence to regain his strength, but he was back in harness by the time the army evacuated Corinth on May 29–30 and continued its retreat to Tupelo. By summer, with Hindman returning to Arkansas to command the Trans-Mississippi District, Govan had a new brigade commander, Brigadier General St. John R. Liddell, a former staff officer for Hardee. It was an all Arkansas unit, including the Second, Fifth, Sixth, Seventh, and Eighth Arkansas Infantry. Brigadier General Simon B. Buckner was Govan's new division commander, and the Second Arkansas remained in Hardee's corps. General Braxton Bragg assumed command of the army, and he wanted to recapture Kentucky. The climactic battle of the resulting campaign, fought at Perryville on October 8, 1862, was Govan's second major action, and he and his regiment performed admirably.[22]

Buckner's division held the center of the Confederate line at Perryville, opposed by the Federal corps of Major General Alexander McCook, a West Point classmate of Hardee. Major General Benjamin F. Cheatham's division opened the attack against McCook on the Union left, but Buckner's men soon joined them, advancing, according to Hardee's instructions, "upon the enemy where the firing is hottest." As Govan moved the Second Arkansas into position, he hurriedly asked Cheatham how the battle was going. "By the Gods, Govan," he replied, "they are giving us particular hell, pitch into them." As Govan later recalled, his regiment proceeded to give the Yanks "what Gen Cheatham said they were giving him." It was a confusing battle, having started in midafternoon and continuing well beyond twilight, "with a bright full moon shining" during the latter phases of the conflict. Govan's regiment suffered its share of the brigade's seventy-one casualties, but it also captured two ambulances, one of them carrying McCook's personal baggage, as well as four Federal flags.[23]

As the army tumbled back out of Kentucky following Bragg's "victory" at Perryville, Murfreesboro, Tennessee, loomed as the next test for Govan, who, based purely on his date of promotion to colonel, found himself the senior regimental commander. The Second Arkansas still formed part of Liddell's brigade, but Pat Cleburne, now a major general, became their division commander. Murfreesboro is of some interest in the study of Govan's military career, for it is the first action for which his battle report survives. Liddell's brigade formed the extreme left of Cleburne's division (indeed, of the entire army) on December 30, the first day of this three-day battle, while the Second Arkansas formed the extreme right of the brigade. Govan's men suffered heavy losses in fighting early that morning when Liddell moved the brigade forward prematurely. The Arkansans found themselves unsupported in fierce fighting against Indiana, Ohio, and Kentucky troops "posted behind a fence and in the woods near a house used by the enemy as a hospital" below Wilkinson Pike. Govan led his regiment to within a hundred yards of the Yankee line, the furthest advance made by the brigade. They held their position until the remaining regiments arrived to help scatter the Federals. Govan then maneuvered his men in a flanking movement to within twenty-five yards of the First Ohio Infantry. After a "closely contested fight," the Federals "gave way in confusion and fled across the field."[24]

Liddell's men rested briefly, but by noon they were up again and had approached the edge of a cedar ridge along the Nashville Turnpike. Cleburne's men swept forward and seemed to have the situation well in hand. Liddell even stopped to chat with some Federal surgeons as his brigade moved ahead easily without his direction. Then things went terribly wrong, and Cleburne's entire division folded. The panic began when two Tennessee brigades, both in advance of Liddell, wavered in the renewed attack. Govan, in Liddell's absence, assumed responsibility for the brigade and made a crucial decision. He and the other regimental commanders had been waiting for additional ammunition to be distributed to their men when the "disorderly and disgraceful" retreat of the Tennesseans swept past them. "I was at first at a loss what course to pursue," admitted Govan. "Our success had been all that we wished." But with his brigade nearly out of ammunition, with no support in front or on his right flank—nearest the enemy—and being exposed in an open field, he conducted an orderly withdrawal. Liddell was mildly critical, noting in his report that "but little firing [was] going on at the time" of the retreat; but he recognized the precarious position the brigade had occupied, and he lavished praise on the Second Arkansas for its performance earlier in the day.[25]

Despite this anticlimactic end to the day's struggle, Bragg's army had gained a victory. Unfortunately, two days of fighting remained. The Second Arkansas engaged only in light skirmishing during this time, but for the entire battle the regiment suffered 118 casualties, including 15 killed, 94 wounded, and 9 missing. The brigade as a whole had 607 losses. During the following months of Bragg's drawn-out retreat toward Chattanooga, Govan continued to gain the admiration of his superiors. Liddell, Cleburne, and Hardee all recommended Govan for promotion to brigadier general in January 1863. Hardee called him "an officer of tried courage of energy & ability," but Bragg had no brigade for him to command. Govan and his regiment performed exceptionally well at Liberty Gap (June 25–26, 1863) during the Tullahoma campaign. By then, however, like many officers in the Army of Tennessee, Govan had lost confidence in Bragg. When Liddell predicted that their commander would make a stand at Tullahoma against the pursuing Federals, Govan quietly but confidently wagered an oyster supper that Bragg would continue to retreat. Govan won.[26]

Bragg again reorganized his army in August 1863, and Govan took command of the entire brigade when Liddell was placed in charge of this and another brigade to form a demi-division in Lieutenant General Daniel H.

Hill's corps. Govan's old regiment had been so depleted by the Murfreesboro fight that it was consolidated with the Fifteenth Arkansas Infantry. In addition, his brigade consisted of an artillery battery, First Louisiana Infantry, Eighth Arkansas, the consolidated Fifth and Thirteenth Arkansas, and the consolidated Sixth and Seventh Arkansas. The noted British military observer Colonel Arthur J. L. Fremantle had been impressed with the brigade after reviewing it earlier in the summer. "General Liddell's brigade was composed of Arkansas troops—five very weak regiments which had suffered severely in the different battles," he reported. "The men were good-sized, healthy, and well clothed, but without any attempt at uniformity in colour or cut." Later, he added, "Most of them were armed with Enfield rifles captured from the enemy. Many, however, had lost or thrown away their bayonets, which they don't appear to value properly, as they assert that they have never met a Yankee who would wait for that weapon. . . . They drilled tolerably well, and an advance in line was remarkably good."[27]

Chickamauga, Georgia, September 18: Bragg thought he had lured Major General William S. Rosecrans into a trap. Govan did not, at first, seem destined to play a very big role in closing that trap, as all of Liddell's division had been assigned to Major General William H. T. Walker's reserve corps. But, in Govan's words, "It proved to be . . . anything but a reserve division." On September 18, Walker ordered Liddell's two brigades to advance by way of Alexander's Bridge, one of the crucial crossings on Chickamauga Creek. They seized the position easily against only light resistance. The Yankees, however, had torn up the planks of the bridge, so Liddell took his men upstream to cross at Byram's Ford that night.[28]

September 19 found the division deployed for battle along the LaFayette Road. Govan's brigade formed the left wing as they moved toward Federal positions along the Brotherton Road at about 10:30 A.M. Liddell cautioned Govan to watch his left flank, which was unprotected. At 11:30 A.M., the line charged and sent the Federals reeling. The men in Govan's front, as so often in the past, came from Ohio and Indiana. It was tough going, even without the fighting, as both sides moved through woods and undergrowth. Visibility was no more than fifty yards. Govan's men rounded up some three hundred prisoners and several pieces of artillery. After pausing only briefly to catch their wind, they then pushed up a small hill against a second Union line, these defenders being from Pennsylvania, Wisconsin, and Indiana. It was no contest. "Steady boys, steady, for God's sake, don't run away," pleaded a Federal officer as his men skedaddled.[29]

The day did not end on this high note, but Govan's personal performance was sterling. As his brigade moved back down the hill with additional trophies of war, Govan was still mindful of Liddell's warning to watch his left flank. He had already deployed the Sixth and Seventh Arkansas as skirmishers on that flank when a long blue column appeared, the vanguard, it turned out, of six full brigades. Govan's single brigade stood no chance, even though it fought stubbornly for a good half hour. When the Sixth and Seventh buckled under the pressure, the rest of the brigade slowly gave way, too. However, as at Murfreesboro, Govan controlled the damage by skillfully withdrawing his men.[30]

As the entire division reformed on a line held by Cheatham's division, most men hoped they had completed their role in the day's fight, but more was to come. Cheatham, desperately needing reinforcements on his right, asked Liddell to send in his division again. The Arkansans pushed ahead briefly, but then fell back, and the hard-luck Sixth and Seventh Arkansas proved to be the weak reed once more. With Govan's men on the right of Brigadier General Edward C. Walthall's brigade, the Sixth and Seventh, forming Govan's own left flank, came under an enfilading fire at about 3:30 P.M. At the same time, one of Govan's own batteries wounded several men by firing short into his ranks. The entire line collapsed, and Govan could do nothing but retreat. He reformed on Walthall's right as night drew near.[31]

The final day of fighting, September 20, proved the least successful for Govan and his men. Ordered against the Federal left at 9:00 A.M. in support of Major General John C. Breckinridge's position, the brigade did not actually advance toward the LaFayette Road, a half mile distant, until noon. One staff officer, aiming his barbs at Polk and Liddell, considered the attack "ill planned and ill conducted." Directed to wheel left and assist the brigade of Brigadier General States Rights Gist, another of Walker's reserve units, Govan and his men played their part successfully. They had a difficult maneuver, made across an open field while Ohio infantrymen fired into them from a woods on the left flank. Yet, in Govan's proud assessment of the day's work, his brigade "pressed gallantly forward and succeeded in driving the enemy from his position." The men continued on enthusiastically southward along the road until they found themselves "almost immediately in the rear of a line of the enemy's strongest breastworks."[32]

But Govan now fell victim to his own success, Gist's failure, and the slowness of the Sixth and Seventh Arkansas. Govan's men had penetrated too far, and Gist's brigade, which Govan had never been able to locate properly in the woods and fields, had already fallen back. Govan's brigade thus found itself "completely isolated." Not only that, but the Sixth and Seventh regiments, unaccountably moving ahead of the main body, had left a considerable gap between themselves and their comrades. The Yankees pounced on the opening with a "vigorous attack" that enfiladed and overlapped the consolidated regiment while the rest of the brigade continued to advance. The Sixth and Seventh fled "in great confusion," the men throwing away arms and accoutrements to speed their flight. The collapse came so rapidly that only the initiative of Govan's regimental commanders on the right saved the day. With Govan unable to communicate with them quickly enough, they coolly organized a coordinated withdrawal in the heat of battle. As they pulled back toward the right, or westward from the field, a cavalry scout intercepted them and led the brigade to safety.[33]

Unhappily, the day's disappointment was not yet spent. At about 5:00 P.M., Liddell reformed both brigades, Govan on the left and Walthall on the right, and moved across LaFayette Road toward Savannah Church and the Mullis House. It was nearly the precise route Govan's men had taken that morning, and again, as in that fighting and on the previous day, the left flank proved the brigade's undoing. Govan grew more and more uneasy as the lines advanced. His left was protected by Breckinridge's men on this afternoon, but Liddell, steadily outpacing Breckinridge, moved ahead too quickly. Govan slowed his own advance at first, trying to maintain contact with the left, but he had been told to dress on Walthall, so forward he went. Brigadier General John B. Turchin's mostly Buckeye brigade lay in wait some three hundred yards to the left. Govan had only a few minutes warning from his skirmishers before the bluecoats came screaming from concealment out of heavy woods. Too late to change front, Govan hurried his men into some undergrowth and ordered them to take cover. The Yanks, supported by two artillery batteries, hit them before they completed the maneuver. The Sixth and Seventh Arkansas, on the brigade's left, broke first. Without this pivot regiment to anchor the line, Govan's remaining regiments melted away. "Under this heavy and galling fire," Govan reported, "no other alternative was left but to withdraw the brigade as speedily as possible to save it from annihilation or capture."[34]

The brigade retreated "in considerable confusion," admitted the colonel, yet he was able to rally and reform the men some four hundred yards to the rear. Miraculously, he also inspired them to advance once more, the third advance of the day. The men surged forward and were making progress along the LaFayette Road, slightly to the right of their previous advance, when the Yanks fell back and the day's fighting ended. It had been a hellish three days, in which Govan lost fifty percent of his command. Yet, given circumstances, he and his men had performed well. Always quick to share credit, Govan praised his regimental commanders for their "coolness and skill" on the two occasions when he had to depend on them for quick, independent action. The brigade, Govan stressed in his report, "never failed to drive the enemy in their front, and advanced each time with a single line unsupported, and with one or the other of my flanks unprotected, and that on no former occasion was the courage and endurance more severely tested, nor in any previous battle did they ever exhibit more determined bravery and gallantry." It was just that damned left flank.[35]

Liddell also seemed satisfied. He appreciated the performances of both Govan and Walthall, especially their "prompt co-operation in every movement and quick apprehension of the constantly recurring necessities that arise on a battle-field." Moreover, Liddell knew that their men, despite being driven back on three occasions, had performed in superior fashion. "Although no brilliant results were directly accomplished," he observed truthfully, "the record for hard fighting cannot be well surpassed." The "best evidence of good soldiers," he maintained, comes "when over-powered by immense numbers on all sides to be able to rally promptly and return again and again to the contest undaunted." Govan and his men had done that on more than one occasion.[36]

As Bragg's army pursued and laid siege to William S. Rosecrans's army in Chattanooga, Govan had much upon which to reflect during the following weeks. His new responsibilities as brigade commander weighed heavily on him. Quiet and introspective by nature, he had always felt the loss of any man under his direction. How much more keenly, then, he must have felt the loss of men who had been killed or maimed under his command in battle. And to lose half his men in three days! . . . well. Then, too, his brother Andrew, fourth youngest of five brothers, and one of three Govan boys present at Chickamauga, had died of wounds suffered in the fighting. Govan grieved and worried about how his mother would take the news. He also had to inform Mary that one of her cousins, Hugh Otey, had been killed.[37]

Despite the success of his men in checking the Federal advance at Chickamauga, another army reorganization required that Liddell's division be broken up in October. Cleburne took command of a new division in Breckinridge's corps with Liddell resuming his place as brigade commander and Govan returning to lead the Second Arkansas. However, when Liddell went on leave in November, Govan once again assumed temporary command of the brigade. Govan himself hoped to take leave soon, for thirty days if he could manage it. Mary was expecting another child, and he wanted to be with her during her "confinement" or shortly after the birth, even if it meant returning to a region swarming with Federals. "I think by the middle of Dec all active military operations will be suspended," he told her, "and my chances are *very* good . . . unless a battle is imminent, in which event I shall feel it my duty to remain."[38]

By September 1863, Govan had reason to be concerned about a pregnant wife in Arkansas. Major General Frederick Steele's Union forces took possession of Little Rock in that month, and turned the capital into a permanent base of operations. More to the point, Helena had been occupied by the Federals since April of 1862, and all Phillips County had suffered at the hands of foragers. An Indiana soldier described Helena as a "miserable . . . looking place," in July 1862, with "the best part of town . . . completely destroyed." "The remaining houses are all deserted by their owners," continued his bleak description, "and are now occupied by Soldiers for different purposes." Major

General Samuel Curtis, for instance, made his headquarters in Hindman's house. The situation was not quite so dangerous on outlying plantations, like the Govan's, but the countryside faced other perils. "Every where, they [Federal soldiers] are devouring and laying waste the labor of man's hands. Our [wives] are not free from their insults—but they walk in armed with pistols & sabres, and thus compell with arms, our wives to cook for them! The surrounding country is, for miles around, swept by foraging parties. Farmers have both negroes and all kinds of stock stolen from them."[39]

The anticipated battle came sooner than Govan would have liked, as his brigade helped shield the retreat of Bragg's army following the fiasco at Missionary Ridge on November 25, 1863. The division had already been evacuated to Chickamauga Station on November 23 in anticipation of joining Lieutenant General James Longstreet's corps in the siege of Knoxville when Cleburne ordered a portion of his men, including Govan's brigade, back to a position on the right of Missionary Ridge. Although Govan's portion of the line never came under direct attack that day, his old regiment helped to defend the army's flank. Between 11:00 A.M. and 3:00 P.M., the Second Arkansas, now consolidated with the Fifteenth and Twenty-fourth Arkansas under Lieutenant Colonel E. Warfield, bore the brunt of some of the heaviest Union assaults. The regiment captured an enemy battle flag and helped to bag five hundred prisoners.[40]

The rest of Bragg's army did not fare so well, and as the precipitous retreat began, Cleburne's division covered Hardee's corps. Cleburne rushed Govan's men to "dispute the enemy's advance" at Shallow Ford Road, which they did to perfection. The brigade's most dramatic moment, however, came on November 27 at Ringgold Gap. At 8:00 A.M., Govan deployed his regiments in a small ravine across the mouth of the gap in Taylor's Ridge to await the enemy's arrival. Two Napoleon guns from Semple's Battery augmented his infantry. When the Federals came to within 150 yards of his line, the two guns opened up with canister shot. The bluecoats "reeled and staggered." They made several attempts that morning to puncture Govan's defenses, but eventually pulled back. The brigade withdrew that afternoon having helped buy the army an entire day to facilitate an orderly retreat. Cleburne lavished praise on Govan and his other brigade commanders—Mark P. Lowrey, Lucius E. Polk, and Hiram B. Granbury—for their day's work. "Four better officers," he proudly asserted, "are not in the service of the Confederacy." Govan, as always, praised the performance of his men. "For this fight I think we deserve especial credit," he maintained, "as it was fought under peculiarly discouraging circumstances, while retreating in the face of overwhelming numbers, and without assistance or support from any quarter."[41]

The army finally settled that winter around Dalton, Georgia, although Cleburne's division, acting as a buffer between the main body and any Federal advance, positioned itself seven miles to the northwest on Tunnel Hill

Ridge. During this encampment, Cleburne showed the "good judgment" to recognize Govan for his services as brigade commander during the autumn campaigns. Liddell had been seeking a transfer to the Trans-Mississippi theater for several months, and he had urged everyone from Cleburne to General Samuel Cooper, adjutant general of the Confederate army, to promote Govan as his replacement. When Liddell finally received his transfer, Cleburne requested a promotion for his Phillips County friend. "This is one of the best Brigades in my Division," Cleburne informed General Cooper, "it is composed exclusively of Arkansans and I feel a peculiar interest in its being well commanded." Govan had already led the brigade in several battles, he emphasized to Cooper, "and always with skill and valour." The promotion finally approved, Govan took permanent command of the brigade on December 29.[42]

Meantime, in a show of mutual confidence, Govan supported his fellow adopted Arkansan in a bold political step by endorsing Cleburne's extraordinary proposal to train slaves as soldiers in the Confederate cause and, once the South gained its independence, to free those slaves who had loyally served the nation. Govan, like Cleburne, was doubtless appalled by the growing thinness of Confederate ranks, and believed that Southern independence could be secured only by sacrificing slavery. With his own slaves probably long since taken by the Federals as "contraband," Govan was the first of thirteen Cleburne brigade and regimental commanders to sign the document. But the controversial proposal was too hot for most Confederate generals and politicians. Jefferson Davis called the proposal "injurious to the public service" and ordered that it be "kept private."[43]

Otherwise, the winter encampment was a mixture of hard work, hard play, and the usual boredom of inaction. Govan and Cleburne's other senior officers spent part of the winter in school studying tactics and the art of war. Cleburne personally listened to the recitations of his four brigade commanders. They, in turn, heard the recitations of their regimental officers, who drilled the troops and tightened discipline considerably. On the lighter side, Govan's and Polk's brigades waged a monster snowball fight following a four-inch snowfall in late March. Thousands of men waged a day-long war for supremacy, and while there is no record of Govan's participation, he probably joined the mob. Cleburne led the ailing Polk's brigade, and Govan would not have missed a chance to lead his men in such a contest. Incidentally, Govan's brigade won.[44]

Govan and his men engaged in rugged action all the next spring and summer during the Atlanta campaign. The men of his Arkansas Brigade, as it was now known, were battle-hardened veterans by this stage of the war. All had "seen the elephant," and their ranks had been thinned considerably. The brigade's eight Arkansas regiments had been consolidated into four: the Second with the Twenty-fourth, Fifth with Thirteenth, Sixth with Seventh, and Eighth with Nineteenth. At Resaca (May 14), Pickett's Mill/New Hope Church (May 27), Kennesaw Mountain (June 27), and Atlanta (July 22), his men fought doggedly and Govan led them expertly. "We had a continuous battle" from Dalton to Atlanta, recalled one of Govan's soldiers. "We moved back slowly, and only when flanked and outnumbered. When we adopted a new line a few miles back, we built breastworks. Thus marching, battling, building works, in rain and mud, with no camp, no tents and but little food, the campaign went on." Govan himself recalled, "It was a campaign that tried the endurance of the soldier to the utmost extent, and tried his soldierly qualities as in a crucible. We slept on our arms, lulled to sleep (if we slept at all) to the sound of the bullet, and awoke in the morning saluted by artillery."[45]

Yet Govan and his men paid a heavy price for the heroics of that spring and summer. In fighting from May 7 to May 20, including the action at Resaca, the Arkansas Brigade lost 3 killed and 31 wounded, the third highest casualties in Cleburne's four brigades. Granbury's Texans played the critical role at Pickett's Mill/New Hope Church, but, as one of Govan's men recalled, the Arkansas Brigade was "in that fight from start to finish." The brigade lost 34 killed and 86 wounded on that day. The tables at last turned on Kennesaw Mountain, where Major General William T. Sherman suffered fearful losses.[46]

In the Battle of Atlanta, Govan's men, exhausted by lack of sleep and two days of endless skirmishing, marching, and entrenching, faced a "murderous fire" yet routed the enemy in hand-to-hand combat and took hundreds of prisoners. "My men carried three distinct works of the enemy without being once repulsed," reported a satisfied Govan, "and held the ground gained until ordered back from the last position, being unsupported." A Confederate artillery officer admired how Govan had maneuvered the brigade deftly in the face of the enemy. "The most successful movement [of the day] was performed by two of General Govan's regiments," he reported, and went on to insist that "Generals Govan, Lowrey, and Smith acted nobly, and all their men fought with the courage of lions."[47]

But Govan's men were also mauled on July 22, and worse was to follow. On the morning of July 22, Govan's brigade counted 1,221 effectives; on the morning of July 23, it numbered just 722. He had lost 86 killed, 322 wounded, and 91 captured. A week later, on September 1, he and his men suffered humiliation and their only real defeat of the war at Jonesboro. General John Bell Hood, now commanding the Army of Tennessee, had decided belatedly that Jonesboro, an important railroad terminus fifteen miles south of Atlanta, must be protected. He ordered troops, including Cleburne's division, to defend it in late August. Corps commander Hardee's report of the ensuing action was terse: "The enemy attacked my whole line fiercely at Jonesborough yesterday, turning my right flank at the same time. The assault was everywhere repulsed, except upon Lewis's and Govan's brigade, which gave way." In point of fact, the Yankees captured Govan and six hundred of his men, along with eight artillery pieces, largely because the brigade had been placed in an indefensible position.[48]

Here is what happened. Early on the morning of September 1, the Arkansas Brigade formed part of a thin defensive line formed by Major General John C. Brown's division and three of Cleburne's brigades. They did not hold an ideal position. Cleburne's men braced themselves behind "an inferior line of entrenchments incomplete and on ground badly selected." Bad as it was along some parts of the line, Govan's men, positioned on the right flank, "found no works at all, but a few fence rails." When Hardee, fearing an attack on the right, hurriedly rushed two more brigades to that flank, Govan's men had to make room for them. They wound up relinquishing what scanty protection they enjoyed and were still working frantically to erect new temporary works when the Federals advanced. The Arkansans turned back the first assault, but when the Yankees brought up reinforcements, and three divisions converged on the brave men of Govan and Brigadier General Joseph H. Lewis, the bulk of both brigades, outmanned and ill protected, surrendered. "Their fire upon the advancing enemy was steady," reported one Confederate officer, "but the immense numbers and overwhelming forces of the Yankees ran upon the works, sweeping over the right of Govan's fortifications." Sergeant Patrick Irwin of the Fourteenth Michigan Infantry captured Govan.[49]

The captivity did not last long. Govan and his men had been transported only as far as Nashville when, on September 9, Union and Confederate authorities arranged for "a special exchange" of 2,000 prisoners. Govan served as the central rebel in the trade, being exchanged for Union general George Stoneman. The entire brigade had returned to duty within a fortnight. Atlanta had been abandoned by then, and the army was preparing to embark on its fateful Tennessee Campaign (November 29–December 27). A series of forced marches in October and November carried Hood's men through Georgia and Alabama, and into Tennessee. Spring Hill, Franklin, and Nashville awaited an army that would soon cease to exist.[50]

A falling out between Hood and Hardee had resulted in the latter being replaced by Cheatham as Govan's corps commander, but Cleburne still led the division. Cleburne's men could have given the Confederacy a signal triumph at Spring Hill (November 29), but miscalculation and miscommunication by his superiors produced an unsatisfying half victory. Govan's brigade formed the center of Cleburne's command as it made a tricky maneuver to reinforce Lowrey's brigade in a rout of the Federal left. Typical of Govan, while his men reformed for another attack—the potentially lethal strike that never came—the general halted with his staff to rescue a family whose house had been set ablaze by Union artillery.[51]

As the Union army slipped away, Hood pursued, and the gods of war gave him another opportunity to seize victory at Franklin (November 30). Blaming his subordinates—particularly Cheatham, Cleburne, and Brown—for his own failures at Spring Hill, Hood ordered a direct assault against a formidable Union line. Incredibly, Hood thought Cleburne's men had lacked spirit and resolve at Spring Hill. He intended to teach them a lesson on this day. As Cleburne gathered his brigade commanders to relate Hood's orders, Govan could see his friend was "greatly depressed." Upon hearing the orders—to carry the enemy works at all hazards—Govan and the others knew they must suffer horrendous casualties. As Govan turned to leave the brief conference, he told Cleburne, "Well, General, few of us will ever return to Arkansas to tell the story of this battle." Cleburne nodded grimly: "Well, Govan, if we are to die, let us die like men."[52]

Cleburne tried to shield his men by forming them on a narrow front into columns of brigades, rather than in lines. The Confederates actually succeeded in breaking through the first Federal line, but they lost much unit cohesion in the process and poured after the fleeing enemy "more like a wild howling mob than an organized army." They fell on the main Union lines in a frenzy of hand-to-hand combat. "I raised my gun to get my man," related one gray-clad survivor, "and was shot in rig[h]t elbow my hat shot off, my gun stock broken with 100 holes thru my blanket. So I laid down in the ditch. They began to cross fire on us and were doing us lots and lots of harm." Govan's men, forming the center of Cleburne's three attacking brigades, were in the thick of it. "General Cleburne then moved forward on foot, waving his cap," recalled Govan, "and I lost sight of him in the smoke and din of battle." Plunging ahead of Govan and his men, Cleburne fell dead eighty yards from the enemy works. Five other Confederate generals died that day at Franklin, including Granbury. Cleburne's division, reported Govan, had been "decimated," "literally ruined," with a loss of fifty percent. Among the wounded was a nephew, Frederick H. Govan, whom the general had recently added to his staff. "Our Brigd and the Ark. Brigd are so badly cut up that we can't move," wrote one of Granbury's men. "Some officers have no men, and some companies have [no] officers—So we have to reorganize and consolidate, a Captain has to command the Brigade."[53]

Govan remained bitter for many years about the Franklin fight. Although he did not use the word, he and many survivors believed that Hood had "betrayed" the army. Two weeks after the battle, he told Mary it had been the "most desperate fight" he had ever seen. He rejoiced in his "safe deliverance from the perils of that bloody conflict the bloodiest by far for the time it lasted of all the battles" in which he had engaged. "As usual," he continued, "Gen Cleburne's and Cheatham Div. bore the brunt of the fight and sustained the heaviest loss. It really seems as if it is intended that we should do the fighting of the Army and where the severest opposition is to be encountered there we are surely to be placed."[54]

With the army still staggering from Franklin, the subsequent Nashville Campaign must have seemed like a nightmare. Hood continued to pursue the Federals as they fell back toward the Tennessee capital, but he had no army. Hood laid siege to Nashville with a mere 23,000 men. The Federals soon had twice that number, and when they assaulted Hood's works on December 15, the once mighty Army of Tennessee disintegrated. Govan held the left of his division's position, now commanded by Brigadier General James A. Smith, which in turn anchored the army's right wing. Govan's brigade, consisting of ten regiments consolidated into three, had 534 men. Still, they shredded a Yankee regiment sent against them during the day's fighting. That night, Hood shifted Cheatham's corps to the left, where Govan's brigade formed the extreme left of the line. Incredibly, the men maintained their sense of humor. As they passed Major General William B. Bate's men, Govan's troops shouted out, "Lie down, Mr. Bate, Mr. Govan is gwine to pop a cap." They needed more caps. They found themselves assigned to defend a sector that required a division, and when the Yanks resumed the attack on December 16, the line snapped. Govan's men, unable to withstand the concentrated artillery fire that poured in on them, fled in disorder. The general fell with a bullet wound to the throat near the Granny White Pike.[55]

Govan returned home to recover from his wound. He could easily have sat out the rest of the war, but he returned to service in April 1865 with a reformed Army of Tennessee near Greensboro, North Carolina. John Brown commanded Cleburne's old division, once again in Hardee's corps. Govan commanded two regiments, the First Arkansas (a consolidation of ten Arkansas regiments and the Third Confederate) and the First Texas (a consolidation of eight Texas regiments). His men had given a spirited performance in fighting at Bentonville, North Carolina, that March, but the fighting was over for Govan. He and his men surrendered as part of General Joseph E. Johnston's army on April 26 and were paroled on May 1.[56]

Govan's postwar career was quiet but active. He resumed planting near Marianna, where he was "very popular and much loved by all." He refused repeated suggestions that he enter public life, although a resident of his neighborhood reported that Govan helped lead Helena's Ku Klux Klan in the 1860s. In 1894, he moved briefly to Washington state to serve as Indian agent for President Grover Cleveland at the Tulalip Agency. Mary died in 1896, and Govan returned to West Tennessee two years later. Thereafter he resided with one or another of his fourteen children in Tennessee, Arkansas, and Mississippi. He died on March 12, 1911, of pulmonary edema and old age at the home of a daughter in Memphis and was buried at Holly Springs, Mississippi.[57]

Like many Civil War veterans, Govan could not let go of the war. In 1878, he wrote a brief history of Cleburne's division, published after his death in 1911. He also corresponded with other veterans—both northern and southern—and tried to set the record straight about events in which he had participated or had special knowledge. He corresponded with St. John Liddell, and Liddell sent him a portion of his reminiscences for comments and criticisms. He remained interested, almost hauntingly so, in the Atlanta Campaign, Spring Hill, and Franklin. He became active in veterans' affairs and attended occasional veterans' reunions, including those of his old brigade. In 1900, he served on the executive committee for the United Confederate Veterans reunion in Memphis.[58]

Govan made an honest confession of the lingering hold the war had on him. In the 1880s, he corresponded with former Union general William W. Belknap concerning the Atlanta Campaign, in which Belknap and his Hawkeyes frequently ran up against Govan's Arkansans. Toward the end of one long letter, in which he recounted his brigade's role in the Battle of Atlanta, Govan paused to say, "I find myself, General, unconsciously fighting over again the battle of the 22nd of July, reciting facts and incidents well known to you and not exactly pertinent to the subject of inquiry of your letters; but you must really excuse it from the well-known weakness of all old soldiers, to fight their battles over again."[59]

The highlight of Govan's veterans' activities came in September 1883, when Belknap invited him to attend a reunion of the Sixteenth Iowa Infantry at Cedar Rapids, Iowa. The men of the Sixteenth had been among the hundreds of prisoners taken by Govan's Arkansans at Atlanta on July 22, 1864. His brigade had also captured the regiment's flag, which Govan had kept for nearly twenty years. He went to Iowa to return the colors. Word spread quickly through the North of Govan's gesture. A month later, at a reunion of the Army of the Tennessee, one of the Union armies against which Govan had waged war so furiously, a member of the Sixteenth Iowa related the Govan tale to the assembled crowd. "The roof of the immense house nearly cracked with the noise of the shouting when I alluded to your name & to your chivalric act," the veteran informed Govan. "You may depend upon it that you will always have a most warm place in our hearts."[60]

Govan had been a steady, patient commander who lacked the charisma of Cleburne, but who became known for his coolness and common sense in battle. He had a knack for surrounding himself with "intelligent, prompt, active and efficient" subordinates who performed "all duties required of them well & with alacrity." No one ever questioned his courage or loyalty. Thus his quiet nature and military talent produced an odd combination of qualities, noted by many. Hardee praised Govan for being as "true and brave as he was courteous and gentle." Liddell described him as "an unpretending and reliable gentleman, zealous in the cause, deliberate and cool in judgment. . . . Brave, yet cautious, unwilling to sacrifice his men without cause." Irving A. Buck, staff officer and chronicler of Cleburne's campaigns, eulogized Govan by saying, "I regard him as one of the best soldiers it was my good fortune to know—a true Christian gentleman, a noble patriot, a loyal and uncompromising friend."[61]

Most importantly, Govan's men had always responded to him, had always done their best for him. He, in turn, drew confidence from their loyal service. During the Atlanta Campaign, "gallant Govan" saw one of his company commanders go down. "The General dismounted," declared a witness, "and, calling him by name, asked if he could do anything for him. Between gasps, as if each would be his last, the brave captain said: 'Give them ——, General.'" Govan rose, ordered that the captain be tended, remounted, and moved forward with his brigade determined "to comply" with the captain's directive. Another of Govan's men recalled "the old days when Cleburne leaned on Govan in peril's utmost stress," the days when there was no better officer in the Confederacy.[62]

Notes

"No Better Officer in the Confederacy: The Wartime Career of Daniel C. Govan" was originally published in *Arkansas Historical Quarterly* 54, no. 1 (Autumn 1995): 269–303, and is reprinted with permission by the Arkansas Historical Society. The author wishes to thank Dr. Anne J. Bailey, Bryan Thomasson of Alma, Charles West of Marianna, Sarah Ward, and Ted Smith of Fayetteville for their assistance in the preparation of the original article.

1. Quoted are Wiley Sword, *The Confederacy's Last Hurrah: Spring Hill, Franklin, and Nashville* (Lawrence, KS, 1993), 127, 221; and John M. Harrell, *Arkansas,* in *Confederate Military History,* 12 vols., ed. Clement A. Evans (Atlanta, 1899), 10:402. As many as twenty-nine Confederate generals have been claimed by Arkansas, but the best analysis of the issue speaks convincingly for sixteen. Seven of the twenty-nine have been claimed by Phillips County, but that number is now generally recognized to be five. See Jonathan P. Morrow Jr., "Confederate Generals From Arkansas," *Arkansas Historical Quarterly* 21 (Autumn 1962): 231–46. The most reliable biographical sketches of all these officers are to be found in William C. Davis, ed., *The Confederate General,* 6 vols. (New York, 1991).
2. Davis, *Confederate General* 3:17–18; *Biographical and Historical Memoir of Eastern Arkansas* (Chicago, 1890), 594–95. Most modern biographical sketches of Govan, including those in the *Dictionary of American Biography, Generals in Gray, The Confederate General, The Encyclopedia of the Confederacy,* and *American National Biography,* list Govan's year of birth as 1829. The correct year is 1827, as given in "D. C.

Govan," *Confederate Veteran* 19 (Sept. 1911): 444, and Dale P. Kirkman, "General Govan," *Phillips County Historical Quarterly* 23 (Dec. 1984–Mar. 1985):94–100.

3. Thomas W. Cutrer, *Ben McCulloch and the Frontier Military Tradition* (Chapel Hill, NC, 1993), 106–8. Cutrer refers to a man named Daniel *Gavan* on the California expedition, but this is almost certainly Govan.

4. *Historical Memoir of Eastern Arkansas,* 594; Cutrer, *Ben McCulloch,* 108–18. Cutrer causes confusion about the marriage and removal to Arkansas of Daniel and Mary Govan by implying that their first child was born in 1852 or 1853, but in the face of other evidence cited below, this early date was clearly impossible.

5. Bureau of the Census, *Population Schedule of the Eighth Census [Free], 1860, Phillips County, Arkansas,* p. 78, Microcopy 653, Roll 47, NA; Bureau of the Census, *Population Schedule of the Eighth Census [Slave], 1860, Phillips County, Arkansas,* p. 63, Microcopy 653, Roll 54, NA; Bureau of the Census, *Eighth Census, 1860, Schedule 4: Productions of Agriculture During the Year Ending June 1, 1860, Phillips County, Arkansas* (Washington, DC, 1864), 25–26; Mary Pugh Govan to Maria Thomson, Feb. 1855, letter in possession of Sarah Ward; Rachel Skoney, "Large Planters and Planter Persistence in Antebellum Arkansas, 1850–1860" (master's thesis, Univ. of Arkansas, 1991), 57–58, 74.

6. Carl H. Moneyhon, *The Impact of the Civil War and Reconstruction on Arkansas* (Baton Rouge, LA, 1994), 16–20; James M. Woods, *Rebellion and Realignment: Arkansas's Road to Secession* (Fayetteville, AR, 1987), 138, 156–57, 190–91; Davis, *Confederate General* 3:17–18; Howell Purdue and Elizabeth Purdue, in Pat Cleburne, *Confederate General: A Definitive Biography* (Hillsboro, TX, 1973), 76; U.S. War Department, *The War of the Rebellion: A Compilation of the Official Records of the Union and Confederate Armies,* 128 vols. (Washington, DC, 1880–1901), series 1, vol. 3:590 (hereinafter cited as *OR;* all references are to series 1 unless otherwise indicated); Daniel C. Govan to wife, July 4, 1861, Daniel Chevilette Govan Papers,1861–1908, Southern Historical Collection, Wilson Library, Univ. of North Carolina at Chapel Hill (hereinafter cited as Govan Papers); Diane Neal and Thomas W. Kremm, *The Lion of the South: General Thomas C. Hindman* (Macon, GA, 1993), 90; Nathaniel C. Hughes Jr., *General William J. Hardee: Old Reliable* (Baton Rouge, LA, 1965), 73–75. A complete roster for Company F appears in "Confederate Soldiers from Phillips County," *Phillips County Historical Quarterly* 2 (Sept. 1963): 32.

7. *OR,* vol. 3:588, 609–10; Daniel C. Govan to wife, July 4, 1861, July 28, 1861, Govan Papers; Hughes, *Hardee,* 75–76; Mark K. Christ, ed., *Rugged and Sublime: The Civil War in Arkansas* (Fayetteville, AR, 1994), 17; St. John Richardson Liddell, *Liddell's Record: St. John Richardson Liddell, Brigadier General . . . ,* ed. Nathaniel C. Hughes Jr. (1866; Dayton, OH, 1985), 121.

8. Hughes, *Hardee,* 76–77; Neal and Kremm, *Lion of the South,* 91; *OR,* vol. 3:609–10, 614–15; Purdue and Purdue, *Cleburne,* 82–84; Daniel C. Govan to wife, July 28, 1861, Govan Papers.

9. Daniel C. Govan to wife, July 28, 1861, Govan Papers.

10. Hughes, *Hardee,* 78–80; Purdue and Purdue, *Cleburne,* 84–88; Daniel C. Govan to wife, Aug. 17, 1861, Govan Papers.

11. Hughes, *Hardee,* 80–83; Neal and Kremm, *Lion of the South,* 93–94; Daniel C. Govan to wife, Oct. 15, 1861, Govan Papers.

12. Hughes, *Hardee,* 83; Daniel C. Govan to wife, Oct. 15, 1861, Govan Papers.

13. Hughes, *Hardee,* 84–85; Daniel C. Govan to wife, Oct. 25, 1861, Govan Papers. For a common soldier's view of the tedium and boisterousness of camp life at Cave City, see *The Autobiography of Sir Henry Morton Stanley,* ed. Dorothy Stanley (Boston, 1909), 179–85.

14. Daniel C. Govan to wife, Oct. 25, 1861, Nov. 4, 1861, Govan Papers; *OR,* vol. 4:507, 555; Thomas L. Connelly, *Army of the Heartland: The Army of Tennessee, 1861–1862* (Baton Rouge, LA, 1967), 70.

15. Bureau of the Census, *Population Schedule of the Eighth Census [Free],* 1, 84, 103; Daniel C. Govan to wife, Oct. 28, 1861, Govan Papers. Several other Govans, mostly Daniel's brothers, served in Arkansas and Mississippi regiments, but at least two of them—William and Jack—are mentioned being in Phillips County at various time during the war. William was usually the closest one available to check on Mary. He served as a major and quartermaster at various times under Hindman and Theophilus Holmes. Mary S. Edmondson,"'This Old Book': The Civil War Diary of Mrs. Mary Sale Edmondson,"ed. Russell P. Baker, *Phillips County Historical Quarterly* 11 (Mar. 1973): 3; Sue Cook,"Diary of Sue Cook (1844–1912), 1864–1865,"ed. Betty M. Faust, *Phillips County Historical Quarterly* 5 (Dec. 1866): 37; *OR,* vol. 13:246, 833; ibid., vol. 47, pt. 3:850.

16. Daniel C. Govan to wife, Oct. 28, 1861, Govan Papers. For official confirmation of Govan's view, see *OR,* vol. 4:481–82.

17. Hughes, *Hardee,* 87; Neal and Kremm, *Lion of the South,* 97–99; *OR,* vol. 7:14–21.

18. Daniel C. Govan to wife, Dec. 3, 22, 1861, Govan Papers; *OR,* vol. 4:555.

19. Davis, *Confederate General* 3:18; Neal and Kremm, *Lion of the South,* 99–102; Hughes, *Hardee,* 89.

20. Daniel C. Govan to C. J. Lincoln, Apr. 9, 1862, Compiled Service Records of Confederate Soldiers Who Served in Organizations from the State of Arkansas, Microcopy 317, Roll 55, Record Group 109, National Archives (hereinafter cited as Compiled Service Records); Fay Hempstead, *Historical Review of Arkansas: Its Commerce, Industry, and Modern Affairs,* 3 vols. (Chicago, 1911), 1:243; Davis, *Confederate General* 3:18; *OR,* vol. 10, pt. 1:383, 575–76; Wiley Sword, *Shiloh: Bloody April* (New York, 1974), 145, 151–55, 311–12, 317–18, 458.

21. Neal and Kremm, *Lion of the South,* 107–9; Sword, *Shiloh,* 281, 293–94; *OR,* vol. 10, pt. 1:576.

22. St. John R. Liddell to Samuel Cooper, Oct. 14, 1863, Compiled Service Records; Neal and Kremm, *Lion of the South,* 113–14; Liddell, *Liddell's Record,* 77–78; Robert U. Johnson and Clarence C. Buel, eds., *Battles and Leaders of the Civil War,* 4 vols. (New York, 1884–88), 3:30; Davis, *Confederate General,* 3:18.

23. Hughes, *Hardee,* 128–31; Liddell, *Liddell's Record,* 91–96; Hempstead, *Historical Review of Arkansas* 1:243; *OR,* vol. 16, pt. 1:1158–60.

24. Peter Cozzens, *No Better Place to Die: The Battle of Stones River* (Urbana, IL, 1990), 74–75, 94–98, 230; *OR,* vol. 20, pt. 1:660, 860–61.

25. Cozzens, *No Better Place to Die,* 147–50; Purdue and Purdue, *Cleburne,* 168–75; Liddell, *Liddell's Record,* 109–13; *OR,* vol. 20, pt. 1:858–59, 861–63.

26. *OR,* vol. 20, pt. 1:680–82; ibid., vol. 23, pt. 1:589–90, 592–93; St. John R. Liddell to Samuel Cooper, Jan. 16, 1863, with endorsements, St. John R. Liddell to Jefferson Davis, Jan. 19, 1863, both in Compiled Service Records; Peter Cozzens, *This Terrible Sound: The Battle of Chickamauga* (Urbana, IL,1992), 19.

27. Davis, *Confederate General* 3:18; *OR,* vol. 30, pt. 1:231; Arthur J. L. Fremantle, *Three Months in the Southern State: April–June 1863* (Edinburgh, 1863), 156–57. Hardee at this time was much impressed by the improvement his Arkansas troops had made since the previous summer. Liddell, *Liddell's Record,* 120–21.

28. Hempstead, *Historical Review of Arkansas* 1:244; *OR,* vol. 30, pt. 2:14, 239, 257–58; Cozzens, *This Terrible Sound,* 108–10.

29. *OR,* vol. 30, pt. 2:258; Cozzens, *This Terrible Sound,* 141–46.

30. *OR,* vol. 30, pt. 2:258–59; Cozzens, *This Terrible Sound,* 149–51.

31. *OR,* vol. 30, pt. 2:258–59; Cozzens, *This Terrible Sound,* 192–95. Liddell blamed the withdrawal on the short artillery rounds, which he called an "unlucky accident." *OR,* vol. 30, pt. 2:252. The continuing poor performance of the Sixth and Seventh, whatever its reason, was becoming an embarrassment to the men. One member of the regiment who kept a detailed diary of his service made entries for September 18, 20, and 21, but not September 19. Rufus W. Daniel, Diary, Sept. 18–21, 1863, U.S. Army Military History Institute, Carlisle Barracks, PA.

32. Russell K. Brown, *To the Manner Born: The Life of General William H. T. Walker* (Athens, GA, 1994), 176–77; Walter B. Cisco, *States Rights Gist: A South Carolina General of the Civil War* (Shippensburg, PA, 1991), 101–2; *OR,* vol. 30, pt. 2:259. Cozzens, *This Terrible Sound,* 353–55, says Govan's brigade performed "poorly" on the previous day, September 19, but this is too harsh a judgment, based solely, it appears, on the final retreat.

33. Cozzens, *This Terrible Sound,* 355–56; *OR,* vol. 30, pt. 2:259. Rufus Daniel fails to mention yet another collapse of his regiment on September 20. His diary entry for that date includes only the regiment's earlier advance.

34. Cozzens, *This Terrible Sound,* 490–91, 493–94; Brown, *To the Manner Born,* 177–78; *OR,* vol. 30, pt. 2:259–60.

35. *OR,* vol. 30, pt. 2:260–61.

36. Ibid., 254.

37. Daniel C. Govan to wife, Nov. 28, 1861, Nov. 16, 1863, Govan Papers; *OR,* vol. 11, pt. 2:753; *OR,* vol. 11, pt. 21:603; Dunbar Rowland, ed., *Mississippi,* 3 vols. (Atlanta, 1907), 1:793–94.

38. Davis, *Confederate General* 3:18; *OR,* vol. 31, pt. 2:660; Daniel C. Govan to wife, Nov.16, 1863, Govan Papers.

39. Christ, *Rugged and Sublime,* 44; Moneyhon, *Civil War and Reconstruction in Arkansas,* 130; Anne J. Bailey, *Between the Enemy and Texas: Parsons's Texas Cavalry in the Civil War* (Fort Worth, TX, 1989), 85–97; "Indiana Troops at Helena: Part III," *Phillips County Historical Quarterly* 17 (Mar. 1979): 11; William H. Barksdale, "William Henry Barksdale Journal," ed. Russell P. Baker, *Phillips County Historical Quarterly* 15 (Dec. 1976): 40–41. See also Carl H. Moneyhon, "The Civil War in Phillips County, Arkansas," *Phillips County Historical Quarterly* 19 (June–Sept. 1981): 18–36.

40. *OR,* vol. 31, pt. 2:745–53.

41. Irving A. Buck, *Cleburne and His Command* (1908; reprint, Jackson, TN, 1959), 175, 182–84; Purdue and Purdue, *Cleburne,* 253–63; *OR,* vol. 31, pt. 2:752–57, 763; Hempstead, *Historical Review of Arkansas* 1:245.

42. Purdue and Purdue, *Cleburne,* 267–68; Liddell, *Liddell's Record,* 150–51, 165–66; St. John R. Liddell to Samuel Cooper, Oct. 14, 1863, and Patrick R. Cleburne to Samuel Cooper, Dec. 3, 1863, Compiled Service Records.

43. Purdue and Purdue, *Cleburne,* 268–78; Liddell, *Liddell's Record,* 150–51, 165–66; *OR,* vol. 31, pt. 3:869; ibid., vol. 52, pt. 2:586–92; Steven E. Woodworth, *Jefferson Davis and His Generals: The Failure of Confederate Command in the West* (Lawrence, KS,1990), 262–63; Robert F. Durden, *The Gray and the Black: The Confederate Debate on Emancipation* (Baton Rouge, LA, 1972), 53–62. For the possible reasoning of Cleburne and Govan, see Thomas J. Key, "The Diary of a Soldier: Thomas J. Key," ed. Dale P. Kirkman, *Phillips County Historical Quarterly* 20 (Dec. 1981–Mar. 1982): 35–36; for a good account of how this episode may have affected Cleburne's military career, see Paul R. Fessler, "The Case of the Missing Promotion: Historians and the Military Career of Major General Patrick Ronayne Cleburne, C.S.A.," *Arkansas Historical Quarterly* 53 (Summer 1994): 211–31.

44. Purdue and Purdue, *Cleburne,* 268; William E. Bevens, *Reminiscences of a Private: William E. Bevens of the First Arkansas Infantry, C.S.A.,* ed. Daniel E. Sutherland (Fayetteville, AR, 1992), 151–55; Ben LaBree, ed., *Camp Fires of the Confederacy* (Louisville, KY, 1898), 48–53; S. R. Watkins, "Snow Battle at Dalton," *Confederate Veteran* 1 (Sept. 1893): 261–62.

45. *OR,* vol. 38, pt. 3:639; Albert Castel, *Decision in the West: The Atlanta Campaign of 1864* (Lawrence, KS, 1992), 404–5; Davis, *Confederate General* 3:18; Sutherland, *Reminiscences of a Private,* 159; Purdue and Purdue, *Cleburne,* 338.

46. *OR,* vol. 38, pt. 3:686–87, 720–26; Purdue and Purdue, *Cleburne,* 322–28, 347–50; Buck, *Cleburne and His Command,* 230–40; Sutherland, *Reminiscences of a Private,* 171–77; Stan C. Hurley,"Govan's Brigade at Pickett's Mill," *Confederate Veteran* 12 (Feb. 1904): 75; Edward Bourne,"Govan's Brigade at New Hope Church," *Confederate Veteran* 31 (Mar. 1923): 89; Philip L. Secrist,"Scenes of Awful Carnage," *Civil War Times* 10 (June 1971): 5–9, 45–48; T. B. Roy,"General Hardee and the Military Operations Around Atlanta," *Southern Historical Society Papers* 8 (Aug.–Sept. 1880): 364.

47. *OR,* vol. 38, pt. 3:737–41; Purdue and Purdue, *Cleburne,* 332–35, 350–62; Wiley A. Washburn,"Reminiscences of Confederate Service by Wiley A. Washburn," ed. James L. Nichols and Frank Abbott, *Arkansas Historical Quarterly* 35 (Spring 1976): 70–71; Thomas J. Key,"The Diary of a Soldier: Thomas J. Key, Part V," ed. Dale P. Kirkman, *Phillips County Historical Quarterly* 22 (June–Sept. 1984): 9–10, 13. This portion of the field could be identified for many years by two Atlanta streets that ran through it: Hardee and Govan. The name of Govan Street was changed not long ago to Caroline. Key,"Diary of a Soldier: Part V," 30n9.

48. Purdue and Purdue, *Cleburne,* 372–80; *OR,* vol. 38, pt. 3:696, 728–29, 741.

49. Castel, *Decision in the West,* 512–18; Nichols and Abbott,"Reminiscences of Confederate Service," 74–76; Thomas J. Key,"The Diary of a Soldier: Part 6," ed. Dale P. Kirkman, *Phillips County Historical Quarterly* 23 (Dec. 1984–Mar. 1985): 13–14; *OR,* vol. 38, pt. 1:676; ibid., pt. 3:728–29. Major Key thought Govan's works "were fairly strong," but he seems to have been alone in this opinion.

50. Davis, *Confederate General* 3:18; Key,"Diary of a Soldier: Part 6," 34n10; Purdue and Purdue, *Cleburne,* 382; Thomas L. Connelly, *Autumn of Glory: The Army of Tennessee, 1862–1865* (Baton Rouge, LA, 1971), 476–93. An interesting anecdote concerning Govan's captivity is in Frank S. Roberts,"C. P. Roberts, Adjutant, Second Georgia Battalion," *Confederate Veteran* 22 (Mar. 1914): 112.

51. Sword, *Confederacy's Last Hurrah,* 126–39.

52. Richard M. McMurry, *John Bell Hood and the War of Southern Independence* (Lexington, KY, 1982), 171–75; Sword, *Confederacy's Last Hurrah,* 177–80; Purdue and Purdue, *Cleburne,* 419–25.

53. Sword, *Confederacy's Last Hurrah,* 183, 190–96, 200, 220–24, 265; Nichols and Abbott,"Reminiscences of Confederate Service," 82; Hempstead, *Historical Review of Arkansas* 1:242; Daniel C. Govan to wife, Dec. 14, 1864, Govan Papers; *Historical Memoir of Eastern Arkansas,* 595; Hempstead, *Historical Review of Arkansas* 1:246; Samuel T. Foster, *One of Cleburne's Command: The Civil War Reminiscences and Diary of Capt. Samuel T. Foster . . . ,* ed. Norman D. Brown (Austin, TX, 1980), 150.

54. Sword, *Confederacy's Last Hurrah,* 268–69; Daniel C. Govan to wife, Dec. 14, 1864, Govan Papers.

55. Sword, *Confederacy's Last Hurrah,* 322–25, 370–73, 377; Christopher Losson, *Tennessee's Forgotten Warriors: Frank Cheatham and His Confederate Division* (Knoxville,

TN, 1989), 232–40; B. L. Ridley,"Last Battles of the War," *Confederate Veteran* 3 (Jan. 1895): 20; *OR*, vol. 45, pt. 1:667, 680, 739–40; Jack D. Welsh, *Medical Histories of Confederate Generals* (Kent, OH, 1995), 85.

56. Davis, *Confederate General* 3:19; *OR*, vol. 47, pt. 1:1061, 1106–9.

57. *Historical Memoir of Eastern Arkansas*, 595;"Reminiscences of the Late Lon Slaughter as Told to Mrs. E. D. Wall, September 3, 1920," *Arkansas Historical Quarterly* 8 (Summer 1949): 169; Frank Armstrong to Daniel C. Govan, Apr. 24, 1894, commission as Indian agent dated May 24, 1894, Govan Papers; Davis, *Confederate General* 3:19; Memphis *Commercial Appeal*, Mar. 13, 1911, p. 4;"D. C. Govan," *Confederate Veteran*, 444; Welsh, *Medical Histories*, 85.

58. Hempstead, *Historical Review of Arkansas* 1:243–46; Liddell, *Liddell's Record*, 7, 129; William W. Belknap to Daniel C. Govan, Sept. 2, 1878, T. B. Roy to Daniel C. Govan, Mar. 8, 1880, J. W. Green to Daniel C. Govan, Nov. 26, 1886, Hugh B. Roland to Daniel C. Govan, May 5, 1894, N. W. Cooper to Daniel C. Govan, Nov. 27, 1895, J. P. Young to Daniel C. Govan, Mar. 23, 1897, and Irving A. Buck to Daniel C. Govan, Aug. 20, 1907, all in Govan Papers; *Veteran: Devoted to the Interests of the Grand Army of the Republic* (hereinafter cited as *Veteran*) 2 (Jan. 1882): 11; "Officers of the Memphis Reunion," *Confederate Veteran* 8 (Sept. 1900): 391.

59. *Veteran* 2 (Jan. 1882): 9–11.

60. William W. Belknap to Daniel C. Govan, Oct. 1, 21, 1883, and J. S. Bosworth to Daniel C. Govan, Sept. 22, 1884, both in Govan Papers.

61. Purdue and Purdue, *Cleburne*, 268, 370–71; St. John R. Liddell to Samuel Cooper, Oct. 14, 1863, Compiled Service Records; Davis, *Confederate General* 3:19.

62. "Humors of Camp Life," *Confederate Veteran* 2 (Dec. 1894): 357; Will H. Thompson to Daniel C. Govan, Nov. 22, 1894, Govan Papers.

Recommended Readings

For those readers wishing to learn more about the individuals and topics discussed in this volume, in addition to the items cited in the bibliography we recommend the following.

Brown, Kent M., ed. *The Civil War in Kentucky: Battle for the Bluegrass State.* Mason City, IA: Savas Publishing, 2000.

Browning, Robert M., Jr. *Forrest: The Confederacy's Relentless Warrior.* Washington, DC: Potomac Books, 2004.

Castel, Albert."Savior of the South? Was Albert Sidney Johnston the'Robert E. Lee of the West'?" *Civil War Times Illustrated* 36, no. 1 (Mar. 1997): 38–40.

Cooper, William J., Jr. *Jeffèrson Davis, American.* New York: Alfred A. Knopf, 2000.

———. *Jefferson Davis and the Civil War Era.* Baton Rouge: Louisiana State Univ. Press, 2008.

Cunningham, O. Edward. *Shiloh and the Western Campaign of 1862.* Edited by Gary D. Joiner and Timothy B. Smith. New York: Savas Beatie, 2007.

Daniel, Larry J. *Shiloh: The Battle that Changed the Civil War.* New York: Simon & Schuster, 1997.

Davis, Stephen."A Reappraisal of the Generalship of John Bell Hood in the Battles for Atlanta." In *The Campaign for Atlanta & Sherman's March to the Sea: Essays on the American Civil War in Georgia, 1864,* edited by Theodore P. Savas and David A. Woodbury, 49–95. Campbell, CA: Savas Publishing, 1994.

Gallagher, Gary W., and Joseph T. Glatthaar, eds. *Leaders of the Lost Cause: New Perspectives on the Confederate High Command.* Mechanicsburg, PA: Stackpole Books, 2004.

Grabau, Warren E. *Ninety-eight Days: A Geographer's View of the Vicksburg Campaign.* Knoxville: Univ. of Tennessee Press, 2000.

Hafendorfer, Kenneth A. *Perryville: Battle for Kentucky.* Louisville, KY: KH Press, 1981.

Hurst, Jack. *Nathan Bedford Forrest: A Biography.* New York: Alfred A. Knopf, 1993.

Lash, Jeffrey N. *Destroyer of the Iron Horse: General Joseph E. Johnston and Confederate Rail Transport, 1861–1865.* Kent, OH: Kent State Univ. Press, 1991.

McDonough, James L. *Shiloh—In Hell Before Night.* Knoxville: Univ. of Tennessee Press, 1977.

———. *War in Kentucky: From Shiloh to Perryville.* Knoxville: Univ. of Tennessee Press, 1994.

McMurry, Richard M. "'The *Enemy* at Richmond': Joseph E. Johnston and the Confederate Government." *Civil War History* 27 (1981): 5–31.

Noe, Kenneth W. *Perryville: This Grand Havoc of Battle.* Lexington: Univ. Press of Kentucky, 2001.

Pemberton, John C. *Compelled to Appear in Print: The Vicksburg Manuscript of General John C. Pemberton.* Edited by David M. Smith. Cincinnati, OH: Ironclad Publishing, 1999.

Robinson, Thomas W. "One Division Saves An Army." *North & South* 10, no. 3 (2007): 28–43.

Stanbury, Jim. "A Failure of Command: The Confederate Loss of Vicksburg." *Civil War Regiments* 2, no. 1 (1992): 36–68.

Symonds, Craig L. "A Fatal Relationship: Davis and Johnston at War." In *Jefferson Davis's Generals,* edited by Gabor Boritt, 3–26. New York: Oxford Univ. Press, 1999.

———. *Joseph E. Johnston: A Civil War Biography.* New York: W. W. Norton, 1992.

Wills, Brian Steel. *A Battle from the Start: The Life of Nathan Bedford Forrest.* New York: HarperCollins, 1992.

Bibliography

Manuscripts

Adjutant and Inspector General's Office, Record of Telegrams Received, 1862–1865. War Department Collection of Confederate Records. Record Group 109. National Archives.

Anderson, George Wayne. Papers, 1758–1896. Southern Historical Collection. Wilson Library, Univ. of North Carolina at Chapel Hill.

Anderson, James Patton. Collection. P. K. Yonge Library of Florida History, Univ. of Florida, Gainesville.

Beatty, Taylor. Diary. Taylor Beatty Papers, 1780-1917. Southern Historical Collection. Wilson Library, Univ. of North Carolina at Chapel Hill.

Beauregard, P. G. T. Papers. Manuscript Division, Library of Congress, Washington, DC.

Beauregard, P. G. T. Papers, 1844–1893. Rare Book, Manuscript, and Special Collections Library, Duke University Library, Durham, NC.

Benson, Berry. Papers, 1845; 1865–1922. Southern Historical Collection. Wilson Library, Univ. of North Carolina at Chapel Hill.

Bragg, Braxton. Papers. William K. Bixby Collection. Missouri Historical Society, St. Louis.

Bragg, Braxton. Papers. William P. Palmer Collection. Western Reserve Historical Society, Cleveland, OH.

Bragg, Braxton. Papers, 1847–1869. Rare Book, Manuscript, and Special Collections Library, Duke Univ., Durham, NC.

Bragg, Thomas. Diary. Thomas Bragg Papers, 1861–1862. Southern Historical Collection. Wilson Library, Univ. of North Carolina at Chapel Hill.

Compiled Service Records of Confederate General and Staff Officers, and Non-Regimental Enlisted Men. Microcopy 331. Record Group 109. National Archives.

Compiled Service Records of Confederate Soldiers Who Served in Organizations from the State of Arkansas. Microcopy 317. Record Group 109. National Archives.

Confederate Miscellany. Manuscript Division, Emory University Library, Atlanta, GA.

Confederate States of America Archives, 1861–1865. Rare Book, Manuscript, and Special Collections Library, Duke University Library, Durham, NC.

Daniel, Rufus W. Diary. September 18–21, 1863. U.S. Army Military History Institute, Carlisle Barracks, PA.

Department of South Carolina, Georgia, and Florida, Letters Sent, 1863–1864. War Department Collection of Confederate Records. Record Group 109. National Archives.

Drermon, William A. Diary. May 30–July 4, 1862. Mississippi Department of Archives and History, Jackson.

Elliott, Habersham. Papers, 1863–1885. Southern Historical Collection. Wilson Library, Univ. of North Carolina at Chapel Hill.

Ferguson, Samuel W. Memoir. Vicksburg National Military Park, Vicksburg, MS.

Ferguson, S. W. Reminiscences. Heyward and Ferguson Family Papers, 1806–1923. Southern Historical Collection. Wilson Library, Univ. of North Carolina at Chapel Hill.

Fleming, Walter L. Papers, 1685–1932. New York Public Library.

Georgia Portfolio. Manuscript Division, Duke University Library, Durham, NC.

Govan, Daniel Chevilette. Papers, 1861–1908. Southern Historical Collection. Wilson Library, Univ. of North Carolina at Chapel Hill.

Govan, Mary Pugh, to Maria Thomson. February 1855. Letter in possession of Sarah Ward.

Hardee, Charles Seton Henry. "Reminiscences and Recollections of Old Savannah." N.d. Southern Historical Collection. Wilson Library, Univ. of North Carolina at Chapel Hill.

Headquarters Book of Albert Sidney Johnston. Louisiana Historical Association Collection. Manuscripts Division, Howard-Tilton Memorial Library, Tulane Univ., New Orleans.

Hyatt, Arthur W. Papers. Louisiana and Lower Mississippi Valley Collection. Hill Memorial Library, Louisiana State Univ., Baton Rouge.

Inventory of Tools and Machines at the Arsenal. Chap. 4, vol. 39. War Department Collection of Confederate Records. Record Group 109. National Archives.

Johnston, Albert Sidney and William Preston. Papers. Mrs. Mason Barret Collection. Manuscripts Division, Howard-Tilton Memorial Library, Tulane Univ., New Orleans, LA.

Johnston, J. Stoddard. Military Papers. Filson Club, Louisville, KY.

Jones, Charles Colcock. Papers, 1763–1893. Manuscript Division, Duke Univ. Library, Durham, NC.

Jones, Sam. Papers, 1861–1864. War Department Collection of Confederate Records. Record Group 109. National Archives.

Lee, Robert Edward. Papers. Virginia Historical Society, Richmond.

Letters and Telegrams Received, Superintendent of Laboratories. Chap. 4, vol. 37. War Department Collection of Confederate Records. Record Group 109. National Archives.

Letters and Telegrams Sent by W. H. McMain, Military Storekeeper of Ordnance, Apr. 1862–Apr. 1865. Chap. 4, vol. 27. War Department Collection of Confederate Records. Record Group 109. National Archives.

Letters Sent and Received by Gen. J. C. Pemberton. Chap. 2, vol. 21. War Department Collection of Confederate Records. Record Group 109. National Archives.

Letters Sent by the Superintendent of the Armory, June 1862–April 1865. Chap. 4, vol. 31. War Department Collection of Confederate Records. Record Group 109. National Archives.

Lovell, Mansfield. Papers. Keeler Collection. Henry E. Huntington Library, San Marino, CA.

Manuscripts Collection. Mississippi Department of Archives and History, Jackson.

McLaws, Lafayette. Papers, 1836–1897. Southern Historical Collection. Wilson Library, Univ. of North Carolina at Chapel Hill.

McLaws, Lafayette. Papers, 1861–1865. War Department Collection of Confederate Records. Record Group 109. National Archives.

Mercer, George Anderson. Diary, 1851–1889. Southern Historical Collection. Wilson Library, Univ. of North Carolina at Chapel Hill.

Moore, Thomas O. Papers. Louisiana and Lower Mississippi Valley Collections. Hill Memorial Library. Louisiana State Univ., Baton Rouge.

Orders and Circulars, 1862–1865. Records of the Department of Mississippi and East Louisiana and of the Department of Alabama, Mississippi, and East Louisiana. Entry 94. War Department Collection of Confederate Records. Record Group 109. National Archives.

Pemberton, Israel. Papers. Pemberton Family Papers. Historical Society of Pennsylvania, Philadelphia.

Pemberton, John C. Papers. Pemberton Family Papers. Historical Society of Pennsylvania, Philadelphia.

Pemberton, John C. Papers. U.S. Naval Academy Archives, Annapolis, MD.

Pemberton, John C. Papers, 1814–1942. Southern Historical Collection. Wilson Library, Univ. of North Carolina at Chapel Hill.

Pemberton, John C. Papers, 1862–1864. Entry 131. War Department Collection of Confederate Records. Record Group 109. National Archives.

Pemberton, John Clifford. Papers. Library of Congress, Washington, DC.

Polk, Leonidas. Papers. Univ. of the South, Sewanee, TN.

Preston, William. Diary. War Department Collection of Confederate Records. Record Group 109. National Archives.

Savannah Squadron Papers, 1862–1865. Manuscript Division, Emory University Library, Atlanta, GA.

Seddon, James A. Papers, 1862–1865. Rare Book, Manuscript, and Special Collections Library, Duke Univ., Durham NC.

Telegrams Sent, Superintendent of Laboratories, Oct. 1863–Apr. 1865. Chap. 4, vol. 52. Records of Ordnance Establishments. War Department Collection of Confederate Records. Record Group 109. National Archives.

Van Duzer, John C. Diary, 1864. Rare Book, Manuscript, and Special Collections Library, Duke University Library, Durham, NC.

Wigfall, Louis T. Family Papers, 1836–1897. Manuscripts Division, Library of Congress, Washington, DC.

Wyckliffe-Preston Papers. Department of Special Collections, University of Kentucky Library, Lexington.

Government Documents

U.S. Bureau of the Census. *Eighth Census, 1860, Schedule 4: Productions of Agriculture During the Year Ending June 1, 1860, Phillips County, Arkansas.* Washington, DC.: Government Printing Office, 1864.

———. *Population Schedule of the Eighth Census [Free], 1860, Phillips County, Arkansas.* Microcopy 653, Roll 47. National Archives.

———. *Population Schedule of the Eighth Census [Slave], 1860, Phillips County, Arkansas.* Microcopy 653, Roll 54. National Archives.

U.S. Naval War Records Office. *Official Records of the Union and Confederate Navies in the War of the Rebellion.* 31 vols. Washington, DC., 1894–1927.

U.S. War Department. *The Official Atlas of the Civil War.* 1892. Reprint. New York: Thomas Yoseloff, 1958.

———. *The War of the Rebellion: A Compilation of the Official Records of the Union and Confederate Armies.* 128 vols. Washington, DC., 1880–1901.

Newspapers and Periodicals

Charleston Daily Courier.

Charleston Mercury.

Harper's Weekly.

Memphis Commercial Appeal.

Memphis Daily Appeal.

Mobile Advertiser and Register.

New Orleans Daily Picayune.

New Orleans Evening Delta.

New York Herald.

New York Times Book Review.

Niles' National Register.

Richmond Examiner.

Richmond Whig.

Savannah Daily Morning News.

The Veteran; Devoted to the Interests of the Grand Army of the Republic.

Printed Primary Sources

"Autobiography of Gen. Patton Anderson, transcribed by Mrs. Anderson." *Southern Historical Society Papers* 24 (1896): 57–72.

Barksdale, William H. "William Henry Barksdale Journal." Edited by Russell P. Baker. *Phillips County Historical Quarterly* 15 (Dec. 1976): 34–46.

Beauregard, P. G. T. *With Beauregard in Mexico: The Mexican War Reminiscences of P. G. T. Beauregard.* Edited by T. Harry Williams. Baton Rouge: Louisiana State Univ. Press, 1956.

Bevens, William E. *Reminiscences of a Private: William E. Bevens of the First Arkansas Infantry, C.S.A.* Edited by Daniel E. Sutherland. Fayetteville: Univ. of Arkansas Press, 1992.

Blackmore, Bettie R. "Behind the Lines in Middle Tennessee, 1863–1865: The Journal of Bettie Ridley Blackmore." Edited by Sarah Ridley Trimble. *Tennessee Historical Quarterly* 12 (1953): 48–80.

Chestnut, Mary B. *A Diary from Dixie.* Edited by Ben Ames Williams. Boston: Houghton, Mifflin, 1949.

———. *Mary Chestnut's Civil War.* Edited by C. Vann Woodward. New Haven: Yale Univ. Press, 1981.

Claiborne, John F. H. *Life and Correspondence of John A. Quitman.* 2 vols. New York: Harper and Brothers, 1860.

Cook, Sue. "Diary of Sue Cook (1844–1912): 1864–1865." Edited by Betty M. Faust. *Phillips County Historical Quarterly* 5 (Dec. 1966): 32–40.

Dahlgren, Madeleine Vinton. *Memoir of John A. Dahlgren, Rear-Admiral United States Navy.* Boston: J. R. Osgood, 1882.

Davis, Jefferson. *Jefferson Davis, Constitutionalist: His Letters, Papers and Speeches.* Edited by Dunbar Rowland. 10 vols. Jackson: Mississippi Department of Archives and History, 1923.

———. *The Papers of Jefferson Davis.* Vol. 2, *June 1841–July 1846.* Edited by James T. McIntosh. Baton Rouge: Louisiana State Univ. Press, 1974.

Dorsey, Sarah A. *Recollections of Henry Watkins Allen: Brigadier-General Confederate States Army, Ex-Governor of Louisiana.* New York: M. Doolady, 1866.

Duncan, Thomas D. *Recollections of Thomas D. Duncan; A Confederate Soldier.* Nashville: McQuiddy Printing, 1922.

Edmondson, Mary S. "'This Old Book': The Civil War Diary of Mrs. Mary Sale Edmondson." Edited by Russell P. Baker. *Phillips County Historical Quarterly* 11 (Mar. 1973): 1–10.

Foster, Samuel T. *One of Cleburne's Command: The Civil War Reminiscences and Diary of Capt. Samuel T. Foster, Texas Brigade, C.S.A.* Edited by Norman D. Brown. Austin: Univ. of Texas Press, 1980.

Frederick II. *Instructions for His Generals.* Translated by Brig. Gen. Thomas R. Phillips. Harrisburg, PA: Military Service, 1944.

Fremantle, Arthur James Lyon. *The Fremantle Diary: Being the Journal of Lieutenant Colonel Arthur James Lyon Fremantle, Coldstream Guards, on His Three Months in the Southern States.* Edited by Walter Lord. Boston: Little, Brown, 1954.

French, Samuel G. *Two Wars: An Autobiography of Gen. Samuel G. French, an Officer in the Armies of the United States and the Confederate States, a Graduate from the U.S. Military Academy, West Point, 1843.* 1901. Reprint. Huntington, WV: Blue Acorn Press, 1999.

Gorgas, Josiah. *The Civil War Diary of General Josiah Gorgas.* Edited by Frank E. Vandiver. University: Univ. of Alabama Press, 1947.

Grant, Ulysses S. *Personal Memoirs of U. S. Grant.* Edited by E. B. Long. 2 vols. 1885. Reprint. Cleveland: World Publishing, 1952.

Gray, John Chipman, and John Codman Ropes. *War Letters 1862–1865 of John Chipman Gray and John Codman Ropes.* New York: Houghton Mifflin, 1927.

Green, John William. *Johnny Green of the Orphan Brigade: The Journal of a Confederate Soldier.* Edited by A. D. Kirwan. Lexington: Univ. of Kentucky Press, 1956.

Hancock, R. R. *Hancock's Diary: or, A History of the Second Tennessee Cavalry.* Nashville: Brandon Printing, 1887.

Henderson, Col. G. F. R. *The Science of War: A Collection of Essays and Lectures, 1891–1903.* Edited by Colonel Neill Malcolm. London: Longmans, Green, 1933.

Holmes, Emma. *The Diary of Miss Emma Holmes, 1861–1866.* Edited by John F. Marszalek. Baton Rouge: Louisiana State Univ. Press, 1979.

Hood, John Bell. *Advance and Retreat: Personal Experiences in the United States and Confederate States Armies.* New Orleans: Published for the Hood Orphan Memorial Fund [by] G. T. Beauregard, 1880.

Hurley, Stan C. "Govan's Brigade at Pickett's Mill." *Confederate Veteran* 12 (Feb. 1904): 75.

Johnson, Robert U., and Clarence C. Buel, eds. *Battles and Leaders of the Civil War: Being for the most part contributions by Union and Confederate officers based upon "The Century War Series" edited by Robert Underwood Johnson and Clarence Clough Buel, of the editorial staff of The Century Magazine.* 4 vols. New York: Century, 1884–88.

Johnston, Joseph E. *A Narrative of Military Operations During the Late War Between the States.* 1874. Reprint. New York: Kraus, 1969.

———. "Some War Letters of General Joseph E. Johnston." Edited by R. M. Hughes. *Journal of the Military Service Institution of the United States* 50 (May–June 1912): 318–28.

Jones, John B. *A Rebel War Clerk's Diary at the Confederate States Capital.* Edited by Howard Swiggett. 2 vols. New York: Old Hickory Bookshop, 1935.

Kean, Robert G. H. *Inside the Confederate Government: The Diary of Robert Garlick Hill Kean.* Edited by Edward Younger. New York: Oxford Univ. Press, 1957.

Key, Thomas J. "The Diary of a Soldier: Part 6." Edited by Dale P. Kirkman. *Phillips County Historical Quarterly* 23 (Dec. 1984–Mar. 1985): 1–36.

———. "The Diary of a Soldier: Thomas J. Key." Edited by Dale P. Kirkman. *Phillips County Historical Quarterly* 20 (Dec. 1981–March 1982): 26–51.

———. "The Diary of a Soldier: Thomas J. Key, Part V." Edited by Dale P. Kirkman. *Phillips County Historical Quarterly* 22 (June–Sept. 1984): 1–32.

Lee, Robert E. *The Wartime Papers of R. E. Lee.* Edited by Clifford Dowdey and Louis H. Manarin. Boston: Little, Brown, 1961.

Liddell, St. John Richardson. *Liddell's Record: St. John Richardson Liddell, Brigadier General, C.S.A., Staff Officer and Brigade Commander, Army of Tennessee.* Edited by Nathaniel Cheairs Hughes Jr. Dayton, OH: Morningside Books, 1985.

Porter, David D. *Incidents and Anecdotes of the Civil War.* New York: D. Appleton, 1885.

"Reminiscences of the Late Lon Slaughter as Told to Mrs. E. D. Wall, September 3, 1920." *Arkansas Historical Quarterly* 8 (Summer 1949): 167–69.

Richardson, James D., ed. *A Compilation of the Messages and Papers of the Confederacy, including the Diplomatic Correspondence, 1861–1865.* 2 vols. Nashville: United States Publishing, 1905.

Ruffin, Thomas. *The Papers of Thomas Ruffin.* Edited by J. G. de Roulhac Hamilton. 4 vols. Raleigh, NC: Edwards & Broughton Printing, 1918–20.

Sherman, William T. *Home Letters of General Sherman.* Edited by M. A. De Wolfe Howe. New York: C. Scribner's Sons, 1909.

———. *Memoirs of General William T. Sherman.* 2 vols. New York: D. Appleton, 1875.

Smith, Gustavus W. *Confederate War Papers: Fairfax Court House, New Orleans, Seven Pines, Richmond and North Carolina.* New York: Atlantic Publishing and Engraving, 1884.

Stanley, Henry M. *The Autobiography of Sir Henry Morton Stanley.* Edited by Dorothy Stanley. Boston: Houghton Mifflin, 1909.

Stevenson, William G. *Thirteen Months in the Rebel Army: Being a Narrative of Personal Adventures in the Infantry, Ordnance, Cavalry, Courier, and Hospital Services; with an Exhibition of the Power, Purposes, Earnestness, Military Despotism, and Demoralization of the South.* New York: A. S. Barnes & Burr, 1862.

Taylor, Richard. *Destruction and Reconstruction: Personal Experiences of the Late War.* Edited by Richard B. Harwell. 1879. Reprint. New York: Longmans, Green, 1955.

Washburn, Wiley A. "Reminiscences of Confederate Service by Wiley A. Washburn." Edited by James L. Nichols and Frank Abbott. *Arkansas Historical Quarterly* 35 (Spring 1976): 47–90.

Watkins, Sam R. *"Co. Aytch," Maury Grays, First Tennessee Regiment, or A Side Show of the Big Show.* Jackson, TN: McCowat-Mercer Press, 1952.

———. "Snow Battle at Dalton." *Confederate Veteran* 1 (Sept. 1893): 261.

Wright, Mrs. D. Giraud. *A Southern Girl in '61.* New York: Doubleday, Page, 1905.

Secondary Sources

Published Secondary Sources

Bailey, Anne J. *Between the Enemy and Texas: Parsons's Texas Cavalry in the Civil War.* Fort Worth: Texas Christian Univ. Press, 1989.

Ballard, Michael B. *The Campaign for Vicksburg.* Conshohocken, PA: Eastern National Park and Monument Association, 1996.

———. *Pemberton: A Biography.* Jackson: Univ. Press of Mississippi, 1991.

Bearss, Edwin C. *Forrest at Brice's Cross Roads and in North Mississippi in 1864.* Dayton, OH: Morningside Books, 1979.

———. *The Vicksburg Campaign.* 3 vols. Dayton, OH: Morningside Books, 1985–86.

Bevier, R. S. *History of the First and Second Missouri Confederate Brigades, 1861–1865.* St. Louis: Bryan, Brand, 1879.

Biographical and Historical Memoirs of Eastern Arkansas. Chicago: Goodspeed Publishing, 1890.

Black, Robert C., III. *The Railroads of the Confederacy.* Chapel Hill: Univ. of North Carolina Press, 1952.

Bourne, Edward."Govan's Brigade at New Hope Church."*Confederate Veteran* 31 (Mar. 1923): 89.

Brown, D. Alexander. *Grierson's Raid.* Urbana: Univ. of Illinois Press, 1962.

Brown, Russell K. *To the Manner Born: The Life of General William H. T. Walker.* Athens: Univ. of Georgia Press, 1994.

Buck, Irving A. *Cleburne and His Command.* 1908. Reprint. Jackson, TN: McCowat-Mercer Press, 1959.

———."Cleburne and His Division at Missionary Ridge and Ringgold Gap."*Southern Historical Society Papers* 8 (1880): 464–75.

Burne, Alfred H."General J. B. Hood."*Army Quarterly and Defence Journal* 79 (Jan. 1960): 220–28.

———. *Lee, Grant and Sherman: A Study in Leadership in the 1864–65 Campaign.* New York: C. Scribner's Sons, 1939.

Burton, Milby. *The Siege of Charleston, 1861–1865.* Columbia: Univ. of South Carolina Press, 1970.

Carman, Ezra A. *General Hardee's Escape from Savannah.* Military Order of the Loyal Legion of the United States. Commandery of the District of Columbia. War Papers. No. 13. Washington, DC, 1893.

Castel, Albert. *Decision in the West: The Atlanta Campaign of 1864.* Lawrence: Univ. Press of Kansas, 1992.

———. *General Sterling Price and the Civil War in the West.* Baton Rouge: Louisiana State Univ. Press, 1968.

Cauthen, Charles E. *South Carolina Goes to War, 1860–1865.* Chapel Hill: Univ. of North Carolina Press, 1950.

Chambers, William N., and W. Dean Burnham, eds. *The American Party Systems: Stages of Development.* New York: Oxford Univ. Press, 1967.

Christ, Mark K., ed. *Rugged and Sublime: The Civil War in Arkansas.* Fayetteville: Univ. of Arkansas Press, 1994.

Cisco, Walter B. *States Rights Gist: A South Carolina General of the Civil War.* Shippensburg, PA: White Mane, 1991.

Clausewitz, Carl von. *On War.* Translated by Colonel J. J. Graham. 3 vols. London: Routledge and Kegan Paul, 1962.

Cleaves, Freeman. *Rock of Chickamauga: The Life of General George H. Thomas.* Norman: Univ. of Oklahoma Press, 1948.

"Confederate Soldiers from Phillips County." *Phillips County Historical Quarterly* 2 (Sept. 1963): 29–32

Connelly, Thomas L. *Army of the Heartland: The Army of Tennessee, 1861–1862.* Baton Rouge: Louisiana State Univ. Press, 1967.

———. *Autumn of Glory: The Army of Tennessee, 1862–1865.* Baton Rouge: Louisiana State Univ. Press, 1971.

Connelly, Thomas L., and Archer Jones. *The Politics of Command: Factions and Ideas in Confederate Strategy.* Baton Rouge: Louisiana State Univ. Press, 1973.

Cooper, William J., Jr. "A Reassessment of Jefferson Davis as War Leader: The Case from Atlanta to Nashville." *Journal of Southern History* 36, no. 2 (May 1970): 189–204.

Cox, Jacob D. *Atlanta.* New York: C. Scribner's Sons, 1882.

———. *The March to the Sea: Franklin and Nashville.* New York: C. Scribner's Sons, 1884.

Cozzens, Peter. *No Better Place to Die: The Battle of Stones River.* Urbana: Univ. of Illinois Press, 1990.

———. *This Terrible Sound: The Battle of Chickamauga.* Urbana: Univ. of Illinois Press, 1992.

Craven, Avery O., ed. *Essays in Honor of William E. Dodd.* Chicago: Univ. of Chicago Press, 1935.

Craven, Avery O., and Frank E. Vandiver. *The American Tragedy: The Civil War in Retrospect.* Hampden-Sydney, VA: Hampden-Sydney College, 1959.

Cullum, G. W. *Biographical Register of Officers and Graduates, U.S. Military Academy.* West Point, NY: U.S. Military Academy, 1891.

Curry, Roy W. "James A. Seddon, a Southern Prototype." *Virginia Magazine of History and Biography* 63 (Apr. 1955): 123–50.

Cutrer, Thomas W. *Ben McCulloch and the Frontier Military Tradition.* Chapel Hill: Univ. of North Carolina Press, 1993.

Davis, Jefferson. *Rise and Fall of the Confederate Government.* 2 vols. New York: D. Appleton, 1881.

Davis, William C., ed. *The Confederate General.* 6 vols. Harrisburg, PA: National Historical Society, 1991.

"D. C. Govan." *Confederate Veteran* 19 (Sept. 1911): 444.

Dillahunty, Albert. *Shiloh.* Washington, DC: Department of the Interior, National Park Service, 1951.

Donald, David, ed. *Why the North Won the Civil War.* Baton Rouge: Louisiana State Univ. Press, 1960.

Donaldson, Gary."'Into Africa': Kirby Smith and Braxton Bragg's Invasion of Kentucky." *Filson Club Historical Quarterly* 61 (Oct. 1987): 444–65.

Dowdey, Clifford. *Experiment in Rebellion.* Garden City, NY: Doubleday, 1946.

———. *The Land They Fought For: The Story of the South as the Confederacy, 1832–1865.* Garden City, NY: Doubleday, 1955.

———. *Lee's Last Campaign: The Story of Lee and His Men Against Grant—1864.* Boston: Little, Brown, 1960.

Dufour, Charles L. *The Night the War Was Lost.* New York: Doubleday, 1960.

———."The Night the War Was Lost. The Fall of New Orleans: Causes, Consequences, Culpability." *Louisiana History* 2 (1961): 157–74.

Duke, Basil W. *Morgan's Cavalry.* New York: Neale, 1909.

Durden, Robert F. *The Gray and the Black: The Confederate Debate on Emancipation.* Baton Rouge: Louisiana State Univ. Press, 1972.

Dyer, Frederick H. *A Compendium of the War of the Rebellion.* 3 vols. 1908. Reprint. New York: T. Yoseloff, 1959.

Dyer, John P. *The Gallant Hood.* Indianapolis: Bobbs-Merrill, 1950.

Eaton, Clement. *Jefferson Davis.* New York: Free Press, 1977

Eisenschiml, Otto. *The Story of Shiloh.* Chicago: Norman Press, 1946.

Evans, Clement A., ed. *Confederate Military History.* 12 vols. Atlanta: Confederate Publishing, 1899.

Fessler, Paul R."The Case of the Missing Promotion: Historians and the Military Career of Major General Patrick Ronayne Cleburne, C.S.A." *Arkansas Historical Quarterly* 53 (Summer 1994): 211–31.

Freeman, Douglas Southall. *Lee's Lieutenants: A Study in Command.* 3 vols. New York: Charles Scribner's Sons, 1942–44.

———. *R. E. Lee: A Biography.* 4 vols. New York: Charles Scribner's Sons, 1934–35.

F. W. M."Career and Fate of Gen. Lloyd Tilghman." *Confederate Veteran* 1 (Sept. 1893): 274–75, 296.

Govan, Gilbert E., and James W. Livingood. *A Different Valor: The Story of General Joseph E. Johnston, C.S.A.* Indianapolis: Bobbs-Merrill, 1956.

Hallock, Judith Lee. *Braxton Bragg and Confederate Defeat.* Vol. 2. Tuscaloosa: Univ. of Alabama Press, 1991.

Hartje, Robert G. *Van Dorn: The Life and Times of a Confederate General.* Nashville: Vanderbilt Univ. Press, 1967.

Hay, Thomas Robson."Braxton Bragg and the Southern Confederacy." *Georgia Historical Quarterly* 9 (1925): 267–316.

———. "Confederate Leadership at Vicksburg." *Mississippi Valley Historical Review* 11 (Mar. 1925): 543–60.

———. "The Davis-Hood-Johnston Controversy of 1864." *Mississippi Valley Historical Review* 11 (June 1924): 54–84.

———. *Hood's Tennessee Campaign.* New York: W. Neale, 1929.

Heitman, Francis B. *Historical Register of the United States Army, from Its Organization, September 29, 1789 to September 29, 1889.* Washington, DC: National Tribune, 1890.

Heleniak, Roman J., and Lawrence L. Hewitt, eds. *Leadership During the Civil War: The 1989 Deep Delta Civil War Symposium; Themes in Honor of T. Harry Williams.* Shippensburg, PA: White Mane, 1992.

Hempstead, Fay. *Historical Review of Arkansas: Its Commerce, Industry, and Modern Affairs.* 3 vols. Chicago: Lewis, 1911.

Hill, Louise B. *Joseph E. Brown and the Confederacy.* Chapel Hill: Univ. of North Carolina Press, 1939.

Horn, Stanley F. *The Army of Tennessee: A Military History.* 1941. Reprint. Norman: Univ. of Oklahoma Press, 1953.

———. "Perryville." *Civil War Times Illustrated* 4 (Feb. 1966): 4–11, 42–47.

Hughes, Nathaniel C., Jr. *General William J. Hardee: Old Reliable.* Baton Rouge: Louisiana State Univ. Press, 1965.

———. "Hardee's Defense of Savannah." *Georgia Historical Quarterly* 47 (Mar. 1963): 43–67.

Hughes, Robert M. *General Johnston.* New York: D. Appleton, 1893.

"Humors of Camp Life." *Confederate Veteran* 2 (Dec. 1894): 357.

"Indiana Troops at Helena: Part III." *Phillips County Historical Quarterly* 17 (Mar. 1979): 11.

Johnston, William Preston. *The Life of General Albert Sidney Johnston.* New York: D. Appleton, 1878.

Jones, Archer. *Confederate Strategy from Shiloh to Vicksburg.* Baton Rouge: Louisiana State Univ. Press, 1961.

———. "Tennessee and Mississippi, Joe Johnston's Strategic Problem." *Tennessee Historical Quarterly* 18 (June 1959): 134–47.

Jones, Charles Colcock, Jr. *The Battle of Honey Hill: An Address Delivered Before the Confederate Survivors' Association in Augusta, Georgia, at Its Seventh Annual Meeting, on Memorial Day, April 27, 1885.* Augusta, GA: Chronicle Printing Establishment, 1885.

———. *The Siege of Savannah in December, 1864, and the Confederate Operations in Georgia and the Third Military District of South Carolina during General Sherman's March from Atlanta to the Sea.* Albany, NY: J. Munsell, 1874.

Jordan, Thomas, and J. P. Pryor. *The Campaigns of Lieut.-General N. B. Forrest, and of Forrest's Cavalry, with Portraits, Maps, and Illustrations.* New Orleans: Blelock, 1868.

Kirkman, Dale P. "General Govan." *Phillips County Historical Quarterly* 23 (Dec. 1984–Mar. 1985): 94–100.

LaBree, Ben, ed. *Camp Fires of the Confederacy.* Louisville: Courier-Journal Job Printing, 1898.

Lewin, Ronald. *Rommel as Military Commander.* Princeton, NJ: Van Nostrand, 1968.

Livermore, Thomas L. *Numbers and Losses in the Civil War in America, 1861–65.* Boston: Houghton, Mifflin, 1901.

Livermore, William Roscoe. *The Story of the Civil War.* New York: G. P. Putnam's Sons, 1913.

Losson, Christopher. *Tennessee's Forgotten Warriors: Frank Cheatham and His Confederate Division.* Knoxville: Univ. of Tennessee Press, 1989.

Malone, Dumas, ed. *Dictionary of American Biography.* 20 vols. New York: C. Scribner's Sons, 1928–44.

May, Robert E. *John A. Quitman: Old South Crusader.* Baton Rouge: Louisiana State Univ. Press, 1985.

McMurry, Richard M. *John Bell Hood and the War of Southern Independence.* Lexington: Univ. of Kentucky Press, 1982.

———. "Patton Anderson: Major General, C.S.A." *Blue & Gray Magazine* 1, no. 2 (Oct.–Nov. 1983): 10–17.

———. *Two Great Rebel Armies: An Essay in Confederate Military History.* Chapel Hill: Univ. of North Carolina Press, 1989.

McWhiney, Grady. *Braxton Bragg and Confederate Defeat.* Vol. 1, *Field Command.* New York: Columbia Univ. Press, 1969.

———. *Confederate Crackers and Cavaliers.* Abilene, TX: McWhiney Foundation Press, 2002.

Memminger, R. W. "The Surrender of Vicksburg—A Defence of General Pemberton." *Southern Historical Society Papers* 12 (1884): 352–60.

Moneyhon, Carl H. "The Civil War in Phillips County, Arkansas." *Phillips County Historical Quarterly* 19 (June–Sept. 1981): 18–36.

———. *The Impact of the Civil War and Reconstruction on Arkansas.* Baton Rouge: Louisiana State Univ. Press, 1994.

Morrow, Jonathan P., Jr. "Confederate Generals From Arkansas." *Arkansas Historical Quarterly* 21 (Autumn 1962): 231–46.

Neal, Diane Neal, and Thomas W. Kremm. *The Lion of the South: General Thomas C. Hindman.* Macon, GA: Mercer Univ. Press, 1993.

"Officers of the Memphis Reunion." *Confederate Veteran* 8 (Sept. 1900): 391.

Parks, Joseph H. *General Edmund Kirby Smith, C.S.A.* Baton Rouge: Louisiana State Univ. Press, 1954.

———. *General Leonidas Polk, C.S.A.: The Fighting Bishop.* Baton Rouge: Louisiana State Univ. Press, 1962.

Patrick, Rembert W. *Jefferson Davis and His Cabinet.* Baton Rouge: Louisiana State Univ. Press, 1944.

Pemberton, John C., III. *Pemberton: Defender of Vicksburg.* Chapel Hill: Univ. of North Carolina Press, 1942.

Perry, William Stevens, ed. *The History of the American Episcopal Church, 1587–1883.* 2 vols. Boston: J. R. Osgood, 1885.

Pickett, William Douglas. *Sketch of the Military Career of William J. Hardee, Lieutenant-General C.S.A.* Lexington, KY: n.p., 1910.

Polk, William M. *Leonidas Polk: Bishop and General.* 2 vols. New York: Longmans, Green, 1893.

Pollard, Edward A. *Life of Jefferson Davis, with a Secret History of the Southern Confederacy.* Philadelphia: National Publishing, 1869.

———. *The Lost Cause: A New Southern History of the War of the Confederates.* New York: E. B. Treat, 1866.

———. *The Second Year of the War.* Richmond: West & Johnston, 1863.

———. *Southern History of the War: The Last Year of the War.* New York: C. B. Richardson, 1866.

Purdue, Howell, and Elizabeth Purdue. *Pat Cleburne, Confederate General: A Definitive Biography.* Hillsboro, TX: Hill Junior College Press, 1973.

Reed, D. W. *The Battle of Shiloh.* Washington, DC: Government Printing Office, 1909.

Ridley, B. L. "Last Battles of the War." *Confederate Veteran* 3 (Jan. 1895): 20.

Roberts, Frank S. "C. P. Roberts, Adjutant, Second Georgia Battalion." *Confederate Veteran* 22 (Mar. 1914): 112.

Roland, Charles P. *Albert Sidney Johnston: Soldier of Three Republics.* Austin: Univ. of Texas Press, 1964.

———. *History Teaches Us to Hope: Reflections on the Civil War and Southern History.* Edited by John David Smith. Lexington: Univ. Press of Kentucky, 2007.

Roman, Alfred. *The Military Operations of General Beauregard in the War between the States, 1861 to 1865.* 2 vols. New York: Harper & Bros., 1884.

Rowland, Dunbar, ed. *Mississippi.* 3 vols. Atlanta: Southern Historical Publishing Association, 1907.

Roy, T. B. "General Hardee and the Military Operations Around Atlanta." *Southern Historical Society Papers* 8 (1880): 337–87.

Scharf, John T. *History of the Confederate States Navy.* 1887. Reprint. New York: Fairfax Press, 1977.

Sears, Stephen W. *George B. McClellan: The Young Napoleon.* New York: Ticknor and Fields, 1988.

Secrist, Philip L. "Scenes of Awful Carnage." *Civil War Times* 10 (June 1971): 5–9, 45–48.

Seitz, Don C. *Braxton Bragg: General of the Confederacy.* Columbia, SC: State, 1924.

Shakespeare, William. *The Tragedy of Julius Caesar.* New York: Washington Square Press, 1966.

Smith, George Winston. "Cotton from Savannah in 1865." *Journal of Southern History* 21 (Nov. 1955): 495–512.

Spiller, Roger J., ed. *Dictionary of American Military Biography.* 3 vols. Westport, CT: Greenwood Press, 1984.

Stevenson, Burton, ed. *The Home Book of Quotations: Classical and Modern.* 10th ed. New York: Dodd, Mead, 1967.

Strode, Hudson. *Jefferson Davis: American Patriot, 1808–1861.* New York: Harcourt, Brace, 1955.

———. *Jefferson Davis: Confederate President.* New York: Harcourt, Brace, 1959.

———. *Jefferson Davis, Tragic Hero: The Last Twenty-five Years, 1864–1889.* New York: Harcourt, Brace, 1964.

Sutherland, Daniel E. "Mansfield Lovell's Quest for Justice: Another Look at the Fall of New Orleans." *Louisiana History* 24 (1983): 233–59.

———. "No Better Officer in the Confederacy: The Wartime Career of Daniel C. Govan." *Arkansas Historical Quarterly* 54, no. 3 (Autumn 1995): 269–303.

Sword, Wiley. *The Confederacy's Last Hurrah: Spring Hill, Franklin, and Nashville.* Lawrence: Univ. Press of Kansas, 1993.

———. *Mountains Touched with Fire: Chattanooga Besieged, 1863.* New York: St. Martin's Press, 1995.

———. *Shiloh: Bloody April.* New York: William Morrow, 1974.

Sykes, Edward Turner. *Walthall's Brigade: A Cursory Sketch, with Personal Experiences of Walthall's Brigade, Army of Tennessee, C.S.A., 1862–1865.* Jackson: Mississippi Historical Society, 1916.

Symonds, Craig L. *Stonewall of the West: Patrick Cleburne and the American Civil War.* Lawrence: Univ. Press of Kansas, 1997.

Tucker, Glenn. *Chickamauga: Bloody Battle in the West.* Indianapolis: Bobbs-Merrill, 1961.

United States Military Academy, Association of Graduates. *Fifteenth Annual Reunion of the Association of the Graduates of the U.S. Military Academy.* East Saginaw, MI: Evening News Printing and Binding House, 1884.

Vandiver, Frank E. "General Hood as Logistician." *Military Affairs* 16 (Spring 1952): 1–11.

———. *Ploughshares into Swords: Josiah Gorgas and Confederate Ordnance.* Austin: Univ. of Texas Press, 1952.

———. *Rebel Brass: The Confederate Command System.* Baton Rouge: Louisiana State Univ. Press, 1956.

Warner, Ezra J. *Generals in Gray: Lives of the Confederate Commanders.* Baton Rouge: Louisiana State Univ. Press, 1959.

Welsh, Jack D. *Medical Histories of Confederate Generals.* Kent, OH: Kent State Univ. Press, 1995.

Williams, Kenneth P. *Lincoln Finds a General.* 5 vols. New York: Macmillan, 1949–59.

Williams, T. Harry. *P. G. T. Beauregard: Napoleon in Gray.* Baton Rouge: Louisiana State Univ. Press, 1955.

———. *Lincoln and His Generals.* New York: Knopf, 1952.

Wolseley, Garnet. "General Forrest." *United Service Magazine* 5 (1892): 1–14, 113–24.

Wood, W. Birkbeck, and James E. Edmonds. *Military History of the Civil War with Special Reference to the Campaigns of 1864 and 1865.* New York: Putnam, 1960.

Woods, James M. *Rebellion and Realignment: Arkansas's Road to Secession.* Fayetteville: Univ. of Arkansas Press, 1987.

Woodworth, Steven E., ed. *The Art of Command in the Civil War.* Lincoln: Univ. of Nebraska Press, 1998.

———, ed. *Civil War Generals in Defeat.* Lawrence: Univ. Press of Kansas, 1999.

———. *Jefferson Davis and His Generals: The Failure of Confederate Command in the West.* Lawrence: Univ. Press of Kansas, 1990.

———. *No Band of Brothers: Problems in the Rebel High Command.* Columbia: Univ. of Missouri Press, 1999.

———. *Six Armies in Tennessee: The Chickamauga and Chattanooga Campaigns.* Lincoln: Univ. of Nebraska Press, 1998.

Wright, Marcus J. *General Officers of the Confederate Army, Officers of the Executive Departments of the Confederate States, Members of the Confederate Congress by States.* New York: Neale, 1911.

Wyeth, John A. *Life of Lieutenant-General Nathan Bedford Forrest.* 1899. Reprint. Dayton, OH: Morningside Press, 1975.

Dissertations and Theses

Hallock, Judith Lee. "Braxton Bragg and Confederate Defeat, Volume II." Ph.D. diss., Stony Brook Univ., New York, December 1989.

Messmer, Charles K. "City in Conflict: A History of Louisville, Kentucky, 1860–1865." Master's thesis, Univ. of Kentucky, 1953.

Skoney, Rachel. "Large Planters and Planter Persistence in Antebellum Arkansas, 1850–1860." Master's thesis, Univ. of Arkansas, 1991.

Oral Interviews

Hughes, Nathaniel C., Jr. Interview with Mrs. Howard Bowen, Aug. 5, 1957, Birmingham, AL.

Contributors

Note: The institution granting the contributor's highest degree is given in parentheses following his name.

Michael B. Ballard (Mississippi State University) is in his twenty-sixth year as university archivist and coordinator of the Congressional and Political Research Center at Mississippi State and in his first year as the associate editor of *The Papers of Ulysses S. Grant.* He has published ten books: seven authored, two coedited, and one coauthored. His books *A Long Shadow: Jefferson Davis and the Final Days of the Confederacy* (1986), *Pemberton: A Biography* (1991), and *Vicksburg: The Campaign that Opened the Mississippi* (2004) were History Book Club selections. He is working on a book tentatively titled *The Civil War in Mississippi: Campaigns and Battles,* a volume in the Mississippi Historical Society's Heritage of Mississippi Series, and he has begun research on the Vicksburg siege.

Edwin C. Bearss (Indiana University) is chief historian emeritus of the National Park Service. Referred to as the "Pied Piper of History," he leads visitors around historic sites throughout the world. He is author or editor of twenty books on the Civil War, including *Rebel Victory at Vicksburg* (1963), *Hardluck Ironclad: The Sinking and Salvage of the Cairo* (1966), and *The Vicksburg Campaign* (3 vols., 1985–86), and he is senior editor of the *Gettysburg Magazine.* Among the many honors he has received are an honorary doctorate from Lincoln College (Illinois); the Alvin Calman, Charles L. Dufour, Bell I. Wiley, T. Harry Williams, Harry S. Truman, Nevins-Freeman, and Frank E. Vandiver Awards; a Distinguished Service Award from the Department of the Interior; and the Civil War Preservation Trust's first Ed Bearss Award.

Arthur W. Bergeron Jr. (Louisiana State University) is a reference historian with the United States Army Military History Institute. He is a past president of the Louisiana Historical Association and of the Richmond and Baton Rouge Civil War Round Tables. His publications include *Guide to Louisiana Confederate Military Units, 1861–1865* (1989), *Confederate Mobile* (1991), *The Civil War Reminiscences of Major Silas T. Grisamore, CSA* (1993), *Louisianians in the Civil War* (2002), *The Civil War in Louisiana: Military Activity* (2002), *The Civil War in Louisiana: The Home Front* (2004), and *A Thrilling Narrative: The Memoir of a Southern Unionist* (2006).

WILLIAM J. COOPER JR. (Johns Hopkins University) is a Boyd Professor at Louisiana State University, where he has spent his entire professional career. He has held several fellowships, including from the Guggenheim Foundation and the National Endowment for the Humanities. He is the author of *The Conservative Regime: South Carolina, 1877–1890* (1968), *The South and the Politics of Slavery, 1828–1856* (1978), *Liberty and Slavery: Southern Politics to 1860* (1983), *The American South: A History* (coauthor, 1991, 4th ed., 2008), *Jefferson Davis, American* (2000), and *Jefferson Davis and the Civil War Era* (2008). He has also edited four books and written numerous articles.

LAWRENCE LEE HEWITT (Louisiana State University) was professor of history at Southeastern Louisiana University. The 1991 recipient of SLU's President's Award for Excellence in Research and the Charles L. Dufour Award for "outstanding achievements in preserving the heritage of the American Civil War," he is a past president of the Baton Rouge Civil War Round Table and former managing editor of *North & South.* His publications include *Port Hudson: Confederate Bastion on the Mississippi* (1987), *The Confederate High Command and Related Topics: The 1988 Deep Delta Civil War Symposium* (1990), *Leadership During the Civil War: The 1989 Deep Delta Civil War Symposium* (1992), *Louisianians in the Civil War* (2002), and *Kentuckians in Gray: Confederate Generals and Field Officers of the Bluegrass State* (2008).

NATHANIEL C. HUGHES JR. (University of North Carolina), a native Tennessean, has had two careers. One has been as a teacher, coach, and headmaster in Bell Buckle, Tennessee, at St. Mary's Episcopal School and at Girls Preparatory School in Memphis, and at the University of Memphis. His second career has been in writing history. He has published twenty-four books, from *General William J. Hardee: Old Reliable* (1965) to his latest, *Yale's Confederates: A Biographical Dictionary* (2008). He also served in the U.S. Marine Corps as leader of an armored amphibian platoon and takes great pride in his three sons becoming Marine officers.

ARCHER JONES (University of Virginia) was professor emeritus of history at North Dakota State University at the time of his death on January 23, 2006. He shared the Bell I. Wiley Prize with his coauthors for *Why the South Lost the Civil War* (1986) and the Jefferson Davis Award for *Politics of Command: Factions and Ideas in Confederate Strategy* (1973) with Thomas Lawrence Connelly. Among his other books are *Confederate Strategy from Shiloh to Vicksburg* (1961), *How the North Won: A Military History of the Civil War* (1983), *The Art of War in the Western World* (1987), and *Civil War Command and Strategy: The Process of Victory and Defeat* (1992).

RICHARD M. MCMURRY (Emory University) is among the most popular lecturers and tour guides dealing with the Civil War. Best known for his *Two Great Rebel Armies: An Essay in Confederate Military History* (1989), he also wrote *John Bell Hood and the War for Southern Independence* (1982), *Virginia Military Institute Alumni in the Civil War: In Bello Praesidium* (1999), *Atlanta 1864: Last Chance for the Confederacy* (2000), and *The Fourth Battle of Winchester* (2002). He coedited *Rank and File: Civil War Essays in Honor of Bell Irvin Wiley* (1976), *This Wilderness of War: The Civil War Letters of George W. Squier, Hoosier Volunteer* (1998), and *An Uncompromising Secessionist: The Civil War of George Knox Miller, Eighth (Wade's) Confederate Cavalry* (2007).

GRADY MCWHINEY (Columbia University) served in the Marine Corps late in World War II. He made his life's work the study of the Civil War era and became an influential and controversial historian in the process. He taught at Troy State University, Millsaps College, the University of California at Berkeley, Northwestern University, the University of British Columbia, Wayne State University, the University of Alabama, Texas Christian University, and, in retirement, the University of Southern Mississippi and McMurry University. His most notable titles include *Attack and Die: Civil War Military Tactics and the Southern Heritage* (1982), which he coauthored with his doctoral student, Perry D. Jamieson; *Braxton Bragg and Confederate Defeat* (1969); and *Cracker Culture: Celtic Ways in the Old South* (1988). Though he died in 2006, his work continues through the programs and initiatives of the Grady McWhiney Research Foundation based in Abilene, Texas.

CHARLES P. ROLAND (Louisiana State University), a native of Tennessee, served as a combat infantry captain in Europe in World War II and received the Bronze Star Medal for meritorious service and the Purple Heart Medal for wounds sustained in action. He served as assistant to the chief historian of the U.S. Army (1951–52), was a member of the History Department at Tulane University (1952–70), and served as the Alumni Professor of History at the University of Kentucky (1970–88). He served as the Harold Keith Johnson Visiting Professor of Military History at the United States Army Military History Institute and Army War College (1981–82) and as the Visiting Professor of Military History at the United States Military Academy (1985–86, 1991–92). He is the author of many books on the Civil War and the American South, including *The Confederacy* (1960), *Albert Sidney Johnston: Soldier of Three Republics* (1964), *An American Iliad: The Story of the Civil War* (1991), and *History Teaches Us to Hope: Reflections on the Civil War and Southern History* (2007).

DANIEL E. SUTHERLAND (Wayne State University) has served as a professor of history at the University of Arkansas since 1989. He has published thirteen books on nineteenth-century U.S. history. His Civil War titles include *Seasons of War: The Ordeal of a Confederate Community, 1861–1865* (1995), *Fredericksburg and Chancellorsville: The Dare Mark Campaign* (1998), *This Terrible War: The Civil War and Its Aftermath* (2003, 2007) with Michael Fellman and Lesley Jill Gordon, and *A Savage Conflict: The Decisive Role of Guerrillas in the American Civil War* (2009).

CRAIG L. SYMONDS (University of Florida) is professor emeritus at the U.S. Naval Academy and chief historian of the USS Monitor Center at the Mariners' Museum in Newport News, Virginia. He is the author of ten books, including *Stonewall of the West: Patrick Cleburne and the Civil War* (1987); *Joseph E. Johnston: A Civil War Biography* (1992); *Confederate Admiral: The Life and Wars of Franklin Buchanan* (1999); *Decision at Sea: Five Naval Battles that Shaped American History* (2005), which won the Theodore and Franklin D. Roosevelt Prize; and most recently, *Lincoln and His Admirals* (2008), cowinner of the Lincoln Prize in 2009.

Frank E. Vandiver (Tulane University) was John H. and Sara Lindsey Professor in Humanities, the director of the Mosher Institute for International Policy Studies, and president emeritus of Texas A&M at the time of his death on January 7, 2005. His long and varied career, coupled with his countless publications on a variety of subjects, ranks him foremost among historians of the Confederacy. Best known for *Mighty Stonewall* (1957), his other Civil War books include *Ploughshares into Swords: Josiah Gorgas and Confederate Ordnance* (1952), *Rebel Brass: The Confederate Command System* (1956), *Jubal's Raid: General Early's Famous Attack on Washington in 1864* (1960), and *Their Tattered Flags: The Epic of the Confederacy* (1970). While teaching at Rice University, he directed several notable Ph.D. candidates, including Civil War historians Thomas Lawrence Connelly, Joseph Harsh, Emory Thomas, Jon Wakelyn, and Richard Sommers.

T. Harry Williams (University of Wisconsin) retired after thirty-eight years at Louisiana State University as a Boyd Professor of History the week before his death on July 6, 1979. His *Huey Long* (1969) won the National Book Award and the Pulitzer Prize, and he served as president of the Southern Historical Association (1958–59) and of the Organization of American Historians (1972–73). His Civil War publications included *Lincoln and the Radicals* (1941), *Lincoln and His Generals* (1952), *P. G. T. Beauregard: Napoleon in Gray* (1955), *McClellan, Sherman, and Grant* (1962), and *Hayes of the Twenty-Third: The Civil War Volunteer Officer* (1965).

Steven E. Woodworth (Rice University) is professor of history at Texas Christian University and author, coauthor, or editor of twenty-six books. He is a two-time winner of the Fletcher Pratt Award of the New York Civil War Round Table for *Jefferson Davis and His Generals: The Failure of Confederate Command in the West* (1992) and *Davis and Lee at War* (1995), a Pulitzer Prize nominee (for the latter), a two-time finalist for the Peter Seaborg Award of the George Tyler Moore Center for the Study of the Civil War for *While God Is Marching On: The Religious World of Civil War Soldiers* (2001) and *Nothing but Victory: The Army of the Tennessee, 1861–1865* (2005), and the 2002 recipient of the Grady McWhiney Award of the Dallas Civil War Round Table for lifetime contributions to the study of Civil War history. Among his other publications are *No Band of Brothers: Problems of the Rebel High Command* (1999) and *Sherman* (2009).

Index

Page numbers in **boldface** refer to illustrations. Units above the regimental level are listed under their commander's name.